# Free/Open Source Software Development

Stefan Koch
Vienna University of Economics and Business Administration, Austria

**IDEA GROUP PUBLISHING**
Hershey • London • Melbourne • Singapore

Acquisitions Editor:       Mehdi Khosrow-Pour
Senior Managing Editor:    Jan Travers
Managing Editor:           Amanda Appicello
Development Editor:        Michele Rossi
Copy Editor:               Jane Conley
Typesetter:                Sara Reed
Cover Design:              Lisa Tosheff
Printed at:                Yurchak Printing Inc.

Published in the United States of America by
    Idea Group Publishing (an imprint of Idea Group Inc.)
    701 E. Chocolate Avenue, Suite 200
    Hershey PA 17033
    Tel: 717-533-8845
    Fax: 717-533-8661
    E-mail: cust@idea-group.com
    Web site: http://www.idea-group.com

and in the United Kingdom by
    Idea Group Publishing (an imprint of Idea Group Inc.)
    3 Henrietta Street
    Covent Garden
    London WC2E 8LU
    Tel: 44 20 7240 0856
    Fax: 44 20 7379 3313
    Web site: http://www.eurospan.co.uk

Copyright © 2005 by Idea Group Inc. All rights reserved. No part of this book may be reproduced in any form or by any means, electronic or mechanical, including photocopying, without written permission from the publisher.

    Library of Congress Cataloging-in-Publication Data

Free/open source software development / Stefan Koch, Editor.
    p. cm.
  ISBN 1-59140-369-3 -- ISBN 1-59140-370-7 (pbk.) -- ISBN 1-59140-371-5 (ebook)
 1. Computer software--Development. 2. Open source software. I. Koch, Stefan.
  QA76.76.S46F74 2004
  005.1--dc22
                                          2004003748

British Cataloguing in Publication Data
A Cataloguing in Publication record for this book is available from the British Library.

All work contributed to this book is new, previously-unpublished material. The views expressed in this book are those of the authors, but not necessarily of the publisher.

# Free/Open Source Software Development

## Table of Contents

Preface .................................................................................................................. vii

### Section I: F/OSS Development - "Intensive Analysis"

**Chapter I**
**Do Not Check in on Red: Control Meets Anarchy in Two Open Source Projects** ............................................................................................................. 1
    *Jesper Holck, Copenhagen Business School, Denmark*
    *Niels Jørgensen, Roskilde University, Denmark*

**Chapter II**
**Analyzing the Anatomy of GNU/Linux Distributions: Methodology and Case Studies (Red Hat and Debian)** ....................................................................... 27
    *Jesús M. González-Barahona, Universidad Rey Juan Carlos, Spain*
    *Gregorio Robles, Universidad Rey Juan Carlos, Spain*
    *Miguel Ortuño-Pérez, Universidad Rey Juan Carlos, Spain*
    *Luis Rodero-Merino, Universidad Rey Juan Carlos, Spain*
    *José Centeno-González, Universidad Rey Juan Carlos, Spain*
    *Vicente Matellán-Olivera, Universidad Rey Juan Carlos, Spain*
    *Eva Castro-Barbero, Universidad Rey Juan Carlos, Spain*
    *Pedro de-las-Heras-Quirós, Universidad Rey Juan Carlos, Spain*

**Chapter III**
**The Co-Evolution of Systems and Communities in Free and Open Source Software Development** .................................................................... 59
    *Yuwen Ye, University of Colorado at Boulder, USA and*
        *SRA Key Technology Lab, Japan*
    *Kumiyo Nakakoji, University of Tokyo, Japan*
    *Yasuhiro Yamamoto, University of Tokyo, Japan*
    *Kouichi Kishida, SRA Key Technology Lab, Japan*

### Section II: F/OSS Development and Software Engineering Practices - "Extensive Analysis"

**Chapter IV**
**The Role of Modularity in Free/Open Source Software Development** ..... 84
    *Alessandro Narduzzo, Universitá di Bologna, Italy*
    *Alessandro Rossi, Universitá di Trento, Italy*

**Chapter V**
**A Quantitative Study of the Adoption of Design Patterns by Open Source Software Developers** ........................................................................ 103
    *Michael Hahsler, Vienna University of Economics and Business*
        *Administration, Austria*

### Section III: F/OSS Projects as Social Constructs

**Chapter VI**
**Coordination and Social Structures in an Open Source Project: VideoLAN** ............................................................................................. 125
    *Thomas Basset, Centre de Sociologie des Organisations, France and*
        *Ecole Normale Superieure de Chachan, France*

**Chapter VII**
**Free Software Development: Cooperation and Conflict in a Virtual Organizational Culture** ........................................................................ 152
    *Margaret S. Elliott, University of California, Irvine, USA*
    *Walt Scacchi, University of California, Irvine, USA*

### Section IV: Simulating F/OSS Development - "Dynamic Swarms"

**Chapter VIII**
**Dynamical Simulation Models of the Open Source Development Process** .................................................................................................. 174

    I.P. Antoniades, Aristotle University of Thessaloniki, Greece
    I. Samoladas, Aristotle University of Thessaloniki, Greece
    I. Stamelos, Aristotle University of Thessaloniki, Greece
    L. Angelis, Aristotle University of Thessaloniki, Greece
    G.L. Bleris, Aristotle University of Thessaloniki, Greece

**Chapter IX**
**Modeling the Free/Open Source Software Community: A Quantitative Investigation** ................................................................................... 203

    Gregory Madey, University of Notre Dame, USA
    Vincent Freeh, North Carolina University, USA
    Renee Tynan, University of Notre Dame, USA

### Section V: F/OSS Development Interacting with Commercial and Public Organizations

**Chapter X**
**Benefits and Pitfalls of Open Source in Commercial Contexts** ................ 222

    Jiayin Hang, Siemens Business Services GmbH & Co. OHG, Germany
    Heidi Hohensohn, Siemens Business Services GmbH & Co. OHG, Germany
    Klaus Mayr, IFS IT GmbH, Germany
    Thomas Wieland, University of Applied Sciences Coburg, Germany

**Chapter XI**
**Experiences Enhancing Open Source Security in the POSSE Project** ................................................................................................. 242

    Jonathan M. Smith, University of Pennsylvania, USA
    Michael B. Greenwald, University of Pennsylvania, USA
    Sotiris Ioannidis, University of Pennsylvania, USA
    Angelos D. Keromytis, Columbia University, USA
    Ben Laurie, AL Digital, Ltd., USA
    Douglas Maughan, Defense Advanced Research Projects Agency, USA
    Dale Rahn, University of Pennsylvania, USA
    Jason Wright, University of Pennsylvania, USA

## Section VI: Implications of the F/OSS Development Model - "The Broad Picture"

**Chapter XII**
**The Impact of Open Source Development on the Social Construction of Intellectual Property** ............................................................... 259
   *Bernd Carsten Stahl, De Montfort University, UK*

**Chapter XIII**
**The Social Production of Ethics in Debian and Free Software Communities: Anthropological Lessons for Vocational Ethics** ................ 273
   *E. Gabriella Coleman, University of Chicago, USA*
   *Benjamin Hill, Debian Project, USA*

**About the Editor** ........................................................... 296

**About the Authors** ......................................................... 297

**Index** ...................................................................... 306

# Preface

In the last few years, free and open source software has gathered increasing interest, both from the business and the academic worlds. As some projects in different application domains, like Linux together with the suite of GNU utilities, GNOME, KDE, Apache, sendmail, bind, and several programming languages, have achieved huge success in their respective markets, they have demonstrated that this new development paradigm can produce output of considerable quality. This has led to massive business interest, has given rise to new corporations like RedHat or VA Software (formerly having achieved a record-breaking IPO 1999 with a 700 percent gain on its first day of trading under the name of VA Linux), and has spurred organizations both small and large (like IBM, Sun Microsystems or Netscape) to invest in one way or the other into this new field.

Academic interest in this new form of collaborative software development has also grown, arising from very different backgrounds including software engineering, sociology, management, and psychology, and has gained increasing prominence, as can be deduced from the number of international journals like *Management Science, Information Systems Journal, Electronic Markets* or *Research Policy*, and conferences like ICSE dedicating special issues, workshops, and tracks to this new field of research. As diverse as the background of researchers are their approaches and the issues tackled. The current research that can be attributed to this field ranges from quantitative analysis of source code or other artifacts of the software development to uncover programming practices and the efficiency of this development model, to sociological field work soliciting information in interviews about the ways in which coordination and communication in these virtual teams are accomplished. This book will try to give an overview of current research. It aims to be an up-to-date inventory of research approaches and outlooks. As yet, an edited volume of academic papers dealing with free and open source software development has not been available, and it is hoped that this book will provide a first step towards attributing this line of research the prominence and credibility it so richly deserves, given the high-quality output produced, as can now be witnessed by any interested reader.

A note is in order about the title of this book. In a first version, the proposed title was *Open Source Software Development*, the announcement of which led to an e-mail response from Richard Stallman, founder of the GNU project, in which he argued that by using this title all relevant and important work of the Free Software community would be subsumed by the Open Source movement (which of course was never intended), and its very existence denied. Following his reasoning, the title was changed, not going further into ideological differences in the outlook of both communities, but explicitly acknowledging the inspiring work done by both. An additional term often used in this context, especially in Europe, *libre software* was not included to maintain readability and because it constitutes more of an artificial term with which no larger group of developers identifies.

# ORGANIZATION OF THE BOOK

The organization of this book is intended to reflect the very different research approaches taken in the field of free and open source software development. Therefore, the chapters have been grouped into no less than six parts, each one dealing with a slightly different focus or outlook. With this, the book provides an overview of this very active field of research, and an interested reader or researcher might be able to identify the approach or focus he or she would like to take in future reading or work. Although there are many different possibilities for classifying the chapters presented here, and even more for all the research currently done, the organization presented here tries to constitute a starting point for discussing and developing a coherent framework for the diverse activities in the field of free and open source software development.

## Section I: F/OSS Development - "Intensive Analysis"

Section I contains three chapters that provide what seems to constitute the heart and most important first step of current research. They all deal with a small number of projects and detail several facets of these. Therefore, the term *intensive analysis* is used here, to be distinguished from the extensive analyses performed in the chapters forming Section II. The first chapter, "Do Not Check in on Red: Control Meets Anarchy in Two Open Source Projects" by Jesper Holck and Niels Jørgensen, is a prototypical example in which the authors describe the most important elements of the software development process in the Mozilla and FreeBSD projects. For each of these elements, the struggle for optimal balance between control—supposedly necessary for producing high-quality software—and anarchy—supposedly necessary for attracting and keeping voluntary developers—is discussed. The authors give a superb picture of the free and open source software development process, documenting and analyzing similarities between free and open source projects and commercial software development.

The second chapter, "Analyzing the Anatomy of GNU/Linux Distributions: Methodology and Case Studies (Red Hat and Debian)" by Jesús M. González-Barahona, Gregorio Robles, and colleagues, presents a quantitative and longitudinal study of one of the most important and visible aspects of free and open source software: GNU/Linux distributions. Using a total of nine different versions of the Red Hat and Debian distributions, the authors base their analysis on the source code itself, and also include a detailed description of the methodology applied. This approach allows the study of the evolution of parameters like total distribution size, size and version of different packages, or usage of different programming languages, thus offering insights into the free and open source projects providing the foundation for these distributions, as well as the process of producing a distribution, which poses an enormous integration task and therefore a large project in itself.

Chapter III is titled "The Co-Evolution of Systems and Communities in Free and Open Source Software Development" and is written by Yunwen Ye, Kumiyo Nakakoji, Yasuhiro Yamamoto, and Kouichi Kishida. In this chapter, the authors see beyond the software product and document its relationship to the development community, analyzing the co-evolution that results. Using four projects in which the authors are involved as case studies, they find that projects co-evolve differently depending on the goal of the system and the structure of the community. This leads to a proposed classification scheme for free and open source projects and practical implications of recognizing the co-evolution and the type of project.

## Section II: F/OSS Development and Software Engineering Practices - "Extensive Analysis"

The two chapters forming Section II have both their focus on the role of a single concept from "traditional" software engineering literature in free and open source software development and their analysis of several projects in common, therefore providing an example of extensive research. The first chapter in this part is by Alessandro Narduzzo and Alessandro Rossi and is titled "The Role of Modularity in Free/Open Source Software Development," which very adequately describes its theme. The authors discuss especially the development of different free and open Unix-systems using the theory of modularity applied to both the software architecture and the organization of the projects.

Chapter V, "A Quantitative Study of the Adoption of Design Patterns by Open Source Software Developers" by Michael Hahsler, describes a large-scale quantitative analysis of the usage of design patterns. The analysis is based on data extracted from the version-control system of the SourceForge hosting site and encompasses almost 1,000 projects. Results indicate that design patterns, as proposed by software engineering literature, are indeed used as a practice to improve communication, as larger projects tend towards increased adoption rates and more productive community members. In addition to these results, this chapter serves as an excellent example of the "testbed" function that free and open source software projects might provide for software engineering researchers because their publicly available data allows for large-scale studies.

## Section III: F/OSS Projects as Social Constructs

Section III of this book, "F/OSS Projects as Social Constructs," contains two chapters that take a distinctly sociological position and foremost view free and open source projects as social constructs. The first of these, "Coordination and Social Structures in an Open Source Project: VideoLAN" by Thomas Basset, presents a mixture of sociological research approaches including social network analysis based on observation and a questionnaire, and quantitative data derived from source code. An analysis of the VideoLAN project is presented, detailing the influence of social relationships between developers on the distribution of work. Face-to-face relationships are found to have great importance as well as friendship that can favor the circulation of advice. In addition to technical expertise, a second kind of expertise—the ability to be aware of who is working on what—determines the hierarchy within the project that is found to be similar to a collegial organization.

The second chapter, Chapter VII, is written by Margaret S. Elliott and Walt Scacchi and has the title "Free Software Development: Cooperation and Conflict in a Virtual Organizational Culture." Using the GNUenterprise project as a case study, this chapter details an ongoing ethnography of this virtual organization using the grounded theory approach with participant-observer techniques. Several examples are used to demonstrate how beliefs, values, and norms interact in this virtual organizational culture to result in community building, resolution of conflicts, and strengthened teamwork.

## Section IV: Simulating F/OSS Development - "Dynamic Swarms"

Section IV of this book combines two chapters that both provide simulation models for free and open source software development. The potential impact of such models can not be understated, as they can be used to predict project variables, for example by project coordinators, or even the evolution of whole project communities. The first chapter, Chapter VIII, "Dynamical Simulation Models of the Open Source Development Process" by I.P. Antoniades, I. Samoladas, I. Stamelos, L. Angelis, and G.L. Bleris, first describes a general framework for dynamical simulation models for free and open source software projects and then introduces a specific simulation model. This model is applied to the Apache Project and the gtk+ module of the GNOME project, and simulation outputs are compared to real data. The second chapter in this part, "Modeling the Free/Open Source Software Community: A Quantitative Investigation," takes a slightly different viewpoint and provides a simulation model for the community level, using data on developers' participation from project hosting sites. The authors, Gregory Madey, Vincent Freeh, and Renee Tynan, then proceed to develop a model of this community as a collection of ad hoc, social networks consisting of heterogeneous agents, self-organizing into projects and clusters of projects. A computer simulation of this model using an agent-based simulation toolkit is presented.

## Section V: F/OSS Development Interacting with Commercial and Public Organizations

While many free and open source software projects work away in splendid isolation, the increased commercial and public interest in this form of software has led to more and more interactions with "traditional" organizations. These interactions are the common theme of the Chapters X and XI found in Section V of this book. The first of these, "Benefits and Pitfalls of Open Source in Commercial Contexts" by Jiayin Hang, Heidi Hohensohn, Klaus Mayr, and Thomas Wieland, deals with commercial ventures and describes a case study in which an open source framework was used to build a commercial product. The process followed leads to some lessons, particularly for companies that offer projects instead of products. The second part of this chapter offers a classification of the different players in the software business such as distributors, system integrators, and software or hardware vendors, their roles, motivations and restraints, partially based on a survey. Chapter XI, "Experiences Enhancing Open Source Security in the POSSE Project" by Jonathan M. Smith, Michael B. Greenwald and colleagues, deals with interactions with public organizations—in this case, the U.S. Department of Defense. The authors describe the project goals, including increasing security in open source systems and more broadly disseminating security knowledge, the organization created to manage this project, and the results produced.

## Section VI: Implications of the F/OSS Development Model - "The Broad Picture"

The last section of this book is devoted to two chapters that deal with the implications that free and open source software development might have on fields other than software development. The first of these is titled "The Impact of Open Source Development on the Social Construction of Intellectual Property," written by Bernd Carsten Stahl, and it argues from the constructionist theoretical foundation that intellectual property is a social construction that is created and legitimized by narratives. After discussing the narratives justifying the creation and protection of intellectual property, the influences of the use of information and communication technology and open source software on these narratives are presented. It is argued that open source software, while partly based on ownership of intellectual artefacts, changes the perception of intellectual property because it offers evidence that some of the classical narratives are simplistic; potential changes to institutions, laws and regulations of intellectual property due to this change are then discussed. The last chapter, "The Social Production of Ethics in Debian and Free Software Communities: Anthropological Lessons for Vocational Ethics" by E. Gabriella Coleman and Benjamin Hill, goes beyond the realm of software and intellectual property and, by describing the Debian project and its new maintainer process as an ethnographic case study, argues that ethics are reinforced through sustained collaborative development in free and open source software communities, and that a similar model of ethical volunteerism based on institutional independence, volunteer labor, and networks of trust might be applicable to a range of vocations, for example medicine.

Stefan Koch
Vienna, October 2003

# Acknowledgments

As is usual for such a large piece of work, many people have contributed to it (in the best tradition of free/open source development). First and foremost among these are the authors themselves and the reviewers who devoted their time to help in selecting and improving the submissions. Overall, 76 distinct authors from 16 countries spanning America, Asia, Australia, and Europe have contributed submissions, and 51 people have provided valuable feedback and assistance in the review process. Contrary to common practice, but in the spirit of free/open source development, we want to list those reviewers (with their permission) to especially recognize their contribution:

Ioannis P. Antoniades (Department of Informatics, Aristotle University of Thessaloniki), Cornelia Boldyreff (University of Durham), Terry Anthony Byrd (Auburn University), Kevin Crowston (Syracuse University School of Information Studies), Margaret S. Elliott (Institute for Software Research, School of Information and Computer Science, University of California, Irvine), Nikolaus Franke and Reinhard Prügl (Department of Entrepreneurship, Vienna University of Economics and BA), Giampaolo Garzarelli (Dipartimento di Teoria Economica e Metodi Quantitativi per le Scelte Politiche, University of Rome, "La Sapienza"), Rishab Aiyer Ghosh (MERIT/Infonomics, University of Maastricht), Jesús M. González-Barahona (Universidad Rey Juan Carlos, Madrid, Spain), Stefan Haefliger (University of St. Gallen), PD Dr. Guido Hertel (University of Kiel), Francis Hunt (University of Cambridge), Dr. Till Jaeger (ifrOSS), Christopher M. Kelty (Rice University), Sandeep Krishnamurthy (University of Washington), Selahattin Kuru (Isik University, Istanbul, Turkey), Karim R. Lakhani (MIT Sloan School of Management), Jan Ljungberg (Dept. of Informatics, Gothenburg University), Greg Madey (University of Notre Dame), Nikhil Metha (Auburn University), Martin Michlmayr (Department of Computer Science and Software Engineering, University of Melbourne), Johann Mitloehner (Vienna University of Economics and BA), Audris Mockus (Avaya Labs), Jae Yun Moon (Hong Kong University of Science and Technology), Mogens Kühn Pedersen (Informatics, Copenhagen Business School), Stephen Rank (University of Durham), Alan I Rea, Jr. (Western Michigan University), Gregorio Robles (Universidad Rey Juan Carlos), Georg Schneider (Vienna University of Economics and BA), Barbara Scozzi (Politecnico di Bari, Italy), Srinarayan Sharma (Oakland University), Bernd Simon (Vienna University of Economics and BA), Jonathan

M. Smith (University of Pennsylvania), Haggen So (Royal Melbourne Institute of Technology), Sebastian Spaeth (University of St. Gallen), Ioannis Stamelos (Dept. of Informatics, Aristotle University of Thessaloniki, Greece), Susanne Strahringer (Department of Information Systems, European Business School), Giancarlo Succi (Center for Applied Software Engineering, Free University of Bolzano-Bozen), Dr. Thomas Wieland (University of Applied Sciences, Coburg, Germany), Yunwen Ye (University of Colorado at Boulder, USA & SRA Key Technology Laboratory, Inc., Japan), Huseyin Yildirim (Duke University), and an additional ten reviewers who prefer to remain anonymous.

Many thanks also go to the people at Idea Group Publishing, especially to Medhi Khosrow-Pour, Michele Rossi, Jan Travers, and Amanda Appicello who have been a great help through the whole process.

Also, last but not least, I would like to thank my family, especially my wife Lale, for support, encouragement, and patience during the time of preparing this book.

Stefan Koch
Vienna, October 2003

# SECTION I:

# F/OSS Development – "Intensive Analysis"

# Chapter I

# Do Not Check in on Red:
## Control Meets Anarchy in Two Open Source Projects

Jesper Holck, Copenhagen Business School, Denmark

Niels Jørgensen, Roskilde University, Denmark

### ABSTRACT

*For two Free/Open Source Software projects, Mozilla and FreeBSD, we describe the central elements in the software development processes: the technological infrastructure, the work organization, and the software process models. For each of these elements we discuss how the projects try to find an optimal balance between control (supposedly necessary for producing high-quality software) and anarchy (supposedly necessary for attracting and keeping voluntary developers). Several important considerations are identified: most importantly, control of access to bug-tracking systems and source code repositories, quality control of both individual contributions and production releases, the importance of the development branch, and control of developers' prioritization of work tasks and availability. The results show that the two projects, even though they produce very different kinds of software (a web-browser suite and an operating system), are similar in many respects. However, they also show how difficult the balance between anarchy and control may be and that it is likely to shift over time.*

Copyright © 2005, Idea Group Inc. Copying or distributing in print or electronic forms without written permission of Idea Group Inc. is prohibited.

# INTRODUCTION

The statement "Do Not Check in on Red" can be found in red, capital letters on the website of Mozilla, a well-known F/OSS (Free/Open Source Software[1]) web browser. Mozilla, like many F/OSS and some commercial projects, employs the principle of *continuous integration*, where a number of developers individually and in parallel add source code to a central repository. The statement is a request for developers *not* to check in new changes to the repository if severe errors have been found in the present source code, a situation announced with unmistakable red boxes on the website. Taken literally, the statement would have the unfortunate consequence of also barring possible fixes to these errors, effectively blocking all further progress; rather, it should be interpreted as a strong request for developers to only check in changes necessary to fix the detected errors while the tree is red. In this way, the statement "Do Not Check in on Red" conveys a fundamental tension in the project: allowing a continuous flow of changes and improvements to the software on one hand, and assuring quality by minimizing the problems resulting from deficient contributions on the other.

As discussed by O'Mahony (2003), it is generally assumed that F/OSS developers ("hackers") don't "embrace centralized modes of governance" and are "less likely to welcome formal organizing mechanisms," instead believing "in the value of challenging work, technical autonomy, self-management, and freedom from a positional basis of power." These assumptions indicate that F/OSS developers would prefer rather loosely controlled projects with a flat hierarchy, relying on individual autonomy, tacit norms, and self-organization rather than commands, control, and explicit rules. This is important for a F/OSS project facing the challenge of attracting and retaining voluntary, competent developers; even though some developers are paid for their work, typically this salary does not come from the project as such, but from companies or organizations volunteering for the project.

It is, however, also generally assumed that software project success is dependent on diligent project management, "planning, organizing, directing, and controlling ... company resources ... to complete specific goals and objectives" (Kerzner, 1989, as cited in Jurison, 1999), and tight control of the software development process is a key element in Software Process Improvement (Paulk, Curtis, Chrissis, & Weber, 1993).

The theme of this chapter will be the tension between these two, apparently conflicting demands on a F/OSS project: how are successful F/OSS projects able to find a reasonable balance between *anarchy* (supposedly necessary for attracting and keeping voluntary developers) and *control* (supposedly necessary for the effective production of high-quality software). In this context we will construe anarchism positively as "holding all forms of government authority to be unnecessary and undesirable and advocating a society based on voluntary cooperation and free association of individuals and groups" rather than "a state of lawlessness or political disorder due to the absence of governmental authority" (*Anarchism*, 2003).

We will illustrate this balance through case studies of Mozilla and FreeBSD, two rather large and well-established F/OSS projects of comparable size, but producing very different software products: the Mozilla web browser suite and the FreeBSD operating system. Both projects have been able to both attract developers and produce reasonably successful and long-lived software, so we believe it is interesting to examine how they have managed to find a middle ground between the two apparent opposites. The focus will be on the production of source code, so we will not go into details regarding how the projects handle tasks like infrastructure maintenance, public relations, and user documentation.

Several previous studies have looked into how F/OSS is developed, often assuming the existence of a generic F/OSS development process. Feller and Fitzgerald (2002) identify a number of characteristics "common to most OSS [open source software] projects": parallel development; large, globally distributed development communities; truly independent peer review; prompt feedback; highly talented, highly motivated developers; actively involved users; and rapid release schedule. This picture is, however, questioned by Healy and Schussman (2003) who, based on analysis of a large number of F/OSS projects, suggest that "generalizations about 'the OSS approach'… may be much too broad." In their view there is a gap between theory and data, as "the typical project has one developer, no discussion or bug reports, and is not downloaded by anyone."

Some case studies of F/OSS projects have also been published. Mockus, Fielding and Herbsleb (2002) describe central aspects of the development processes used in the Apache and Mozilla projects and present many quantitative data from these (size of communities, distribution of work, defect density, etc.). The evolution of Linux (Godfrey & Tu, 2000), configuration management in KDE, Mozilla, and Linux (Asklund & Bendix, 2002), and support in Apache (Lakhani & von Hippel, 2003) have also been studied. Finally, we have previously published studies of coordination and motivation in FreeBSD and Mozilla (Holck & Jørgensen, 2003a; Holck & Jørgensen, 2003b; Jørgensen, 2001).

We believe that there is still a need for detailed case studies of the development of F/OSS, and even though this chapter neither establishes new theory nor tests existing theories, we hope it can contribute to the understanding of the "inner workings" of F/OSS projects. The results should be interesting, not only in relation to the F/OSS communities, but also in relation to commercial software companies, as these increasingly are becoming engaged in F/OSS development (Henkel, 2003). FreeBSD and Mozilla have been chosen as our objects of study because they both have quite open organizations, where a large group of developers have the rights to update and change files in the repositories, unlike, e.g., Linux, where Linus Torvalds often is referred to as a "benevolent dictator," having the final decision in all questions.

Our study is a multiple, explorative case study in the sense of Yin (1998). We have pulled statistical data from software repositories, studied mailing lists and newsgroups, read through much of the documentation available on the projects' websites, and finally drawn on a survey of 72 FreeBSD developers performed in November 2000 that was also the basis for Jørgensen (2001).

# SHORT DESCRIPTION OF THE TWO PROJECTS

Before we go any further, we believe some general knowledge of the two projects will be relevant, so we will briefly mention some data for each (see Table 1 for basic facts).

*Mozilla* is a web browser, or rather a web suite consisting of a browser, a mail and newsreader client, a HTML editor, and a chat client. The F/OSS project was created when Netscape decided to free the source of its proprietary Navigator web browser in 1998 (Hamerly, Paquin, & Walton, 1999). New versions of Netscape Navigator are almost entirely based on Mozilla, which is developed for a number of different operating system platforms, including Windows, MacOS X, Linux, FreeBSD, and AIX (*Mozilla Release FAQ*, 2002).

*Table 1: Key data for FreeBSD and Mozilla.*

| Name | FreeBSD | Mozilla |
|---|---|---|
| Main product | Operating system | Web browser suite |
| Major product qualities | Robustness, security | Platform independence, open interfaces, user interface |
| Major platforms | Intel x86, Alpha, SPARC64, PC-98 | Windows, Linux, MacOS |
| Approximate size of development version | 29,000 files, 11 million lines | 40,000 files, 6 million lines |
| Activity on development version in October 2002 | 118 persons committed 2,063 changes | 107 persons committed 2,856 changes |
| 50% of these commits made by | 12 developers | 7 developers |
| Project management | 9 person elected Core Team | 11 person Mozilla.org Staff |

*FreeBSD* is an operating system of the Unix-family (like GNU/Linux), but with a special focus on robustness and security, and so it is mainly used as a server operating system. In addition to the operating system itself, the project also includes *ports*, a large number of application programs ported to the FreeBSD operating system. The project was born in 1993 as one of a long line of descendants from the legendary Berkeley Software Distribution from 1977. FreeBSD runs on a number of different hardware platforms, most importantly Intel x86 processors, but also Alpha, PowerPC, Sparc64, NEC PC-98, and IA-64 (*Supported Platforms*, 2003).

The two projects are of comparable size, but differ in most other aspects. The Mozilla suite is a typical end-user product; the FreeBSD operating system is of most interest to server administrators and advanced, professional users. Backwards compatibility is not a big issue for Mozilla, but is certainly important for FreeBSD. Mozilla has a relatively short history as F/OSS, compared to FreeBSD's 25 years. The Mozilla project was created by and for a long time maintained tight connections with Netscape, who supported the project with both technology and developers. FreeBSD has had no similar relation to a single company. On the basis of these differences we find it surprising that – as we shall later see – the two projects have remarkably similar infrastructures and process models.

# THE TECHNOLOGICAL INFRASTRUCTURE

In this section we will present major elements from the two projects' technological infrastructures. We will focus on the elements that are most important in relation to the software development processes, described in more detail later in this chapter. Along with the major human responsibilities, these elements are illustrated in Figure 1.

For each element in the infrastructure there are a number of rules regarding its use in the software development process. These rules may either be implemented in the technology (e.g., as password-controlled access to the repository) or be publicized regulations (e.g., disallowing changes to the repository at certain times). We will describe the most important of these along with the elements.

## Repository

Both projects feature central source code repositories under CVS control (Fogel & Bar, 2001) and reachable via the Internet. Developers contribute to the projects by continually and in parallel updating (changing, adding, and deleting) repository files. The CVS-controlled source code repositories guarantee that:

- Each file change (called a commit or check in) creates a new file version. All previous versions remain accessible, thereby making it possible to go back to any older file version if problems or errors are introduced.

*Figure 1: Key elements in the software development processes.*

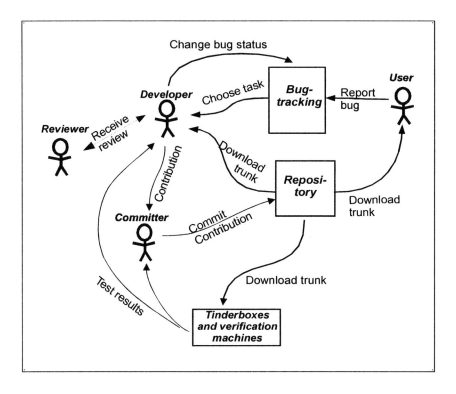

- Each file change is logged, making it possible to see who made the commit, when it was made, which lines were changed, and what comments the committer supplied.
- Conflicting changes to a file (as when two developers want to change the same lines) are handled gracefully; the last committer will have to explicitly accept or reject the changes made by earlier committers.
- For each file version, there is always only one official copy – the one in the repository.

Everyone is allowed to download files from the repositories. This can be done one file at a time, but is more often done by using the command *CVSup*, causing all files on the local machine to be updated to the newest versions. The software product that results from *building* with the newest versions of all files is called the *development version*. Because files are updated continuously, the development version will also be constantly evolving. The aggregation of all files related to the development version is called the *trunk* (or *CURRENT* in FreeBSD). In addition to the trunk, the repository contains *branches* for production releases, making it possible to isolate new development on a file from file versions part of (previous) production releases.

Because it is only superficially tested, there are obvious risks involved when building and running the development version, but nonetheless many people do this frequently. *Developers* do this to test their own changes, not yet committed, with the files from the development version, in order to be sure that no problems arise once the changes are committed. Also, many *users* want to test-drive the newest version, weighing the chance of trying new features and avoiding old bugs over the risks involved in running virtually untested software. For both projects this *community testing* is essential for finding and removing bugs.

Only developers with special privileges, *committers,* are allowed to make changes to the files in the repositories. A developer without these privileges will have to go through a committer in order to get contributions added to the repository.

## Bug-Tracking System

Very much in line with what Raymond (2000) has entitled Linus' Law, "Given enough eyeballs, all bugs are shallow," in both projects everyone can report bugs, and everyone is actively encouraged to do this. A "bug" doesn't have to be an error; rather, it should be understood as a pending work task. FreeBSD uses the term "problem report":

*A FreeBSD problem report (PR) is not necessarily a bug with FreeBSD itself. In some cases it may be reporting a mistake in the documentation ... In other cases it may be a "wish list" item that the submitter would like to see incorporated in to FreeBSD (Support, n.d.).*

Mozilla goes a step further in this direction. All on-going work should be related to one or more bugs (bugs, 2003):

*Not all "bugs" are bugs. Some items in the database are known as Enhancement Requests or Requests for Enhancement (RFE for short). An RFE is a bug whose severity field is set to "enhancement" ... Enter the tasks you're planning to work on as enhancement requests and Bugzilla will help you track them and allow others to see what you plan to work on.*

Also, when committing changes to the repository in Mozilla, the committer must assign the commit to one or more bugs.

Because of the large number of reported bugs[2], both projects rely heavily on central defect-tracking systems, which register all information in relation to every reported bug. Mozilla uses its own Bugzilla system for this; the FreeBSD project uses Gnu Problem Report Management System (GNATS) instead.

These systems make it possible to record and browse information for every bug regarding[3]: *Description* (synopsis, affected product and version, possible fix ...), *state* (open, analyzed, suspended, closed ...), *severity* (critical, trivial, enhancement …), *priority* (high, medium, low), *originator* (person reporting this bug), and *responsible* (person assigned to this bug). It is also possible to associate a bug with a mailing list, causing people on this list to automatically be notified when information regarding the bug is updated. The Bugzilla system has some additional features, most importantly:

- Bugs can be marked as blocking an upcoming release; i.e., they have to be closed before release.
- It is possible to organize bugs hierarchically. For each bug, it is possible to point to other bugs that this bug:
  - Depends on: "sub-bugs" that have to be closed in order for this bug to be closed; or
  - Blocks: "super-bugs" that cannot be closed before this bug is closed.

The hierarchy of bugs in Mozilla leads to "meta bugs," e.g., bug 163993 "tracking those bugs that appear to be of substantially greater interest to the Mozilla community than the 'average' bug."

The two bug-tracking systems can be seen as dynamic requirement specifications, each bug identifying a requirement to a new version of the software. Further, the hierarchical organization of bugs and the option of assigning bugs to releases make Mozilla's Bugzilla system also function as a dynamic work breakdown structure, which, according to Jurison (1999), in a software project is "the basis for all planning activities … it decomposes the project into hierarchically structured well-defined, manageable tasks or activities."

## Tinderboxes

In order to be able to discover errors in newly committed code as fast as possible, Mozilla has around 20 machines running as *tinderboxes*. These are computers with different hardware and operating systems that constantly and automatically are repeating the following cycle:

1. Download newest version from repository.
2. Build.
3. Perform tests.
4. Report results.

The cycle may take from 20 minutes to several hours, depending on the number of tests and the hardware and operating system of the tinderbox. All machines perform a rudimentary run-test of the newly built browser; some machines also run various performance tests, measuring startup time, window open time, memory usage and leak, etc. The results are automatically shown on a number of tinderbox web pages. If the build fails, it is called a *broken-build* situation; this is marked with red color on the tinderbox pages, which also in capital letters shows the message "DO NOT CHECK IN ON RED" (*tinderbox*, n.d.).

Tinderboxes are also used in FreeBSD, but not as heavily as in Mozilla. They run twice a day on three different systems (one for *STABLE*, one for *CURRENT* except PowerPC, and one for *CURRENT* for PowerPC), and the results are shown on a dedicated web page (*FreeBSD tinderbox logs*, n.d.).

## Verification Machines

Supplementing the automatic and continuous tests performed by the tinderboxes, Mozilla follows a rather strict procedure in which the source code in the trunk is tested on a daily basis. The tests follow this scheme, according to Yeh (1999) and our own observations:

- At 8 AM (PST) each working day, the tree *closes,* which means that further updates to the source code are forbidden. Technically it is still possible to update the tree, but doing so is strongly denounced.
- At 8:05, a number of verification machines (12 as of October 2002) begin to download the trunk source code, build it, and test it. These *smoke tests* are similar to the tests performed by the tinderboxes.
- If one of these tests fail and the appointed release engineer is not able to determine and fix the problem, he or she will contact developers who have checked in source code changes since the last successful build, and together they will try to solve the problem; even though the tree is officially closed, changes directly related to solving the problems are allowed.
- If the failed tests relate to one of the "reference platforms (Linux, Macintosh, Win32)" (Yeh, 1999), the tree remains closed until all tests have passed on each of these platforms, or "it is determined that the regressions have a workaround or are not critical features." (Yeh, 1999)
- When all tests are passed successfully, the tree is opened; this is announced in the netscape.public.mozilla.builds newsgroup.

This schedule causes the tree to be closed for check-ins for quite long periods; in the first three weeks of October 2002 (arbitrarily chosen), the tree was closed for periods ranging from one to 10 hours, adding up to a total of 94 hours or almost 20% of the time.

FreeBSD has no verification machines like Mozilla, and does not have a daily routine for opening and closing the tree.

## Mailing Lists and Newsgroups

In addition to the communication taking place through bug reports and commits to the repository, the two projects make heavy use of more informal communication channels, including mailing lists and newsgroups.

Although there are several newsgroups related to discussion of FreeBSD, mailing lists are the recommended communication forums in FreeBSD (*FreeBSD Handbook*, 2003):

*By addressing your questions to the appropriate mailing list you will reach both us and a concentrated FreeBSD audience, invariably assuring a better (or at least faster) response [than by posting to one of the comp.unix.bsd.freebsd.\* groups] .*

There are around 70 of these mailing lists with very different purposes, from announcing job opportunities to discussing the SCSI subsystem. For developers working on *CURRENT*, the mailing list freebsd-current is particularly important, as this is where all announcements of important changes to *CURRENT* will be given. Also, problems in building or running *CURRENT* will be posted to and discussed in this forum; these seem to account for around 75% of the list threads.

Most lists are public and can be both searched and browsed on the FreeBSD website, but several closed lists exist, including lists for the core team security team, release engineer, etc.

Mozilla relies on a number of newsgroups (netscape.public.mozilla.\*, e.g., netscape.public.mozilla.builds), but these don't seem to be as active as FreeBSD's mailing lists. Most of these newsgroups are public, though some are moderated.

## Website

The projects' websites have several important functions. They present both the projects and the products (the software) to the outside world, and they act as portals, making it possible for developers (and everyone else) to locate and download all existing information (project documentation, manuals, and news) related to the projects.

## Balancing Anarchy with Control

The infrastructures described above show only few and minor differences between the two projects. It seems apparent that, for both projects, openness is a very important principle; only very little information seems to be kept secret, as illustrated by the following:

- Everyone can download every version of every file;
- Everyone can monitor each file for who made which changes;
- Test results are free for everyone to see; and
- Nearly all newsgroups and mailing lists can be read by everyone.

But there is also a large amount of openness in contributing to the projects:

- Everyone can report bugs;

- Only few newsgroups and mailing lists are moderated; and
- Once you have commit rights, it is technically possible to update any file in the repository.

The disadvantage of this openness may be the risk that contributions are not always of a sufficient quality: bug reports may be erroneous, newsgroup postings may be misleading or irrelevant, and file changes may introduce new bugs. As an illustrative example, during the week from April 2 to April 9, 2003 (arbitrarily chosen), a total of 828 bugs related to the Mozilla browser changed their resolution status as follows:

- 282 (34%) were marked as *duplicate*, i.e., identical to previously reported bugs;
- 110 (13%) were marked as *invalid*, i.e., not a bug;
- 104 (13%) were marked as *works for me*, i.e., not reproducible; and
- 330 (40%) were marked as *fixed*.

So, during this week, as many as 60% of the resolved bugs appeared to have been mis-reported.

But the projects have taken several measures to reduce the risks associated with poor quality contributions:

- Every change to the repository is logged and reversible; and
- A considerable effort is put into detecting and correcting broken-build situations as quickly as possible.

In the next sections we will see how the projects' work organization and process models also influence the balance between openness and quality control.

## WORK ORGANIZATION

Most of the work in both projects takes place in one-man projects; i.e., individual developers working on bug-fixes or improvements to the source code in the repositories. In effect, anyone can contribute to the projects with bug reports or suggestions for changes; they can do this just once or on a regular basis, and thus it is not obvious whether these contributors should be considered to belong to the organization. But both projects also employ several staff and management functions, i.e., people with more permanent positions, including formal authority and obligations. According to Saers (2003), the FreeBSD project has 18 of these "official hats," covering everything from Public Relations to Standards.

In this section, we will only outline the positions that are most important in relation to our focus on the development models in the two projects.

## Top-Level Management

Both Mozilla and FreeBSD are headed by groups comparable to the board of directors found in traditional organizations. In Mozilla, this group is the *Mozilla.org staff* of 11 persons (*Mozilla Roles and Responsibilities*, 2002): "Mozilla.org staff members provide the overall guidance for the project. This includes the development of Mozilla itself, ... maintaining a development infrastructure, building community, assisting potential new developers and creating overall policies and procedures for the project." You don't apply to join the staff, you are asked to join (*Mozilla Roles and Responsibilities)*: "We anticipate that future staff members will have been Staff Associate members and probably a Module Owner or Activity Owner as well before being asked to join mozilla.org staff."

The FreeBSD project is headed by a 9 person *Core Team (The FreeBSD Core Team*, n.d.): "The FreeBSD core team constitutes the project's 'Board of Directors,' responsible for deciding the project's overall goals and direction as well as managing specific areas of the FreeBSD project landscape." In contrast to the Mozilla.org staff, the FreeBSD Core Team is democratically elected. Election is conducted every second year by and from the active committers (the ones who have committed to the tree within 12 months of the election start date), and there is considerable competition. For example, for the 2002 election, 22 candidates were running for the nine places.

## Release Management

Focusing on the software development process models, the next lower level in the hierarchy of the two projects is the group of people managing the release processes.

In Mozilla, this is a team of 13 *Drivers* that "provide project management for mozilla.org milestone releases." Before milestone releases, all check-ins should be reviewed by one of the drivers, focusing "on the importance of that particular fix to the milestone release" (*Mozilla Roles and Responsibilities*, 2002).

In FreeBSD the *Release Engineering Team* (six persons) is "responsible for setting release deadlines and controlling the release process" (*Committer's Guide*, 2003).

## Module Owners

In the Mozilla project, *module owners* are also part of the formal organizational hierarchy. A module owner is "... someone to whom mozilla.org staff delegates leadership of the module's development ...A module owner's OK is required to check code into that module" (*Mozilla Modules and Module Ownership*, 2003). It can be hard to find a qualified module owner, especially for modules with a low level of activity, and so these modules may for some time be without an owner. Module ownership can be shared among several people.

FreeBSD employs *module maintainers*, in many ways comparable to the Mozilla module owners (*FreeBSD Developers' Handbook*, 2003):

*"The maintainer ... is responsible for fixing bugs and answering problem reports pertaining to that piece of the code ... Changes to directories which have a maintainer defined shall be sent to the maintainer for review before being committed."*

FreeBSD seems, however, to have a more relaxed attitude to module ownership. There is an accepted risk that the maintainer may not be active (*FreeBSD Developers' Handbook*):

*"Only if the maintainer does not respond for an unacceptable period of time, to several emails, will it be acceptable to commit changes without review by the maintainer."*

And it seems to be more like a duty and less like a privilege, compared to Mozilla (*FreeBSD Developers' Handbook*):

*"It is of course not acceptable to add a person or group as maintainer unless they agree to assume this duty. On the other hand it does not have to be a committer and it can easily be a group of people."*

However, it is suggested that you have the changes reviewed by someone else if at all possible.

## Reviewers

In Mozilla, 29 persons are appointed as *super reviewers*; one of these must review most code in-process, i.e., "built with the Mozilla browser/mail/news/editor application suite" (Eich & Baker, 2002), before it is committed.

In FreeBSD, there seems to be no specially appointed reviewers. Usually the maintainer, if present, is expected to review code before check-in. As an alternative, it is recommended to ask for review on the mailing lists.

## Committers

A *committer* is a developer with the right to add or change code in the repository. In both projects, you have to demonstrate your competence first, typically by adding high-quality contributions for some time, before being given CVS write access. Mozilla has a formal bureaucratic procedure for this, involving a formal application, acceptance from a *voucher* (person who already has CVS write access), and written acceptance from three super reviewers (*Getting CVS Write Access to Mozilla*, 2003).

In FreeBSD, the right to commit to the source code repository is given by the Core Team. The team does this on the initiative from one or more committers, but the procedure for this doesn't seem to be as formal as for Mozilla. A new committer will be appointed a *mentor*, i.e., "an experienced committer … responsible for everything you do in the FreeBSD project … answers your questions, reviews your patches …" (Lucas, 2002).

## Contributors

In contrast to the categories above, there are no formal procedures involved in becoming a contributor. A contributor is simply someone who in some way has contributed to the project. You don't "resign" as contributor, you simply stop contributing. As only committers have access rights to the repositories, a committer must always approve the work of a contributor and perform the actual changes to the repository. Historically, more than 1,700 persons have contributed to FreeBSD (*Contributors to FreeBSD*, 2003), and approximately 850 persons to Mozilla (*Our Contributors*, 2003).

A contributor is free to choose the tasks (i.e., bugs) that seem most attractive to work on, and websites and documentation in the two projects actively encourage people to do this. It is also emphasized that you can contribute in many other ways than through programming, e.g., reporting bugs or writing documentation.

## Balancing Anarchy with Control

As Siobhán O'Mahony recently put it (Stark, 2003):

*Several projects … are experimenting with this tension now - 'How much structure can we impose on volunteers?' People are intimately aware of the fact that too much structure will disenfranchise the very people who make the most successful open source projects possible.*

Mozilla and FreeBSD also face this tension, and both have chosen organizations of work that make it easy for new developers to join the projects. No authorization is needed in order to read documents and source code, read and post to most mailing lists and newsgroups, submit bugs, or suggest changes to the repositories.

But both projects also feature a "core" of more formal work organization, including several levels of hierarchy, a number of staff functions, and a certain amount of bureaucracy. As a new developer, you have to "work your way up" in the hierarchy; for example, before becoming a committer or a maintainer, you first have to show your interest and qualifications by working for some time as a contributor. The degree of formalization seems highest in Mozilla, with more detailed and extensive rules in relation to review, module ownership, and obtaining committer status. An important difference between the two organizations can also be found in their top-level management: the Mozilla.org staff is self-elective, but the

FreeBSD Core Team is democratically elected by and from the committers. These differences may be related to the Mozilla project's tight connection with Netscape; if most developers are employed by the same company, it is probably easier to mandate strict methods and procedures, and less attractive to perform democratic election of the top-level management.

# RELEASE LEVEL PROCESS MODELS

After having described some of the most important structural elements in the two projects, namely the technological infrastructure and the organizational hierarchy, we will now focus on the software process models employed in Mozilla and FreeBSD. Even though these models are different from, and in some ways more relaxed than, traditional process models, we will see that the projects are in no way just letting things go. They put a lot of effort into carefully planning and controlling the development processes.

In accordance with Sommerville (2001), we will define a software process as a "set of activities and associated results which lead to the production of a software product," and a software process model as an "abstract representation of a software process." We will focus on the guidelines, routines, rules, etc., that the projects have for software development, but also supplement our description with examples from actual software processes.

In our case studies of FreeBSD and Mozilla, we have been able to identify process models on two different levels: a *release level* and a *contribution level*. The release level model is used to control and coordinate the process of producing a new software release on a general level, not describing the work of the individual developers. The contribution level model, in contrast, is the rather sketchy model for the development process leading to a new contribution to the software repository.

We will first look at the process models for the "release projects," i.e., the projects of producing a new release of the two software products.

## Release Model in Mozilla

In Mozilla, these release projects are expected to go through a set of distinct phases, as illustrated in Figure 2.

1. A six-week period, where even risky changes to the source code are allowed, leading to an Alpha version of the coming release.
2. A four-week period of stabilization, leading to the Beta version.
3. A three-week period in which only "stop-ship" bugs are found and fixed. From the start of this period, all changes must be accepted by *drivers*. After two weeks, a release branch is made, separating final work on the coming release from work on the trunk, and effectively marking the start of the next release

*Figure 2: Mozilla's Spring 2003 milestone schedule (Eich & Hyatt, 2003).*

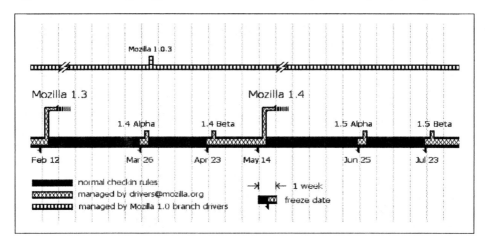

project. If *drivers* don't think the release is of satisfactory quality, this branch may be postponed.
4. It is possible to make changes to a release even after the release date, but this seldom happens. When it does happen, it is most often when a bug-fix from the trunk leads to a similar bug-fix in the release branch (called *merging*). These changes must still be accepted by *drivers*.

Figure 2 indicates that Mozilla has only two long-lived branches: the trunk (also called *HEAD*) and the 1.0 branch. According to the Mozilla 1.0 manifesto (Eich, 2002), the idea behind the 1.0 branch is to provide a "stable, long-lived, *branded* branch," preferably with near-zero bugs. But, effectively, not much work is going on in this branch. At present (April 2003), no commits have been made to this branch during the last three months, so all work is now focused on the trunk.

## Release Model in FreeBSD

In FreeBSD, major releases (e.g., going from 4.x to 5.0) are treated differently from minor releases (e.g., going from 5.0 to 5.1). Major releases are expected to go through the following timetable, also illustrated in Figure 3.

1. Some months before the expected release date the head branch (*CURRENT*) enters a state of *slush* or *feature freeze*. From this point, all significant new features should be discussed with the release engineers before being committed.
2. One month later, *CURRENT* enters a state of *code freeze*. From here on, every commit must be accepted by the release engineers.

*Figure 3: Model for major releases in FreeBSD.*

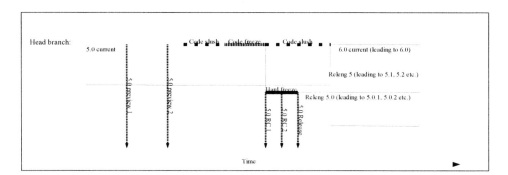

3. After one more month, several important things happen, almost simultaneously:
   a. A new branch, eventually leading to the new major release, is made from *CURRENT*. Source code in this branch is frozen even harder than before.
   b. The first release candidate is made as a snapshot of *CURRENT*.
   c. The source code in *CURRENT* is thawed, but is still in *slush* state where significant changes must be accepted by the release engineers.
4. During the following month, several more release candidates may be released.
5. Finally, the major release is made as a snapshot of the release branch. After this release, the source code in the release branch will be somewhat thawed, but still under control from the release engineers.
6. Eventually, a new *RELENG* branch is made from *CURRENT*. This will cause the source code in *CURRENT* to become unfrozen and on its way to the next major release. The new branch, also known as *STABLE* will lead to the next minor releases.

Minor releases follow a similar, but simpler timetable.

## Comparing the Two Release Models

The project models in the two projects are very similar. One of the key principles seems to be that new releases should be made as simple snapshots of development branches. But, in order to do this, it is necessary to put several restrictions on changes to the development branch in the weeks or months before the release.

During its lifetime, the Mozilla-project has only made one major release (1.0), but produces minor releases on a regular basis. Each minor release is given its own branch, but development in this branch is sporadic to non-existent after the release.

This is in contrast to FreeBSD, which regularly releases both major and minor (and even sub-minor, e.g., 4.6.2) releases, and where development in the release branches will go on for years, merging bug-fixes from *CURRENT* back into the release branches. The difference between Mozilla and FreeBSD in this respect is almost certainly due to the different products. Most users will be able to update their browsers to a new version very easily and with very little risk, but updating the operating system on a crucial server is a major, risk-prone task. Therefore, it is important for system administrators to be able to make, e.g., security updates also to older versions of the operating system.

## Balancing Anarchy with Control

The release models followed in the two projects are based on several years' experience and try to establish an optimal balance between opposing interests. Tight control of the source code in the trunk (long and hard freeze periods) will severely slow down new development and will probably also make contributors lose motivation, because:

- According to Jørgensen (2001), a key motivational factor for contributors is to quickly see the results of their work; a frozen tree will inhibit this.
- When the trunk is finally thawed, there will be a large, accumulated need for commits to the trunk. This will make integration more difficult because many people will be committing many and large changes in a very short time, thus resembling "big bang integration," generally deprecated in software engineering (Pressman, 1992).

On the other hand, very relaxed control of the source code (short or no freeze periods) may lead to bug-filled releases, because new and scarcely tested code can be added to the trunk just before the release.

One way to avoid long freeze periods on the trunk, while still being able to stabilize and test new releases, would be to let releases branch off early. But, according to Walrad and Strom (2002), this "branch-by-release model" has several drawbacks, including "unnecessary complexity in managing post-release code fixes and unnecessarily orphaning changes in progress;" drawbacks that may be especially important in F/OSS projects.

It is important to note that the release plans are the only plans in the two projects. They define *when* the different releases shall be made and, to some degree, the bug-tracking systems define *what* shall be part of each release. But there are no plans on more detailed levels, no timetable for the various contributions that eventually will lead to the next release.

# CONTRIBUTION LEVEL PROCESS MODELS

Each contribution to the repository can be seen as the result of a (probably small) software engineering project. The requirements for this project will often come from the bug-tracking system, but developers can also specify their own requirements—"scratching their personal itch" (Raymond, 2000). Most developers seem to work more or less alone (Jørgensen, 2001), but each developer may be engaged in several projects. At any one time, a large number of these "contribution projects" will go on in paraliel, as can be seen by the more than 50 commits being made each day to each repository. In this section we will describe the process models used in Mozilla and FreeBSD for these contribution projects.

## Contribution Model in Mozilla

The description following is based on information from the Mozilla website, most importantly *The Seamonkey Engineering Bible* (2003)[4] introduced as "Here are the things you must do to be an effective contributor and to keep other team members productive on the project."

1. Assignment – choosing the task to work on. There is a general recommendation to "stay focused on the most important problems," but no explicit, given task assignments.
2. Development – producing the source code changes. There are no explicit guidelines for this process, except for a number of required pre-conditions before actually committing the source code changes to the repository:
   a. Accept may be needed by module owners, super-reviewers, drivers, etc.; and
   b. Coding and documentation standards should be respected.
3. Local test – testing the changes with the newest version of the branch, to which the changes are going to be committed. The required minimum of testing is very simple. Apart from being able to build, it is only required to test the most basic functionality: visiting two websites, changing preferences, and sending and reading an email message. Further, it is recommended to also test ftp and be able to view the HTML source. The limited amount of testing is explained by Duddi (1999): "These tests need to be very short so that developers don't flinch at the list ... the ultimate goal is to prevent the tree from being closed for verification for long periods."
4. Check-in – committing the source code changes to the repository. Except for certain "emergency situations," commits should only be made when the tree is open and green.
5. Tinderbox test – being highly available (*on the hook*) and helping to fix eventual errors encountered in the tinderbox tests. When on the hook, a developer should be "highly available until the tree has a full cycle of green after you[r] check-ins are completed." After the first, successful tinderbox build test

("green tree"), the developer is only expected to be "semi-available" until next morning's verification build.
6. Verification test – again be highly available and help to fix eventual problems encountered in the daily verification tests.
7. Community test – after a successful verification test, the developer is expected to still "watch talkback data reports … follow the newsgroup for patches and suggestions to make your module better."

## Contribution Model in FreeBSD

The main aspects of the contribution model in FreeBSD are very similar to Mozilla's, but in some ways FreeBSD takes a more relaxed approach:

- The rules regarding review are more relaxed. There are no super-reviewers in FreeBSD, only requirements to "discuss any significant change *before* committing [author's emphasis]", "when in doubt, ask for review," and "respect existing maintainers" (The FreeBSD Committers' Big List of Rules, n.d.).
- There is no daily closing of the tree due to failed tinderbox or verification testing. Tinderbox testing is only performed twice each day, and there are no separate verification tests.
- There are no formal rules regarding being available after a commit. Performing "drive-by commits" is, however, considered very bad behavior.

## Comparing the Two Contribution Models

Supplementing the models described above, both Mozilla and FreeBSD have a number of general guidelines and recommendations for developers, covering aspects ranging from rules of good conduct, e.g., "Don't fight in public … it looks bad" (*The FreeBSD Committers' Big List of Rules*, n.d.) to coding style guides.

As can be seen, neither Mozilla nor FreeBSD has any rules regarding the method or model the developer should follow in his or her contribution project. There are a number of general guidelines and recommendations, but basically the only control that can be exercised are the rules for committing to the repository. For example, the Mozilla project can require all commits to be assigned to a bug, but the project has no influence on *when* the developer chooses to assign his or her contribution to a bug; it may be at the very start of the contribution project or just before committing.

Neither project requires that design should precede coding, in the sense of writing, discussing, or approving written design documents prior to coding. Indeed, in practice, for the individual change there is typically no design document. Thirty-one of the 72 committers surveyed in FreeBSD responded that they had never distributed a design document (defined as a separate document, distinct from a source file).

## Balancing Anarchy with Control

In traditional software engineering, the quality assurance process involves defining or selecting standards for the software development process and the software product (Sommerville, 2001), but several factors limit the number and rigidity of the standards Mozilla and FreeBSD can use. The use of rigid process standards would be difficult or impossible:

- Many projects may be unknown to everyone but the developer himself, until the result is finally reviewed and committed.
- The large number of parallel projects would make it a huge task to tightly control each one.
- If the bureaucratic procedures involved in making contributions become too complicated and time-consuming, there will be a significant risk of losing developers, especially if they, as Raymond (2000) puts it, are "self-directed egoists."

Reducing the potential problem of lacking or relaxed standards may be the fact that most contribution projects are performed by very small groups (one to three persons) (Jørgensen, 2001), as this may make "the quality of the development team … more important than the development process used" (Sommerville, 2001). But a complete lack of standards and procedures would also not be convenient:

- Too few rules might lead to poor quality code, in the long run making both users and other developers lose interest and motivation.
- Too little coordination between different contribution projects might lead to parallel, and eventually wasted, work (two developers working on the same bug), also making developers lose motivation.

Both projects have chosen a mixture of quite relaxed and quite strict standards. As examples of the relaxed standards, we will mention:

- No requirement for separate design documentation;
- No requirement to follow a specific model or method;
- Developers can choose their own task; and
- Only very basic tests are required before committing.

On the other hand, the projects also have several, highly enforced rules, e.g.:

- All contributions must go to the trunk: "merge to the tip, merge to the tip, merge to the tip, and when there are lots of people checking in code, merge to the tip" (*The Seamonkey Engineering Bible*, 2003);
- Mozilla in particular has detailed rules for how a "broken-build" situation shall be handled, including both detailed time schedules and responsibilities.

Instead of relying on a large number of strict quality standards, both projects have chosen to focus on relatively few, but highly enforced rules for how contributions shall be made, and for how to handle "broken-build" situations in the repository. As a substitution for the missing standards, both projects seem to rely more on the informal communication and coordination taking place in mailing lists, newsgroups, and chat forums, and in this way develop a "quality culture" where all developers are committed to achieving a high level of product quality. The openness of F/OSS projects may contribute to this, as illustrated by Jørgensen's study (2001) where 86% of the developers agreed that "knowing that my contributions may be read by highly competent developers has encouraged me to improve my coding skill," and one developer added, "Embarrassment is a powerful thing."

# CONCLUSION

In this section, we will summarize important aspects of how the two projects balance anarchy (avoiding centralized modes of governance and formal organization mechanisms, preferring technical autonomy, self-management, and freedom) with control (planning, organizing, directing, and controlling company resources to complete specific goals and objectives).

As seen in this chapter both projects have pretty advanced, well-functioning and well-supported technical infrastructures, including:

- CVS controlled repositories;
- Websites with large amounts of information being continuously updated with contributions and status information;
- Newsgroups and mailing lists; and
- A number of tinderboxes and verification machines (Mozilla).

It is obvious that developing and maintaining this infrastructure requires some amount of organization and control over the organization's resources.

We have also seen that both projects not only rely on self-management, but also have cores of formal hierarchy and bureaucracy, including top-level management, release management, module owners, super-reviewers (Mozilla), and several staff functions. Finding a place in this organization is not up to the individual developer: only after showing your worth as a contributor will you be promoted to committer, and the Mozilla organization also formally appoints super-reviewers and members of *Mozilla.org staff*. FreeBSD seems to be somewhat less hierarchical than Mozilla, with fewer formal appointments and a democratically elected management group, the *Core Team*.

Around these formal cores, both projects have peripheries with only a few and simple control mechanisms. Most information is public (source code, test results, bug reports, discussions on newsgroups and mailing lists) and no or little formal authorization is needed to report bugs or post to newsgroups and mailing lists. Also, there appears to be only quite limited control of the contribution projects: developers can choose whichever task (bug) they want to work on; there are only a few requirements to follow specific models or methods; design documents are not mandated; and only rudimentary testing is necessary before changes are committed. Control of these contribution projects is mostly limited to the step of committing: a contribution must be reviewed before being committed (especially in Mozilla); only authorized developers can commit; developers are required to be "on the hook" and help in solving potential problems with their contributions; and access to commit is restricted during slush or freeze periods (before releases) and in broken-build situations, where the tree is closed or red.

Both projects plan and control the processes leading to new software releases; in particular, FreeBSD has detailed plans for new releases, including well-documented models, time schedules, and task assignments. Planning and controlling these processes imply important decisions regarding duration and strictness of freeze periods, and if and when to make new branches in the repository.

Both projects seem to rely more on requesting contributors to follow "rules of good conduct," (often documented on the websites) than technical mechanisms for control. For example, once you have commit privileges, it appears to be technically possible to commit changes to any file at any time (even when the tree is closed or red) without previous testing and review, even though doing this is, of course, strongly discouraged.

Finding the optimal balance between anarchy and control is a continuous process. Recently, long-time members of Mozilla.org staff, Brendan Eich and David Hyatt (2003), announced a major shift in strategy for the Mozilla project, including changes in the development model. They explain the reason for the change as:

*Simply put: great applications cannot be managed as common land, with whoever is most motivated in a particular area, or just the last to check in, determining the piecewise look and feel of the application. ... great software is originated by one or a few hackers building up and leading a larger team of people who test, clean up, extend, and grow to join or replace the first few.*

What they propose is to:

*Continue the move away from an ownership model involving a large cloud of hackers with unlimited CVS access, to a model ... of vigorously defended modules with strong leadership and clear delegation ... The faux-egalitarian model of CVS access and pan-tree hacking that evolved from the earliest days of Mozilla is coming to an end.*

So, in Mozilla, the pendulum now seems to be shifting away from anarchy and towards control.

## ACKNOWLEDGMENTS

We wish to thank Dag-Erling Smørgrav, Niklas Saers, Stefan Koch, and the anonymous reviewers for their valuable comments and suggestions for improvements to previous versions of this chapter.

## REFERENCES

*Anarchism* (2003). Britannica Concise Encyclopedia. Retrieved July 18, 2003, from: http://search.eb.com/ebc/article?eu=380585.

Asklund, U. & Bendix, L. (2002). A study of configuration management in open source software projects. *IEE Proceedings-Software, 149*(1), 40-46.

*bugs.* (2003). Retrieved April 8, 2003, from: http://www.mozilla.org/bugs/.

*Committer's guide.* (2003). Retrieved April 21, 2003, from: http://www.freebsd.org/doc/en/articles/committers-guide/.

*Contributors to FreeBSD.* (2003). Retrieved May 9, 2003, from: http://www.freebsd.org/doc/en/articles/contributors/article.html.

Duddi, S. (1999). *Pre-checkin tests*. Retrieved May 1, 2003, from: http://www.mozilla.org/quality/precheckin-tests.html.

Eich, B. (2002). *Mozilla 1.0 manifesto*. Retrieved Nov. 15, 2002, from: http://www.mozilla.org/roadmap/mozilla-1.0.html.

Eich, B. & Baker, M. (2002). *Mozilla "super-review."* Retrieved Dec. 1, 2002, from: http://www.mozilla.org/hacking/reviewers.html.

Eich, B. & Hyatt, D. (2003). *Mozilla development roadmap*. Retrieved April 23, 2003, from: http://www.mozilla.org/roadmap.html.

Feller, J. & Fitzgerald, B. (2002). *Understanding open source software development*. London: Pearson Education.

Fogel, K. & Bar, M. (2001). *Open source development with CVS*. Scottsdale, AZ: Coriolis Group.

Free Software Foundation. (2003). *The free software definition*. Retrieved July 18, 2003, from: http://www.fsf.org/philosophy/free-sw.html.

*The FreeBSD committers' big list of rules*. Retrieved Dec. 1, 2002, from: http://www.freebsd.org/doc/en/articles/committers-guide/rules.html.

*The FreeBSD core team*. Retrieved April 21, 2003, from: http://www.freebsd.org/doc/en/articles/contributors/staff-core.html.

*FreeBSD developers' handbook.* (2003). Retrieved April 21, 2003, from: http://www.freebsd.org/doc/en/books/developers-handbook/index.html.

*FreeBSD handbook.* (2003). Retrieved April 11, 2003, from: http://www.freebsd.org/doc/en/books/handbook/index.html.

*FreeBSD tinderbox logs*. Retrieved May 1, 2003, from: http://www.rtp.freebsd.org/~des/.

*Getting CVS write access to Mozilla.* (2003). Retrieved April 22, 2003, from: http://www.mozilla.org/hacking/getting-cvs-write-access.html.

Godfrey, M. W. & Tu, Q. (2000). *Evolution in open source software: A case study.* Paper presented at the International Conference on Software Maintenance (ICSM'00), San José, California.

Hamerly, J., Paquin, T., & Walton, S. (1999). Freeing the source: The story of Mozilla. In M. Stone (Ed.), *Open sources: Voices from the open source revolution.* Sebastopol, CA: O'Reilly & Associates.

Healy, K. & Schussman, A. (2003). *The ecology of open-source software development.* Retrieved May 19, 2003, from: http://opensource.mit.edu/papers/healy-schussman.pdf.

Henkel, J. (2003). *Open source software from commercial firms – Tools, complements, and collective invention.* Unpublished manuscript.

Holck, J. & Jørgensen, N. (2003a). *Continuous integration as a means of coordination: A case study of two open source projects.* Paper presented at the 12th International Conference on Information Systems Development (ISD 2003), Melbourne, Australia.

Holck, J., & Jørgensen, N. (2003b). *Overloading the development branch? A view of motivation and incremental development in FreeBSD.* Position paper presented at the 4th International Conference on Extreme Programming and Agile Methods (XP 2003), Genoa, Italy.

Jørgensen, N. (2001). Putting it all in the trunk: Incremental software development in the FreeBSD open source project. *Information Systems Journal, 11*(4), 321-336.

Jurison, J. (1999). Software project management: The manager's view. *Communications of the Association for Information Systems, 2*(17).

Kerzner, H. (1989). *Project management* (3rd ed.). Newbury Park, CA: Sage Publications.

Lakhani, K. R. & von Hippel, E. (2003). How open source software works: "Free" user-to-user assistance. *Research Policy, 32*(6), 923-943.

Lucas, M. (2002). *How to become a FreeBSD committer.* Retrieved April 22, 2003, from: http://www.onlamp.com/lpt/a/1492.

Mockus, A., Fielding, R. T., & Herbsleb, J. D. (2002). Two case studies of open source software development: Apache and Mozilla. *ACM Transactions on Software Engineering and Methodology, 11*(3), 309-346.

*Mozilla modules and module ownership.* (2003). Retrieved April 21, 2003, from: http://www.mozilla.org/hacking/module-ownership.html.

*Mozilla release FAQ.* (2002). Retrieved April 7, 2003, from: http://www.mozilla.org/docs/mozilla-faq.html.

*Mozilla roles and responsibilities.*(2002). Retrieved April 18, 2003, from: http://www.mozilla.org/about/roles.html.

O'Mahony, S. (2003). Non-profit foundations and their role in community-firm software collaboration. In J. Feller, B. Fitzgerald, S. Hissam, & K. Lakhani (Eds.), *Making sense of the bazaar: Perspectives on open source and free software*, Sebastopol, CA: O'Reilly.

Open Source Initiative (2003). *The open source definition*. Retrieved July 18, 2003, from: http://opensource.org/docs/definition.php.

*Our contributors.* (2003). Retrieved May 9, 2003, from: http://www.mozilla.org/credits/.

Paulk, M. C., Curtis, B., Chrissis, M. B., & Weber, C. V. (1993). Capability maturity model, version 1.1. *IEEE Software, 10*(4), 18-27.

Pressman, R. S. (1992). *Software engineering - A practitioner's approach* (3rd international ed.). New York: McGraw-Hill.

Raymond, E. S. (2000). *The cathedral and the bazaar, v. 3*. Retrieved April 8, 2003, from: http://www.catb.org/~esr/writings/cathedral-bazaar/cathedral-bazaar/.

Saers, N. (2003). *A project model for the FreeBSD project*. Retrieved May 5, 2003, from: http://niklas.saers.com/freebsd-model/freebsd-model.html.

*The Seamonkey engineering bible.* (2003). Retrieved April 21, 2003, from: http://www.mozilla.org/projects/seamonkey/rules/bible.html.

Sommerville, I. (2001). *Software engineering* (6th ed.). Harlow, UK: Addison-Wesley.

Stark, M. (2003). The organizational model for open source. *HBS Working Knowledge, July 7*. Retrieved August 15, 2003, from: http://workingknowledge.hbs.edu/pubitem.jhtml?id=3582&t=technology.

*Support.* Retrieved April 8, 2003, from: http://www.freebsd.org/support.html.

*Supported platforms.*(2003). Retrieved April 7, 2003, from: http://www.freebsd.org/platforms/.

*tinderbox.* Retrieved April 22, 2003, from: http://tinderbox.mozilla.org/showbuilds.cgi?tree=SeaMonkey.

Walrad, C. & Strom, D. (2002). The importance of branching models in SCM. *IEEE Computer, 35*(9), 31-38.

Yeh, C. (1999). *Mozilla tree verification process*. Retrieved April 11, 2003, from: http://www.mozilla.org/build/verification.html.

Yin, R. K. (1998). *Case study research: Design and methods*. Newbury Park, CA: Sage Publications.

# ENDNOTES

[1] Free/Open Source Software. We will use this term to cover both Free Software as defined by the Free Software Foundation (2003), and Open Source Software as defined by the Open Source Initiative (2003).

[2] April 8, 2003: 17,811 active (i.e., new, assigned, or reopened) bugs related to the Mozilla browser trunk. 2,747 active (i.e., open, analyzed, feedback, patched, or suspended) bugs for all FreeBSD versions, including obsolete releases.

[3] The two systems use slightly different terms and categories.

[4] Seamonkey is the codename used in Mozilla for the current Mozilla browser.

## Chapter II

# Analyzing the Anatomy of GNU/Linux Distributions:
## Methodology and Case Studies (Red Hat and Debian)

Jesús M. González-Barahona, Universidad Rey Juan Carlos, Spain

Gregorio Robles, Universidad Rey Juan Carlos, Spain

Miguel Ortuño-Pérez, Universidad Rey Juan Carlos, Spain

Luis Rodero-Merino, Universidad Rey Juan Carlos, Spain

José Centeno-González, Universidad Rey Juan Carlos, Spain

Vicente Matellán-Olivera, Universidad Rey Juan Carlos, Spain

Eva Castro-Barbero, Universidad Rey Juan Carlos, Spain

Pedro de-las-Heras-Quirós, Universidad Rey Juan Carlos, Spain

## ABSTRACT

*GNU/Linux distributions are probably the largest coordinated pieces of software ever put together. Each one is in some sense a snapshot of a large fraction of the libre software development landscape at the time of the release and, therefore, its study is important to understand the appearance of that landscape. They are also the working proof of the possibility of releasing reliable software systems in the range of 50-100 millions of lines of code, even when the components of such systems are built by hundreds of independent groups of developers, with no formal connection to the group releasing the whole system. In this chapter, we provide some quantitative information about the software included in two such distributions: Red Hat and Debian. Differences in policy and organization of both distributions will show up in the results, but some common patterns will also arise. For instance, both are doubling their size every two years, and both present similar patterns in programming language usage and package size distributions. All in all, this study pretends to show how GNU/Linux distributions are with respect to their source code, and how they evolve over time. A methodology of how to make comparable and automated studies on this kind of distributions is also presented.*

## INTRODUCTION

Libre software[1] provides software engineering with a unique opportunity to make detailed characterizations of software projects that can be complete, detailed, and reproducible, since the source code is available for anyone to read. This makes it possible to build complete models based on public and repeatable studies. Based on this idea, it seems reasonable to collect data from libre software projects, to start building up a castle of numbers that can later be used to sustain theories about how libre software is developed.

In this respect, we have found GNU/Linux[2] distributions to be a perfect example of what to study. During the second half of the 1990s, GNU/Linux distributions evolved and grew, to the point that at the beginning of the 2000s they include the most comprehensive, coordinated compilations of libre software. Therefore, when we study the most representative distributions, we are in fact analyzing a very important, and representative, subset of the mature libre software available at the time of the release of such distributions. Answering questions like which languages are more usual in these distributions, or how is the mean package size evolving, tells us about how the libre software community is working, and may help us in making predictions for the future (for instance, "when, if ever, will C++ surpass C as the most popular language in libre software distributions?").

What is more important in terms of libre software engineering is the huge size of these distributions that makes them the state-of-the-art in terms of management of software aggregates (libre or not). It is really difficult to find coordinated collections of software of the size usual in GNU/Linux distributions, with complex interdependences, composed by the results of hundreds of libre software projects (sometimes coordinated by volunteers, sometimes by companies, or, in many cases, by a mixture of both), which, when delivered, satisfy the requirements of literally millions of users worldwide. Understanding with some detail how these distributions are and how they evolve may help us to understand how this delivery of 50-100 millions of lines of source code (MLSOC) is possible, and maybe to determine whether the growth seen during recent years is going to be maintainable. Lessons that could be useful in other very large software systems (libre or not) could also be learned.

Therefore, when we were developing some methodological approaches to the problem of studying libre software projects, it seemed natural to study some of the most representative GNU/Linux distributions and their evolution over time. A lot has been said about Linux, for example, the huge amount of programs that you get when you get a Linux CD-set, the extraordinary cumulative effort of thousands of developers, etc. But we wanted to go a step further—to get actual figures that can help us to talk in more precise terms. How large are GNU/Linux distributions, and which packages do they include? Which languages are used for their development? How large is the estimated effort needed to develop all the included software? How do all those data evolve over time, and from distribution to distribution? In other words, the question we wanted to answer is "How are GNU/Linux distributions in terms of source code analysis?"

While doing such an analyses, we also determined a methodology of how to perform it, which can be used to make comparative studies. It is based on the measurement of the source lines of code (SLOC) of the files in the distribution being studied, for several releases, and the data mining on those measurements looking for the size of the packages, the relative use of several programming languages, Constructive Cost Model (COCOMO) estimations, etc. In this chapter, we present such a methodology, complete with some case studies of its application to several releases of two of the most representative GNU/Linux distributions, Red Hat Linux and Debian GNU/Linux.

Other work has previously been done on related topics. David A. Wheeler (2000) was the first to make public a detailed study of a GNU/Linux distribution, Red Hat Linux 6.2 (later completed with the study of release 7.1 [Wheeler, 2001]), while González-Barahona et al. (2001; 2003) have studied two releases of Debian. Most of the details of the methodology proposed in this chapter can be found in those studies, although they lack most of the evolutionary approach and the formalization of the methodology itself. The data presented here, related to the releases studied in those papers, has been recalculated (to use exactly the same methods in all analysis), but obviously is based on them.

Other authors have also studied the details of some libre software projects. For instance, Mockus, Fielding, and Herbsleb (2002) studied Mozilla and Apache; Koch and Schneider (2000) studied GNOME; Schach, Jin, Wright, Heller, and Offut (2002) the Linux kernel; and Godfrey and Tu (2000) looked at the evolution of libre software projects over time. However, all of them refer to isolated libre software projects, in the range of 1-5 million SLOC. On the contrary, our study is focused on collections of software that include the products of hundreds (or thousands) of libre software projects, with an aggregated size in the tens of millions of SLOC.

The organization of the rest of the chapter is as follows. The next section describes the fundamentals of Red Hat and Debian distributions. The section

"Counting Lines of Code" shows the results of counting the code of the studied releases as a whole, while the following section does the same but for individual packages, and the section "Counting Languages" for programming languages. The section "Estimations of Effort" offers some estimations for the releases, using the COCOMO model, and the section "Comparison with other Systems" compares the results obtained in our studies to other systems. Then, the chaper presents in detail the methodology used for this study and ends with a discussion of some conclusions and possible new lines of research.

# DEBIAN AND RED HAT: TWO KINDS OF GNU/LINUX DISTRIBUTIONS

Red Hat Linux was one of the first commercial GNU/Linux distributions. Today, it is probably the most well known, and for sure is the one considered as the "canonical" among the commercial ones. On the other hand, Debian GNU/Linux was one of the first GNU/Linux distributions put together by a group of volunteers, and today it is probably the most widely known "non-commercial" distribution. Both represent very different ways of collecting the software that Linux users want. Both have very different stories, development models, technical details, goals, and funding methods. Because of this, it seems reasonable to study at least both cases when trying to show how GNU/Linux distributions are.

Despite their differences, both Debian GNU/Linux and Red Hat Linux are distributions in the strict sense of the term; the work done by Red Hat and Debian developers is mainly related to integration tasks and not to software development. Of course, Red Hat and Debian may have software developers (on their staff in the former case or among its contributors in the latter), but this is only secondary for the goals of the distributions and, of course, for our study. We assume that the work done by Red Hat and Debian as integrators was just to take the source code packages (usually the files released by the software authors themselves) and package them in a way that fulfills certain criteria (both technical and organizational). The output of this process is a distribution—a set of packages conveniently organized that will enable the user to install, uninstall, and update them easily.

Distribution makers are also responsible for quality assurance, a very important matter considering that many libre software projects are lead by volunteers (Michlmayr & Hill, 2003). In this respect, they are accountable to their users for the stability and security of the resulting distribution.

Although Debian and Red Hat are both distribution makers and share the above characteristics, their respective goals and policies are very different. For instance, while Red Hat Inc. takes into account marketing parameters to decide what enters (and in what condition) its distribution, the Debian Project is proud of taking into account mainly technical excellence. In this respect, while Red Hat distributions are announced in advance and released on a fairly regular basis (usually around six

*Table 1: Versions of several packages in Red Hat 6.2, Debian 2.2 and Red Hat 7.1.*

| Package | Version in Red Hat 6.2 | Version in Debian 2.2 | Version in Red Hat 7.1 |
|---|---|---|---|
| Linux | 2.2.14 | 2.2.19.1 | 2.4.2 |
| xfree86 | 3.3.6 | 3.3.6 | 4.0.3 |
| Gdb | 1991004 | 1990928 | 20010316 |
| Python | 1.5.2 | 1.5.2 | 1.5.2 |
| Perl | 5.005 | 5.005 | 5.6.0 |
| gnome-libs | 1.0.55 | 1.0.56 | 1.2.8 |
| Apache | 1.3.12 | 1.3.9 | 1.3.19 |
| Glibc | 2.1.3 | 2.1.3 | 2.2.2 |

months), Debian delivers a release "when it is ready" (which means, among other things, that critical bugs are below a very low threshold) and announces no explicit plans about dates for future releases.

From the point of view of their contents, both distributions are also different. As it will become clear later, Debian distributions are larger (in fact, much larger), both in number of packages and in total number of lines of source code. This is reasonable if we consider two factors:

- First is that while Red Hat has to make economic calculations to decide the effort devoted to a new distribution (which is, of course, dependent on the number of packages), Debian counts on volunteers who make no such calculus. The limiting factor in Debian distributions is not economical resources, but developer's (volunteer) time and coordination possibilities.
- Second is that Red Hat is targeted to a certain kind of users, and it shows little interest in maintaining clearly minority pieces of software, while in Debian, it is usually enough that a given package interests one skilled developer to enter the distribution.

With respect to the policy on what versions of packages are included, Debian and Red Hat also display some differences. As a rule of thumb, Debian usually includes older releases of packages in its releases. In this respect, it is very interesting to compare Red Hat 6.2, Debian 2.2, and Red Hat 7.1 (see the versions of some packages included in these distributions in Table 1). The interval between both Red Hat releases is about one year, while Debian 2.2 was released in the middle of this interval. Interestingly enough, many of the Red Hat 6.2 packages match the version in Debian 2.2 (released about five months later). There are even some cases where Debian 2.2 contains versions older than those in Red Hat 6.2. An explanation for this situation should be researched with detail, but one can argue that it could be related to the different nature of both distributions[3]; Red Hat being more commercially-oriented has a bigger motivation for having the latest of the latest, while the pressure in Debian for such a policy is not that high. In fact, the Debian release process includes a phase of freezing. During its duration, no new versions enter the pool of packages that will finally be the stable release; only bug fixes to packages already in the pool are allowed. This phase may take several months, which means that newer versions of the packages will not make it into the final release.

All in all, it could be said safely that in a spectrum of GNU/Linux distributions Red Hat is on one extreme (the commercial one) while Debian is close to the other (the one of the independent projects). In the following subsection the reader can find some more details about both distributions.

## Red Hat Linux and Red Hat Software, Inc.

Red Hat Software, Inc. was founded in 1994 by Bob Young and Marc Ewing. Their first goal was to compile and market a GNU/Linux distribution, which was named (and still is) Red Hat Linux (Young, 1999). It was basically a packaged version of what was available on the Net at that time, including documentation and support. During the Summer of 1995, Red Hat Linux 1.0 saw the light. Some months later, during Fall 1995, version 2.0 was released, including the Red Hat Package Manager (RPM) technology that later became a de facto standard for GNU/Linux packages[4]. In 1998, Red Hat released version 5.2, which was probably the first GNU/Linux distribution that reached the mass market. For a complete history of Red Hat Linux releases, see Smoogen (2003).

Before the RPM system came into existence, almost any GNU/Linux distribution could be installed through a menu-driven procedure, but modifications and, in particular, additions of software packages after the first installation were not easy. RPM went a step beyond the state of the art by providing users with "package management" (Bailey, 1998), which made it simple to remove, install, or upgrade any of the software packages available in the distribution. The RPM is still the most used package managing system among GNU/Linux distributions. The statistics found at DistroWatch (www.distrowatch.com), a site containing information about the best known distributions, show that in May 2003 a majority of its 118 recorded databases used the RPM, exactly 65 (55%). On the other hand, the Debian package manager (known as deb) is only used by 16 distributions, which make up only 14% of the total.

However, Red Hat Software, Inc. is not only known for its software distribution based on Linux. In August 1999, Red Hat went public, and its shares had the eighth biggest first-day gain in Wall Street history. In 2003, the value of Red Hat's shares is about one-hundredth of the peak value achieved before the dot-com crash, but the successful start in the stock market made Red Hat front-page news in journals and magazines not directly related to the computer world. In any case, Red Hat has managed to overcome the problems that other companies face in the libre software business, and it announced "black numbers" for the first time in its history for the last quarter of the year 2002.

## Debian GNU/Linux and the Debian Project

Debian is a free (libre) operating system, which currently uses the Linux kernel to put together the Debian GNU/Linux software distribution (although other distributions based on other kernels, like the Hurd, are expected in the near future). The distribution is available for several architectures, including Intel x86, ARM, Motorola 680x0, PowerPC, Alpha, and SPARC.

The core of the Debian distribution (called section "main," which accounts for the vast majority of the packages) is composed only of libre software, according to the Debian Free Software Guidelines (DFSG) (Debian Project, 2003). It is available on the Net for download, and many redistributors sell it on CDs or other media. The Debian distribution is put together by the Debian Project, a group of over 900 volunteer developers (Debian Database, 2003; Robles, Scheider, Tretkowski & Weber, 2001) spread around the world, collaborating via the Internet. The work done by those developers includes adapting and packaging all the software included in the distribution, maintaining several Internet-based services (web site, on-line archive, bug- tracking system, support and development mail lists, etc.), several translation and internationalization efforts, development of tools specific to Debian, and, in a broad sense, all the infrastructure that makes the Debian distribution possible.

Debian also takes special care to take advantage of the freedom that libre software provides to users, i.e., the availability of source code. Because of that, source packages are carefully crafted for easy compilation and reconstruction of original (upstream) sources. This makes them also convenient for measuring and, in general, for getting statistics about them.

More details about the Debian history and Debian distributions can be found in Debian History (2003) and in Lameter (2002).

# COUNTING LINES OF CODE

The number of physical lines of source code (SLOC) is one of the most simple and widely used techniques for comparing pieces of software. Some well-established methods for estimating effort and optimum scheduling (notably COCOMO) use this metric as the basis for their calculations. Therefore, it seems to be a good metric to apply when estimating the size of GNU/Linux distributions.

For the analysis presented in this chapter, we have measured the SLOC of five Red Hat Linux distributions (5.2, released in October 1998; 6.0, April 1999; 6.2, March 2000; 7.1, April 2001; and 8.0, September 2002) and four Debian GNU/Linux distributions (2.0, released in July 1998; 2.1, March 1999; 2.2, August 2000; and 3.0, July 2002). Those releases represent the main milestones of both distributions

*Figure 1: Size, in SLOC and number of source packages, of the studied distributions.*

SLOC for each release

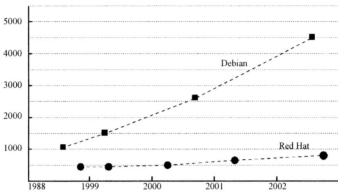

Number of source packages for each release

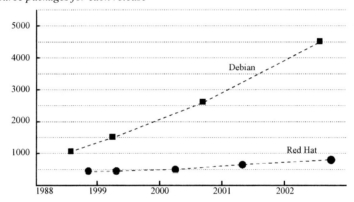

*Table 2: Size of the Red Hat Linux and Debian GNU/Linux Distributions studied.*

| Name | Release date | Source packages | Size (MSLOC) | Mean size of packages (SLOC) |
|---|---|---|---|---|
| Red Hat 5.2 | 1998-10 | 403 | 12 | 31,200 |
| Red Hat 6.0 | 1999-04 | 433 | 15 | 35,500 |
| Red Hat 6.2 | 2000-03 | 470 | 18 | 39,900 |
| Red Hat 7.1 | 2001-04 | 605 | 32 | 52,800 |
| Red Hat 8.0 | 2002-09 | 792 | 50 | 63,000 |
| Debian 2.0 | 1998-07 | 1,096 | 25 | 22,850 |
| Debian 2.1 | 1999-03 | 1,551 | 37 | 23,910 |
| Debian 2.2 | 2000-08 | 2,611 | 59 | 22,650 |
| Debian 3.0 | 2002-07 | 4,579 | 105 | 22,860 |

during a period of about four years (including all stable releases of Debian and the most representative releases of Red Hat Linux during that period of time). Therefore, we can study how Red Hat and Debian have evolved in parallel and make some comparisons at given points during this period.

In the following sections we analyze and show the SLOC counts for all these distributions, starting with the aggregated data shown in Table 2 (and briefly discussed below), and the graphs in Figure 1, which show the evolution of all studied distributions over time, both in SLOC and in number of source packages.

## Counts on Red Hat Distributions

Red Hat Linux 5.2, released in late 1998, included 403 source packages, with about 12.5 millions physical source lines of code (MSLOC). Red Hat 6.0, released about six months later, had 30 packages more and about 15 MSLOC. Release 6.2 (about a year later) included 470 source packages, with about 18 MSLOC. Red Hat 7.1, released another year later, included 605 source packages, with more than 32 MSLOC. Finally, release 8.0 (about a year and a half later) has almost 800 source packages and about 50 MSLOC. All in all, the SLOC found in year 2002 in Red Hat Linux are roughly four times those found in 1998, which means that it is doubling in size approximately every two years. The number of packages, on the contrary, has only doubled after the whole period of four years (which also means that the mean package size in Red Hat has doubled during that time).

## Counts on Debian Distributions

Debian 2.0 included 1,096 source packages, with more than 25 MSLOC. The next Debian release, 2.1 (delivered by the project about nine months later), had more than 37 MSLOC in 1,551 different packages. Debian 2.2 (released about 15 months later than Debian 2.1) contained over 59 MSLOC distributed in 2,611 packages. The latest stable release of the Debian GNU/Linux distribution, Debian 3.0 (delivered almost two years after Debian 2.2) grouped 4,579 source code packages with almost 105 MSLOC. This means that in four years, the size (in SLOC) of Debian GNU/Linux has increased more than four-fold (doubling every 24 months), while the number of packages in the distribution follows in general the same growth pattern. In other words, every two years the Debian Project is incorporating in its distribution as much code as it had incorporated during its whole previous history. As will be shown later, it may be also of interest how the mean package size remains stable during the whole period of the study, close to 23,000 SLOC.

## Comparing Debian with Red Hat

The first data that surprised us when we compared Debian GNU/Linux and Red Hat Linux was how large Debian distributions were compared to those of Red Hat. For instance, as the reader has probably noticed, Debian 2.2 is roughly twice the size of Red Hat 7.1 (released about eight months later) and more than three times the size of Red Hat 6.2 (released five months earlier). The comparison with Debian 3.0 is even more interesting. It is more than three times larger (in lines of code) than Red Hat 7.1 (released about 15 months earlier), and about twice the size of Red Hat 8.0 (two months later). In addition, to put these comparisons into context, it is important to notice the long freezing times of Debian, which in the case of Debian 3.0 accounted for about six months, and which in fact means that the release date is well delayed with respect to the time when the packages in the distribution were "frozen."

It is difficult to know the underlying reasons for these differences, but our guess is that they reflect differences in the policies of the two distributions. While Debian usually includes in the distribution any program that meets some quality and usability criteria, the policy of Red Hat seems to be more restrictive (for instance, targeted to produce a given number of CDs in the final distribution).

In any case, what is clearly the main factor causing these differences is the number of packages included in each distribution. In the case of Debian 2.2, for instance, we have considered more than 2,600 source packages (with a mean of about 21,300 SLOC per package), while Red Hat 7.1 includes only 605 source packages (almost 53,000 SLOC per package).

## Evolution and Trends

It is interesting to note that both Red Hat and Debian are doubling their size every two years. This means that both distributions included at least as much new code between 1998 and 2000 than during its whole previous history (in terms of lines of code). And the same can be said for the period between 2000 and 2002.

In fact, it is actually more than "as much", since when, for instance, Debian doubles from less than 60 MSLOC in Debian 2.2 to more than 105 MSLOC two years later, we have to count as new code not only the obvious difference between 60 and 105, which corresponds to new packages or increments of code size in packages already present in Debian 2.2, but also all the code in "old" packages that has changed (i.e., bug fixes, restructuring of code, sometimes whole rewritings of parts of applications). The changes in code from release to release should of course be carefully measured (which is not a target of this study), but they certainly add another good quantity to this doubling rule.

In addition, it could be assumed that at any given time a very large fraction of all the libre software available for Unix-like systems that is mature and useful to more than a handful of users is available in the Debian release of that time[5]. If this were the case, it could be said, extrapolating the above comments, that every two years the libre software community is delivering at least as much mature and useful code as it delivered during its whole previous history.

It should also be considered that the patterns observed in the past could be maintained in the future. Four years is little time to make extrapolations, but if the future trend could be extrapolated, the quantity of libre software code in a Debian release around 2006 would be about 400 MSLOC, reaching the GigaSLOC (GSLOC) around 2009. In the case of Red Hat, it would include 200 MSLOC around 2006, and wouldn't reach the GSLOC until about 2011. But of course, predicting the future is a risky business, and only detailed future analysis can determine whether these predictions will become true or not. Maybe libre software development will peak at a maximum in the near future, or maybe Debian or Red Hat will fail to scale to these really large sizes, due to coordination (or other) problems.

# SIZE OF PACKAGES

Distributions are composed of packages. The team building them decides when to consider a piece of software as a package, but in most cases they just follow the practices of the original development project (in the sense that when they deliver a certain software as a "unit," the distribution considers it as a package). In the particular case of Debian, there are cases where documentation is set apart in another package, but that fact has little impact on the size of packages in terms of SLOC, since those documentation packages usually contain little or no code. Of course, in this section (and, in general, in the rest of the chapter), we refer to source packages, which include the source code from which binary programs and libraries can be built. On the contrary, users are interested usually in binary packages, which are those found on the installation CDs, for instance. It is important to note that in many distributions (for instance, in Debian), a given source package can be used to produce several binary packages. For instance, in Debian 3.0 there are about 4,500 source packages but more than 10,000 binary packages.

## Distribution of the Size of Packages

In Figure 2, the size (as aggregated SLOC count for all the source code files included in it) of all packages in each of the studied distributions is shown.

From these figures, it is obvious that the "shape" of the curve showing the size of the packages is similar, and that great differences do exist between the larger packages (in the order of the millions of lines of source code) and the smaller ones (with roughly hundreds or tens of SLOC, or even zero). The reader should pay special attention to the fact that the Y axis (representing the size of each package) is in logarithmic scale.

However, there is an obvious difference between the Red Hat shapes and the Debian shapes in these figures. While both distributions show a quick decline to zero in the curves once the count has gone down to under 500 SLOC, the Debian curve in this area is smoother than Red Hat (which presents almost a step at about 300-400 lines of code (see this effect with more detail in the next figure). The gap in packages of that size in Red Hat distributions can also be seen in the histograms presented below. The reasons for this difference can be explained by looking at the list of packages in both distributions. While Debian includes a good quantity of small

*Figure 2: Distribution of the size of packages in Debian (a) and Red Hat (b) releases. Packages are ordered by size in the X axis, SLOC count for each is represented in the Y axis (logarithmic scale).*

*(a) Debian: Each line corresponds to a release (starting from left: 5.2, 6.2, 7.1, 8.0).*

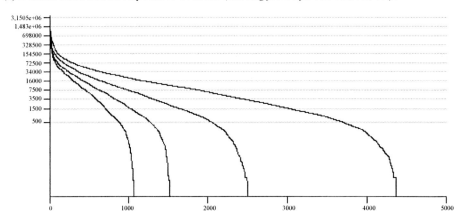

*(b) RedHat: Each line corresponds to a release (starting from left: 2.0, 2.1, 2.2, 3.0).*

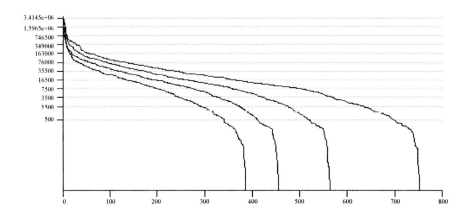

packages (fewer than 1,000 SLOC), containing scripts, small plug-ins, tiny scripting libraries, documentation, etc., Red Hat seems to miss that kind of packages. In the end, probably we are looking here at the effects of the different policies of both distributions for selecting packages to include. While in Debian any developer can include in the distribution his or her petty scripts (if he or she finds them useful for the users), in Red Hat there are other considerations to be taken into account when selecting packages.

Copyright © 2005, Idea Group Inc. Copying or distributing in print or electronic forms without written permission of Idea Group Inc. is prohibited.

*Figure 3: Histogram with the SLOC distribution for Red Hat packages.*

*(a) Red Hat 5.2*

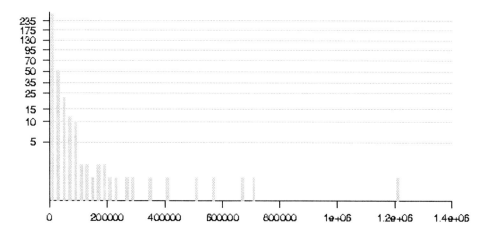

*(b) Red Hat 8.0*

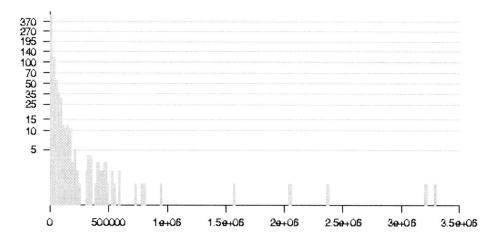

The variations in the size of the packages are possibly more explicit in the histogram representing the distributions of package size (see Figure 3 and Figure 4). In these figures it is clearer how almost all package sizes are below a certain threshold. It is also clear (both from these figures and from the former ones) how as time passes this threshold grows up. In general, as we will see below for a selected collection of packages, there is a certain tendency of any individual package to become larger and larger, following the Continuing Growth Law of Software Evolution as stated in Lehman, Ramil, Wenick, Perry, and Turski (1997).

*Figure 4: Histogram with the SLOC distribution for Debian packages.*

*(a) Debian 2.0*

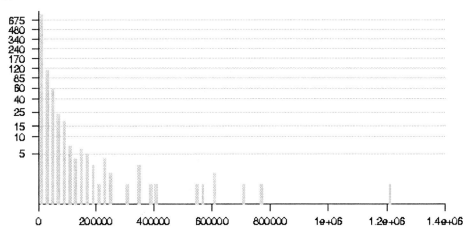

*(b) Debian 3.0*

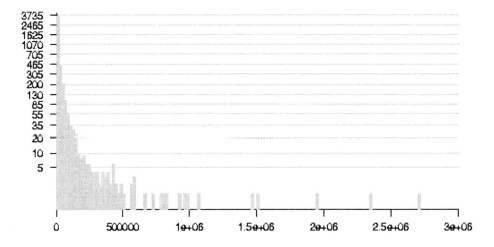

## Mean Size of Packages

When studying the mean size of source packages (total number of SLOC in the distribution divided by the number of source packages), there are two meaningful differences between Debian and Red Hat (see data in Table 2).

•   While the mean size of packages in Debian distribution is astoundingly regular (around 23,000 SLOC for Debian 2.0, 2.1, 2.2, and 3.0), during the studied period of about four years, the variation in Red Hat was considerable and monotonically growing (from about 31,000 SLOC in Red Hat 5.2 to 62,000 SLOC in Red Hat 8.0), in roughly the same period. It is difficult to explain this notable difference, especially the regularity of the Debian mean package, but a theory can be ventured. Maybe the Debian "ecosystem" is richer, and while many packages in it are growing, new smaller ones are also entering, with the neat effect that the mean remains constant. Meanwhile, Red Hat, being much more selective with regard to what enters the distribution or not, could be focusing on a given set of packages that become larger over time.

- The mean of Red Hat packages is larger. And not only "a bit" larger; rather, the mean of the Red Hat package is roughly twice the mean size of the Debian package. Even in 1998, when the mean Red Hat package was much smaller than four years later, it was about 40% larger than the mean Debian 2.0 package. In 2002 this ratio is about one to three. Again, it is difficult to venture an explanation, and certainly it is related to the previous item, but this also certainly reflects deep differences in the policies for selecting the packages that enter the distribution.

## Visiting the Largest Packages

It is also insightful to have a look at the largest packages of the distributions. Most of them are well-known packages with a large history of development and, in some cases, their developmental model is even documented in several papers. Studying how they evolve in the studied distributions, their size, etc., can therefore be meaningful. For this matter, see data offered in Table 3 and Table 4 (only data related to Debian is included so that evolution over time can be easily followed). Looking at these tables, some facts can be highlighted:

- There is a lot of movement in the top 10 (the 10 largest packages) in Debian distributions. Debian 3.0 includes among its 10 largest packages only three of those included in the same list for Debian 2.0 (released about four years earlier). Some of this movement is due to newcomers entering directly the list (as is the case of Mozilla, not present until Debian 3.0). In other cases it is due to packages that are really aggregations of other packages (as is the case, for instance, with mingw32, a cross compiler for C/C++ targeted to Win32 executables). However, the ever-growing lower limit of this list is noticeable. While in Debian 2.0, gcc entered it with not much more than 460,000 SLOC, the smallest package in the list for Debian 3.0, ncbi-tools (a set of libraries for biology applications), has more than 700,000 SLOC.
- The largest packages do not only have more source code, they also tend to have larger source files. While the mean ratio "SLOC per file" for packages in the top 10 ranges from 332 to 359 (332.69 for Debian 2.0, 351.70 for 2.1, 333.63 for 2.2, and 359.05 for 3.0), the same mean ratio for all packages is

*Table 3: Ten largest packages in Debian 2.0.*

| Rank | Package name | Version | SLOC | Files | SLOC/file |
|---|---|---|---|---|---|
| 1 | xfree86 | 3.3.2.3 | 1189621 | 4100 | 290.15 |
| 2 | xemacs20 | 20.4 | 777350 | 1794 | 433.31 |
| 3 | egcs | 1.0.3a | 705802 | 4437 | 159.07 |
| 4 | gnat | 3.10p | 599311 | 1939 | 309.08 |
| 5 | kernel-source | 2.0.34 | 572855 | 1827 | 313.55 |
| 6 | gdb | 4.17 | 569865 | 1845 | 308.87 |
| 7 | emacs20 | 20.2 | 557285 | 1061 | 525.25 |
| 8 | lapack | 2.0.1 | 395011 | 2387 | 165.48 |
| 9 | binutils | 2.9.1 | 392538 | 1105 | 355.24 |
| 10 | gcc | 2.7.2.3 | 351580 | 753 | 466.91 |

*Table 4: Ten largest packages in Debian 3.0.*

| Rank | Package name | Version | SLOC | Files | SLOC/file |
|---|---|---|---|---|---|
| 1 | kernel-source | 2.4.18 | 2574266 | 8527 | 301.9 |
| 2 | mozilla | 1.0.0 | 2362285 | 11095 | 212.91 |
| 3 | xfree86 | 4.1.0 | 1927810 | 6493 | 296.91 |
| 4 | pm3 | 1.1.15 | 1501446 | 7382 | 203.39 |
| 5 | mingw32 | 2.95.3.7 | 1291194 | 6840 | 188.77 |
| 6 | bigloo | 2.4b | 1064509 | 1320 | 806.45 |
| 7 | gdb | 5.2.cvs20020401 | 986101 | 2767 | 356.38 |
| 8 | crash | 3.3 | 969036 | 2740 | 353.66 |
| 9 | oskit | 0.97.20020317 | 921194 | 5584 | 164.97 |
| 10 | ncbi-tools6 | 6.1.20011220a | 830659 | 1178 | 705.14 |

in the range between 228 and 243 for (228.49 for Debian 2.0, 229.92 for 2.1, 229.46 for 2.2, and 243.35 for Debian 3.0). However, there is a large variance in this number, ranges from the 138 SLOC per file of egcs (a derivative of the GNU compiler gcc), version 1.1.2, to the 806 SLOC per file of bigloo (a Scheme compiling system), version 2.4b.

- There is a general increase in size of the packages in the list of the largest packages as time passes. Only the largest package of Debian 2.0 would be in the top 10 for Debian 3.0. This growth is especially curious when considering, as was shown above, that the mean size of packages in Debian distributions remains almost constant over time. To balance the larger "largest" packages, a lot of small ones have to enter each distribution.

From the point of view of application domains, there are no changes in the landscape from distribution to distribution. System tools (compilers, debuggers), graphical systems, editors, special-purpose libraries, and web browsers (Mozilla) are the kind of applications found in the lists of largest packages.

# COUNTING LANGUAGES

In the process of measuring the SLOC count of a given source file, the language in which the file is written is identified. With this identification, very useful statistics on the use of programming languages in the distribution can be compiled. In this section, some of these statistics are presented. However, before discussing them, it is important to note that languages are detected by using heuristics, and therefore some mistakes can be found if packages are carefully inspected. In any case, after the inspection of a good number of randomly selected files, it can be said that the number of errors in the identification of languages is completely negligible for the purposes of this study.

## Most Used Languages

Figure 5 and Figure 6 show pie graphs with the fractions of the most used languages (according to their absolute SLOC count) in some Red Hat Linux and Debian GNU/Linux releases. It is important to note how C is the most widely used language, with percentages between 60% and 85%. Although its evolution in time is decreasing (as data and figures will show later), its importance within GNU/Linux distributions cannot be underestimated. This tendency of C to decrease in SLOC share is clearer in the case of Red Hat, which over the four years of the study shrinks from a 85% to a mere 62% in Red Hat 8.0 (for Debian, the decrease is from 77% in 2.0 to 63% in 3.0).

Another fact that can be highlighted is that the diversity in languages used in the distributions is increasing. With time, the fraction for "other" languages gets larger and larger, and includes more and more languages. As an example, the list of languages found in Debian 3.0 with more than 1% of presence is shown in Table 5. Below the 1% threshold the languages found (in decreasing order) are: PHP, Ada, Modula3, Objective C, Java, Yacc, and ML (all with percentages between 0.30% and 0.60%).

The presence of a small quantity of large packages written mainly in a "minority" language may sometimes explain the language's relatively high position in the ranking. For instance, taking numbers from Debian 3.0, in the case of Ada, three

*Figure 5: Pie graph with the count for main languages in some Red Hat distributions.*

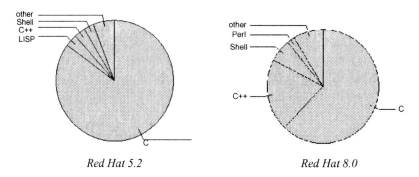

Red Hat 5.2                    Red Hat 8.0

*Figure 6: Pie graph with the count for main languages in some Debian distributions.*

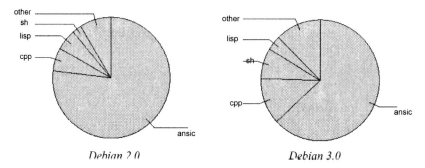

Debian 2.0                     Debian 3.0

*Table 5: Most used languages in Debian 3.0.*

| Language | SLOC | Percentage |
|---|---|---|
| C | 66,549,696 | 63.08% |
| C++ | 13,066,952 | 12.39% |
| Shell | 8,635,781 | 8.19% |
| LISP | 4,087,269 | 3.87% |
| Perl | 3,199,436 | 3.03% |
| Fortran | 1,939,027 | 1.84% |
| Python | 1,458,783 | 1.38% |
| Assembler | 1,367,085 | 1.30% |
| Tcl | 1,080,897 | 1.02% |

packages (gnat, an Ada compiler; libgtkada, an Ada binding to the Gtk library; and Asis, a system to manage Ada sources) account for more than 430,000 SLOC of the total of about 576,000 Ada SLOC for the whole distribution. In the case of LISP, emacs and xemacs alone account for more than 1,200,000 SLOC of the total of about 4,000,000 SLOC for the whole distribution.

## Evolution of Language Use Over Time

The graphs in Figure 7 and Figure 8 show the relative growth of some programming languages for Debian. Figure 7 represents the absolute number of SLOC for the most represented languages, while the latter shows the relative growth of some languages (taking as a base their SLOC in Debian 2.0).

In Figure 8, it can be seen how, while C quadruplicates the number of SLOC from Debian 2.0 to Debian 3.0 (which is roughly the growth ratio for the whole Debian SLOC count during that period), there are other languages whose growth is more spectacular (mainly scripting languages, such as Perl or Python), while more traditional languages (LISP, Fortran, Ada) usually show a minor growth rate (with the notable exception of C++). Not included in this figure are Java and PHP. Especially for the latter, the growth rate is enormous, but in part due to the fact that their appearance in Debian 2.0 is rather testimonial (PHP was not so significant in 1998, while Java had not entered the libre software world by that date). The case of Shell (the language with more growth in the figure), has to be considered having in mind that most packages have some Shell code, be it for configuration, for helper scripts, etc. This means that, in addition to the growth of its use, there is a component of its evolution directly linked to the number of packages in the distribution.

*Figure 7: Evolution of the four mose used languages in Debian distributions.*

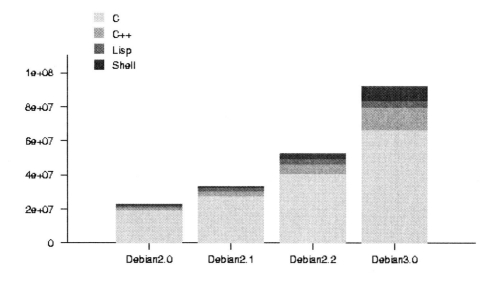

*Figure 8: Relative growth of some programming languages in Debian.*

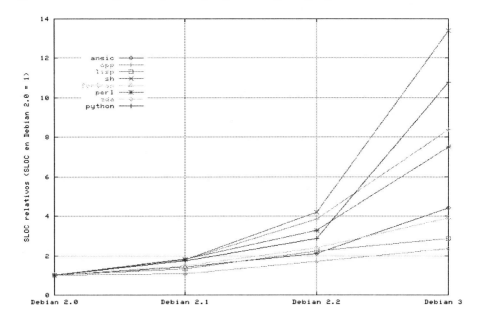

When looking at the evolution of main languages, it is clear how C, despite its growth in absolute terms, is decreasing slowly but continuously, from a percentage of 76% in Debian 2.0 to 63% in Debian 3.0, four years later. However, C remains, with great difference, the most used language in GNU/Linux distributions, still far away from C++ (currently the second language), which in Debian 3.0 amounts for 12% of the SLOC in the distribution. Therefore, we can safely say that, still today, C is the name of the game when we come to languages used in GNU/Linux distributions.

## Mean File Size for Different Languages

Table 6 gives us the mean number of lines of source code in a file for several programming languages. With the exception of Shell, the mean SLOC per file for a given language has little variations through the different releases, even with high relative growth rates for all languages. This fact suggests that somehow, common uses for a given language cause source files to be, usually, of a certain size.

On the other hand, the modularity of a given language can in part be inferred from these numbers. For instance, C++ has a smaller mean SLOC per file value than C, probably due to its object-oriented programming paradigm, which could lead to smaller files. On the other hand, languages such as LISP and Yacc tend to present very large files. The mean size of SLOC per file also remains pretty stable, i.e., around 229 for Debian 2.0, 2.1, and 2.2, climbing only to 243 SLOC per file in Debian 3.0 (in part due to the growth of the mean size of Shell files). Of course,

*Table 6: Mean file size for several languages.*

| Language | Debian 2.0 | Debian 2.1 | Debian 2.2 | Debian 3.0 |
|---|---|---|---|---|
| C | 262.88 | 268.42 | 268.64 | 283.33 |
| C++ | 142.5 | 158.62 | 169.22 | 184.22 |
| LISP | 394.82 | 393.99 | 394.19 | 383.60 |
| Shell | 98.65 | 116.06 | 163.66 | 288.75 |
| Yacc | 789.43 | 743.79 | 762.24 | 619.30 |
| Mean | 228.49 | 229.92 | 229.46 | 243.35 |

much more detailed studies should be done before attributing this circumstances to accident or to any non-casual reason.

# ESTIMATIONS OF EFFORT

The Constructive Cost or COCOMO model (Boehm, 1981) gives an estimation of the effort (human and financial) needed to build software of a given size. In fact, basic COCOMO uses the number of SLOC in the system to estimate the minimum resources that are needed to build such a system. However, the model used for this estimation assumes a proprietary (and "classical") development environment, so it has to be considered with some care. In any case, COCOMO estimations can give us at least a good idea of the order of magnitude of the minimum effort that would have been necessary if a proprietary development model had been used.

Using the SLOC count for the various distributions considered in this chapter, the results provided by the basic COCOMO model are shown in Table 7. These numbers assume that all projects were developed independently from the others, something we could almost state as true in all cases. In the calculation of cost estimation, the mean salary of full-time system programmers in the year 2000 in the U.S.A. (according to Computer World, 2000) was $56,286 USD per year. The overhead factor used for the COCOMO calculations was 2.4, and the effort estimation used has been the one for organic projects (which probably gives an underestimation in a lot of packages, since organic projects are usually relatively small, with a small number of developers who are already familiar with each other and usually working at the same location). Wheeler (2001) provides an explanation for why 2.4 was selected as the overhead factor and some other details on the estimation model used.

A quick look at the numbers shows how, while the total estimated effort for distributions grows quickly with distribution size, the estimated minimum schedule grows at a much slower pace. Without going into detail, this can be attributed to the assumption made that the development is done in parallel for all packages, which

*Table 7: Development effort and total cost estimations for each release.*

| Name | MSLOC | Effort (person-years) | Schedule (years) | Cost (Million USD) |
|---|---|---|---|---|
| Red Hat 5.2 | 12 | 3,216 | 4.93 | 434 |
| Red Hat 6.0 | 15 | 3,951 | 5.08 | 533 |
| Red Hat 6.2 | 18 | 4,830 | 5.45 | 652 |
| Debian 2.0 | 25 | 6,360 | 4.93 | 860 |
| Red Hat 7.1 | 32 | 8,427 | 6.53 | 1,138 |
| Debian 2.1 | 37 | 9,425 | 4.99 | 1,275 |
| Red Hat 8.0 | 50 | 13,313 | 7.35 | 1,798 |
| Debian 2.2 | 59 | 14,950 | 6.04 | 2,020 |
| Debian 3.0 | 105 | 26,835 | 6.81 | 3,625 |

means that merely adding more packages does not increment the minimum schedule. Of course, this assumption ignores the integration problems of a distribution with thousands of packages (which in the case of Debian leads to freezing periods of several months, devoted mainly to coordination and stabilization of the distribution as a whole). In addition, it can be observed how some Red Hat releases have a larger schedule estimation that Debian releases with a much larger size. This is also because of the parallel development assumption, which means that the schedule for a release is as large as the schedule of its largest package. And it happens that there are some packages in some Red Hat releases that are larger than their counterparts in Debian releases.

From another point of view, more focused on the estimation of economic resources, it can be seen how the estimated cost of both Debian and Red Hat distributions is doubling, approximately, every two years. As was said before, this estimation does not take into account the coordination costs that certainly are needed to put together such distributions, nor the applicability of COCOMO to the development models used in libre software. However, the estimation of more than $3.6 billion USD for the development of Debian 3.0 could be a valid indicator of the economic effort that the libre software community has made to build and compile such a collection of software.

## COMPARISON WITH OTHER SYSTEMS

We can put into context the figures offered for Red Hat and Debian by comparing them with some estimations of the size given for several well-known and significant operating systems and office suites.

The cited numbers have been reported in Lucovsky (2000) for Windows 2000, Sun Press Release (2002) for StarOffice 5.2, McGraw (2000) for Windows XP, and Schneier (2000) for the rest of the systems. The entries in the table are shown ordered by increasing number of source lines of code. Most of these estimations (in fact, all of them) are not detailed and cannot be confirmed because of their proprietary nature. It is also difficult to know what is being considered as a line of code (although in most cases, it seems that raw lines of source files, including comments, are considered). However, the estimations should be similar enough to SLOC counting methods to be suitable for comparison, at the very least in order of magnitude.

Note also that Red Hat and Debian include many applications that in many cases can be classified under the same category and share the same goals, while Microsoft and Sun operating systems and office suites are much more limited in this way. If the usual applications used in those environments were counted together, their size would be surely much larger. However, it is also true that all those applications are neither developed nor put together by the same team of developers, as is the case in Linux-based distributions such as the ones we have seen in this chapter.

From these numbers, it can be seen that Linux-based distributions in general, and Debian 3.0 in particular, are some of the largest pieces of software ever put together by a group of developers, during the whole history of software development.

*Table 8: Comparison with proprietary systems.*

| System | Release Date | Lines of Code (millions) |
|---|---|---|
| Microsoft Windows 3.1 | 1992-04 | 3 |
| Sun Solaris 7 | 1998-10 | 7.5 |
| Sun StarOffice 5.2 | 2000-06 | 7.6 |
| Red Hat Linux 5.2 | 1998-10 | 12 |
| Red Hat Linux 6.0 | 1999-04 | 15 |
| Microsoft Windows 95 | 1995-08 | 15 |
| Red Hat Linux 6.2 | 2000-03 | 18 |
| Debian 2.0 | 1998-07 | 25 |
| Microsoft Windows 2000 | 2000-02 | 29 |
| Red Hat Linux 7.1 | 2001-04 | 32 |
| Debian 2.1 | 1999-03 | 37 |
| Windows NT 4.0 | 1996-07 | 40 |
| Red Hat Linux 8.0 | 2002-09 | 50 |
| Debian 2.2 | 2000-08 | 55 |
| Debian 3.0 | 2002-07 | 105 |

## SOME NOTES ABOUT THE METHODOLOGY

After showing the most relevant results of how we have applied the SLOC counting methodology to several Debian and Red Hat releases, in this section we now summarize the methodology itself. First, the reader can find its general description, with the issues common to all the software measured. Later, the specifics of how it was applied to Red Hat and Debian are discussed. At the end of the chapter, some aspects of the methodology are discussed. Some more details (especially on how we have applied it to Debian 2.2 and Debian 3.0) can be found in González-Barahona et al. (2001, 2003).

## GENERAL DESCRIPTION

In all cases, the source packages corresponding to the release to be studied are identified and downloaded, and the source directories used by the developers to create the binary packages are recreated. These processes depend on some peculiarities of the distribution and are detailed below for the cases of Debian and Red Hat.

Once the source directory is available, the final analysis consists on the collection of SLOC data and the elaboration of some statistics regarding the total number of physical SLOC of the release, the SLOC for each package, the SLOC for each of several programming languages considered, etc. Most of these data have been calculated using SLOCCount (Wheeler, 2000), completed with some scripts done by the authors. Therefore, we are providing some details about how it works, which could be of interest:

- SLOCCount starts by analyzing the source code files, using heuristics to identify the programming languages used in each of them. With this information, and for each package, lists are made including the files written in each language.
- For each file, a parser is run, capable of counting the physical lines of code for the language in which the file is written. With this information, the SLOC count for each file in the package is completed.
- Once the package is completely measured, aggregated figures are compiled, including the SLOC count for each language present in the package and for the whole package. In addition, statistics for a collection of packages can also be obtained (by package and by language).
- Based on the SLOC counts calculated above, estimations of effort and value, using the COCOMO model, are computed.
- SLOCCount also allows for the use of several mechanisms to improve the counting process. For example, MD5 checksums on each file are used to identify duplicates within a package, or several heuristics are followed to determine whether a source file was automatically generated by some tool (and therefore should not be counted as code directly written by a human).

- The code in Makefiles and the specifications of packages in RPM packages are not counted. SLOCCount can count them, but since Makefiles are more configuration files than actual source code, it seemed more natural not to consider them.

It is also important to specify what SLOCCount counts as lines of code. The definition of physical line of code (SLOC) in this context is *a line ending in a newline or end-of-file marker, and which contains at least one non-whitespace non-comment character.* Therefore, comments are not counted. However, in languages where the identification of a comment is not trivial, heuristics are used (since SLOCCount does not include complete parsers for these languages), which in some (but anecdotal) cases may lead to bad identifications.

With the aim of measuring only mature code, we have tried not to consider beta packages when included in the releases. This is never the case in Debian (since beta packages do not enter stable releases, by Debian policy), but can happen in Red Hat. In this case, we have not considered all packages that included the string "beta" in its name[6]. Also, with the idea of not measuring any package twice, we have examined, both in the case of Debian and Red Hat, the list of packages in each release and when several versions of the same package were found, only the latest one was counted.

All in all, when referring to the aggregated numbers for the whole release of a distribution, the counting methodology leads to the size of the mature source code directly written by the developers in libre software projects contributing their programs to Red Hat Linux. When referring to packages, the numbers provide information about the size of the source code of the package as it is within the distribution (that is, including any patch or modification that may have applied by the maintainer of the package within the distribution).

To ease the study of the data produced by SLOCCount, a tool has been built to parse the raw data produced by it and dump it in an XML-formatted file. The final data mining (including graph and table generation) has also been automated by a set of scripts that work on the XML files with the data for distributions (sometimes using them to feed an SQL database that is later queried to get some statistics). The whole process—from downloading to final data mining—has been automated by a set of scripts.

## Specifics in the Case of Red Hat Releases

The peculiarities when collecting data related to the Red Hat Linux releases studied in this chapter are summarized by the following:

1. Which source code makes a Red Hat release? For the purposes of this study, we have used source packages (src.rpm) of Red Hat Linux releases found in mirrors of ftp.redhat.com, the official repository of Red Hat software, that in-

clude only the libre software found in Red Hat releases (some Red Hat releases included also proprietary software). For each release, we have considered only the packages found in the directory en/os/i386/SRPMS. This basically means that we have only considered the source packages for the i386 architecture. Since most source packages are architecture independent, and when they are not, they are very similar for all supported architectures, this seems a fair simplification. In addition, it is consistent with the case of Debian, for which we have not considered source packages specific for a non-i386 architecture (though this is a really rare case).

2. Downloading and collecting data: Once the repository with the source packages is chosen, all the files are downloaded and the source directories corresponding to each are recreated using a two-step process. First, the packages within the src.rpm file are extracted, including the spec file, with instructions of how to build the source. Second, using that spec file, the original directory is recreated, some patches are applied to it (if needed), and some scripts may be run. As a result, we get the source package that, when compiled, would result in the binary package one can find in the corresponding Red Hat binary distribution. This process is done with the help of rpm, the Red Hat packaging tool, in a Debian machine. After the recreation of the source directory, the SLOC count of each source file is obtained.

When comparing our criteria on Red Hat to those used by David A. Wheeler (described in detail in Wheeler, 2000 and 2001), the main difference is that we use MD5 checksums only to detect duplicates within packages, while Wheeler also uses them to detect duplicate files in different packages (he counts unique files only once across the whole distribution). The reasons for this change are both pragmatic (it was a bit difficult to do when Debian releases were analyzed, due to the huge quantity of disk space needed) and methodological (we feel that a file in two different and unrelated packages should be counted twice). In any case, the difference in the counts due to this different consideration is negligible for the purposes of our study (usually well below 3%, although there are some packages, like GDB, where it can be as high as 20%).

## Details of Debian Data Collection

The particular issues related to the collection of the data related to the Debian releases are summarized as follows:

1. Which source code makes a Debian release? Source code for current and past Debian releases is archived and available to everyone on the Internet. For every release, they reside in the source directory of the Debian archive[7]. The only problem is to determine the list of source packages for any given release and where to access them, which for releases from Debian 2.0 on is not difficult

since a Sources.gz file is present in the source directory, with information about the source packages for the release, including the files that compose each package.
2. Downloading and collecting data: Once we know what files to download, we have to download all of them before being able to gather data. Since the size of the unpackaged sources for a Debian release is very large, we chose to work on a per-package basis, gathering all the relevant data from it (mainly, the number of lines of code) before deleting it and proceeding with the download of the next one.

## Comments on the Methodology

In this study, there are several sources of inaccuracy that can cause some errors when looking into the details, although they should not alter the main picture. Therefore, remember that the numbers offered in this chapter are just estimations, especially when they refer to aggregates of packages. The main sources of inaccuracy are:

- Some SLOC counts on files could be wrong. Although SLOCCount includes carefully designed heuristics to detect source files to distinguish source lines from comments and to avoid automatically generated files, those heuristics do not always work as expected. For instance, analyzing Debian 2.0 with a recent version of SLOCCount that supports detection and measuring of C# code throws 352 lines of code for that language, which is impossible taking into account that C# was not in use at that time. The error in this case is due to an erroneous identification of C# files and not to a wrong way of counting SLOCs. In any case, the number of such errors can be considered irrelevant in the context of many millions of source lines of code.
- Count of physical SLOC and not logical SLOC. The count of physical lines of code is usually considered as a minimum for estimating the size of a software package. However, in general this count is also considered as worse than logical SLOC (which counts sentences of the programming language using, for instance, terminating semicolons in C). In any case, the decision to measure physical SLOC mandates the selection of COCOMO instead of COCOMO II for effort estimation, since the latter needs logical lines of code as input.

In addition to the inaccuracy in data collection, there is another more serious concern. All the study is based on SLOC counting. However, SLOC is a very simple measure of the size and complexity of a system. We selected it as the central measure for our study since it is easy to calculate (even for source code written in many different languages), well known to the software engineering community, and can be used directly to estimate efforts using the COCOMO model (simple but well accepted as an estimator, at least, of the order of magnitude). But SLOC also has several problems derived from its simplicity: it may have very different meanings for different languages, shows no clear relationship to the complexity of the code (which certainly impacts on effort), or pays no attention to design documents, documentation, etc., to mention just a few. More research is needed to apply other source code metrics to GNU/Linux distributions and to try to correlate the data they may provide with the one shown here.

## CONCLUSIONS AND FURTHER WORK

In this chapter, we have presented some results of our work on counting the lines of code in several Debian and Red Hat releases. They can be used to infer the evolution of Red Hat Linux and Debian GNU/Linux distributions for a period starting in 1998 and ending in 2002. The main similarities and differences between both distributions have been shown, and a detailed analysis from several points of view (total size, package by package data, relevance of different languages, estimations of effort, etc.), has been provided.

In addition to the static analysis of the releases, we have presented a dynamic study, showing the evolution of several parameters (total distribution size, packages size, languages used, etc.) over time. Although four years (the observation period for this study) can be considered a short time, we venture that in the future (or at least in the near future) the data offered in this study will follow the same pattern and trends found during this period. If this assumption were true, this study could also be used for making estimations of how GNU/Linux distributions will be in the coming years. By extension, they could also be used to predict the development of the whole ecosystem of libre software for GNU/Linux and Unix-like systems.

Among the more relevant data presented in this chapter is the analysis of total size of the releases. In the case of Debian 3.0, the largest analyzed (maybe the largest GNU/Linux distribution to date), we found that it is composed of more than 105 millions of SLOC. Using the COCOMO model, this would imply a cost (using traditional, proprietary software development models) of more than $3.6 billion USD and an effort of almost 27,000 person-years. We venture that these numbers are close to reflecting the state of mature and widely used libre software for Linux and other Unix-like systems around 2002. We have also found that the size of Red Hat 8.0, the most well-known Red Hat Linux release of 2002, is about half the size of Debian 3.0, and that the characteristics (e.g., mean size) of the packages included in both distributions are rather different (although the larger packages are almost the same). Also, the size of both Red Hat and Debian releases is doubling every two years.

With respect to the comparison of Red Hat and Debian, the most obvious fact is that the latter is much larger. It has also been observed how the different policies for inclusion of packages cause interesting effects in the pattern of the size of packages and in the mean package size. With respect to this latter number, it is noticeable how all Debian distributions maintain an almost constant mean value of about 23,000 SLOC per package, despite their great differences in total size, languages used, number of packages, etc. Whether this is just a coincidence or there is some kind of constant on the size of packages is something to be researched with more detail.

We have also shown the evolution of the largest packages in the Debian distribution. These top 10 sets have been shown to varying greatly in time but, in any case, include larger and larger packages. Research on future distributions are needed to clarify whether the larger packages are in the system applications domain (operating system, applications oriented to the software developer, etc.) or whether, as data in this study may suggest, shift to more end-user targeted applications is happening.

When looking at the programming languages, the evolution shows a gradual decline of C. However, it is still the most used language (and it will surely remain as such in the near future). Scripting languages have a tendency to grow quickly and are gaining a share in the languages pie, while most traditional, compiled programming languages show a smaller growth rate, even smaller than C. Java and C++ are the exception, both experiencing an important growth. We also point out how, with some minor exceptions, files written in a given programming language have an almost constant mean size over time, measured as the number of SLOC per file. This could be explained through the different characteristics of each language, but further studies should be done to confirm or deny this hypothesis.

This chapter also includes the description of the methodology used to obtain all of these measures, which allows for future studies of new releases of the studied and other distributions. Should those studies conform to the methodology, the resulting numbers could be compared, which will certainly help in the analysis of the differences and similarities between GNU/Linux and other Unix-like libre software distributions. In addition, most of the methodology is automated, which will help to make extensive studies on distributions in the future. Finally, we find it important to repeat once more that all the data offered are only estimations. However, we believe they are accurate enough to draw conclusions and to compare with other systems.

# REFERENCES

Bailey, E. C. (1998). Maximum RPM - Taking the Red Hat Package Manager to the limit. Retrieved on January 19, 2004 from: http://rikers.org/rpmbook/.

Bodnar, L. (2003). Linux distributions - Facts and figures. Retrieved on January 19, 2004 from: http://www.distrowatch.com/stats.php?section=packagemanagement.

Boehm, B. W. (1981) *Software engineering economics*. Englewood Hall, NJ: Prentice Hall.

Computer World (2000) Salary Survey 2000. In Wheeler, D. A. (2001). More than a gigabuck: Estimating GNU/Linux's size. Retrieved on January 19, 2004 from: http://www.dwheeler.com/sloc.

Debian Documentation Team. A brief history of Debian. Retrieved on January 19, 2004 from: http://www.debian.org/doc/manuals/project-history/.

Debian Project. Debian developers database. Retrieved on January 19, 2004 from: http://db.debian.org.

Debian Project. Debian free software guidelines (part of the Debian social contract). Retrieved on January 19, 2004 from: http://www.debian.org/social_contract.

Debian Project. *Debian GNU/Linux 2.2 release information.* Retrieved on January 19, 2004 from: http://www.debian.org/releases/2.2/.

Debian Project. Debian GNU/Linux 2.2, the "Joel 'Espy' Klecker" release, is officially released. Retrieved on January 19, 2004 from: http://www.debian.org/News/2000/20000815.

Debian Project. Debian policy manual. Retrieved on January 19, 2004 from: http://www.debian.org/doc/debian-policy/.

Godfrey, M.W. & Tu, Q. (2000). Evolution in open source software: A case study. In *Proceedings of the Intl Conference on Software Maintenance*, San José, California, October, pp. 131-142. Retrieved on January 19, 2004 from: http://plg.uwaterloo.ca/~migod/papers/icsm00.pdf.

González-Barahona, J.M., Ortuño-Pérez, M. A., de-las-Heras-Quirós, P., Centeno-González, J., & Matellán-Olivera, V. (2001). Counting potatoes: The size of Debian 2.2. *Upgrade Magazine.* 2(6). Retrieved on January 19, 2004 from: http://upgrade-cepis.org/issues/2001/6/up2-6Gonzalez.pdf.

González-Barahona, J.M., Robles, G., Ortuño-Pérez, M., Rodero-Merino, L., Centeno-González, J., Matellán-Olivera, V., Castro-Barbero, E., & de las-Heras-Quirós, P. (2003). *Measuring Woody: The size of Debian 3.0.* Unpublished. Will be available at http://people.debian.org/~jgb/debian-counting/.

Koch, S. & Schneider, G. (2000). Results from software engineering research into open source development projects using public data. Diskussionspapiere zum Tätigkeitsfeld Informationsverarbeitung und Informationswirtschaft, no. 22. Wirtschaftsuniversität. Wien, Austria. Retrieved from: http://wwwai.wu-wien.ac.at/~koch/forschung/sw-eng/wp22.pdf.

Lameter, C. (2002). Debian GNU/Linux: The past, the present and the future. *Free Software Symposium 2002*, October 22, 2002. Japan Education Center, Tokyo, Japan. Retrieved on January 19, 2004 from: http://u-os.org/tokyo/.

Lehman, M.M., Ramil, J.F., Wernick, P.D., Perry, D.E., & Turski, W. (1997). Metrics and laws of software evolution - The nineties view. *4th International Symposium on Software Metrics.* Albuquerque, NM, November 5-7, 1997. Retrieved on January 19, 2004 from: http://www.ece.utexas.edu/~perry/work/papers/feast1.pdf.

Lucovsky, M. (2000). From NT OS/2 to Windows 2000 and beyond - A software-engineering odyssey. In *4th USENIX Windows Systems Symposium*, Seattle, WA, August 3-4, 2000. Retrieved on January 19, 2004 from: http://www.usenix.org/events/usenix-win2000/invitedtalks/lucovsky_html/.

McGraw, G. (2000). Building secure software: How to avoid security problems the right way. In Wheeler, D.A. (2001). More than a gigabuck: Estimating GNU/Linux's size. Retrieved on January 19, 2004 from: http://www.dwheeler.com/sloc/.

Michlmayr, M. & Hill, B.M. (2003), Quality and the reliance on individuals in free software projects. In *Proceedings of the 3rd Workshop on Open Source Software Engineering*, Portland, OR, May 3-10, 2003. Retrieved on January 19, 2004 from: http://opensource.ucc.ie/icse2003/3rd-WS-on-OSS-Engineering.pdf.

Mockus, A., Fielding, R.T., & Herbsleb, J.D. (2002). Two case studies of open source software development: Apache and Mozilla. *ACM Transactions on Software Engineering and Methodology (TOSEM)*, 11(3), 309-346.

Robles, G., Scheider, H., Tretkowski, I., & Weber, N. (2001). WIDI - Who is doing it? A research on Libre Software developers. Retrieved on January 19, 2004 from: http://widi.berlios.de/paper/study.pdf.

Schach, S., Jin, B., Wright, D., Heller, G., & Offut, A. (2002). Maintainability of the Linux kernel. *IEE Proceedings - Software*, 149(1), 18-23. Retrieved on January 19, 2004 from: http://opensource.ucc.ie/icse2002/SchachOffutt.pdf.

Schneier, B. (2000). Software complexity and security. *Crypto-Gram Newsletter*, March 15, 2000. Retrieved on January 19, 2004 from: http://www.schneier.com/crypto-gram-0003.html.

Smoogen, S.J. (2003). The truth behind Red Hat Names. Retrieved on January 19, 2004 from: http://www.smoogespace.com/documents/behind_the_names.html.

Sun Microsystems (2000). Sun Microsystems announces availability of StarOffice(TM) source code on OpenOffice.org. October 16. Retrieved on January 19, 2004 from: http://www.collab.net/news/press/2000/openoffice_live.html.

Wheeler, D.A. (2000). Estimating Linux's size. Retrieved on January 19, 2004 from: http://www.dwheeler.com/sloc/.

Wheeler, D.A. (2001). More than a gigabuck: Estimating GNU/Linux's size. Retrieved on January 19, 2004 from: http://www.dwheeler.com/sloc/.

Young, R. (1999). Giving it away. How Red Hat Software stumbled across a new economic model and helped improve an industry. Retrieved on January 19, 2004 from: http://www.oreilly.com/catalog/opensources/book/young.html.

## ENDNOTES

1. Throughout this chapter, we use "libre software" as a way of referring both to free software and open source software. Though open source software and free software communities are very different, the software is not, since almost all licenses considered to be "free" are also considered "open source," and the other way around.
2. Throughout this chapter, we use the term "Linux" to refer to the kernel, while software distributions based on it will usually be referred as "GNU/Linux."
3. Of course, there are other possible explanations that should be taken into account, the simplest of them being that Red Hat is better at quickly updating the versions of the packages in its distribution.
4. Since version 1.1 of the Linux Standard base (a specification with the goal of achieving binary compatibility among GNU/Linux distributions, released by the Free Standards Group), RPM has been chosen as the standard package format. The Debian project continues with its own package format, as well as many other Debian-package-dependent distributions, and complies with the standardized format by means of a conversion tool called alien.
5. This is of course an unproved assumption that should be backed by more empirical data than is available today. But given the list of packages in Debian releases and the way they are selected for inclusion (basically, because a Debian developer finds them useful), it may not be too risky an assumption.
6. Of course, this simple mechanism is not the more strict way of detecting software still in beta, but it works pretty well for Red Hat releases.
7. ftp://archive.debian.org and mirrors.

## Chapter III

# The Co-Evolution of Systems and Communities in Free and Open Source Software Development

Yunwen Ye, University of Colorado at Boulder, USA
and SRA Key Technology Lab, Japan

Kumiyo Nakakoji, University of Tokyo, Japan

Yasuhiro Yamamoto, University of Tokyo, Japan

Kouichi Kishida, SRA Key Technology Lab, Japan

## ABSTRACT

*Because a Free and Open Source Software (F/OSS) project is unlikely to sustain a long-term success unless there is an associated community that provides the platform for developers, users, and user-turned-developers to collaborate with each other, understanding the well-observed phenomenon that F/OSS systems experience "natural product evolution" cannot be complete without understanding the structure and evolution of their associated communities. This chapter examines the structure of F/OSS communities and the co-evolution of F/OSS systems and communities based on a case study. Although F/OSS systems and communities generally co-evolve, they co-evolve differently depending on the goal of the system and the structure of the community. A systematic analysis of the differences leads us to propose a classification of F/OSS projects into three types: Exploration-Oriented,*

*Utility-Oriented, and Services-Oriented. Practical implications of realizing the co-evolution and recognizing the different types of F/OSS projects are discussed to provide guidance for F/OSS practitioners.*

# INTRODUCTION

Many definitions exist regarding Free and Open Source Software (F/OSS) (DiBona, Ockman, & Stone, 1999). The major difference in those definitions comes from the difference in distribution and re-distribution rights. In this chapter, we use the term Free and Open Source Software inclusively to refer to those systems that give users free access to source code, as well as the right to modify it. F/OSS grants the right to run, read, and change its source code not only to the developers of a system but to all users—who, in fact, are potential developers. Developers, users, and users-turned-developers form a *community of practice* (Lave & Wenger, 1991). A community of practice is a group of people who are informally bonded by their common interest and shared practice in a specific domain. Community members regularly interact with each other for knowledge sharing and collaboration in pursuit of solutions to a common class of problems. A F/OSS project is unlikely to sustain long-term success unless there is an associated community that provides the platform for developers, users, and users-turned-developers to collaborate with each other.

By allowing users to become co-developers, F/OSS encourages *natural product evolution* (O'Reilly, 1999). To understand how this natural product evolution happens, we conducted a case study of four F/OSS projects and systematically examined their similarities and differences. Our study examines not only the evolution of F/OSS systems, but also the evolution of the associated F/OSS communities and the relationship between these two types of evolution.

The case study leads us to the proposition that a strong correlation exists between the evolution of a F/OSS system and that of its associated community. F/OSS systems evolve through the contributions made by its community members, and the contributions made by any member change the role that the member plays in the community, thus resulting in the evolution of the community by reshaping community structure and dynamics. Although F/OSS systems and communities generally co-evolve, they co-evolve differently depending on the goal of the system and the structure of the community. The difference results in different evolution patterns of F/OSS systems and communities. To treat such differences systematically, we propose to classify F/OSS projects into three types: Exploration-Oriented, Utility-Oriented, and Service-Oriented. Such a classification provides the basis for finding better technological and managerial support for a particular F/OSS project.

## RELATED WORK

Since the publication of Raymond's seminal paper on F/OSS (Raymond, 2001), F/OSS has received enormous attention. Researchers have started analyzing F/OSS from a variety of perspectives, including its evolutionary path, its development process, its collaborative nature, its developer profiles, and its community dynamics.

Godfrey and Tu (2000) analyzed the evolutionary growth of Linux both at the system level and within the major subsystems. They found that, over the years, Linux had been experiencing surprisingly super-linear growth, which contradicted the inverse square growth rate hypothesis (Turski, 1996) that resulted from previous software evolution research based on the analysis of proprietary software systems. Most of the growth of Linux, however, came from the addition of new features and support for new hardware architecture rather than defect fixing.

Dempsey, Weiss, Jones, and Greenberg (2002) have examined the demographic composition of Linux contributors and found that most contributions come from only a handful (2.2%) of developers, although thousands of developers worldwide have made contributions. The set of core contributors shifts over time, enabling the sustainable development of Linux.

Koch and Schneider (2002) confirmed that a small "inner circle" of developers was responsible for the development of GNOME, and those developers were also active participants in the associated mailing lists. The number of active developers rises quickly at the start-up of the system, but levels off at the point when the system reaches stability. Parallel programming on a same file is rare, indicating a high degree of division of labor in F/OSS development.

By examining the development processes of Apache and Mozilla, Mockus, Fielding, and Herbsleb (2002) have formed several partially verified hypotheses that are essential for the success of F/OSS. One of their major hypotheses is that a successful F/OSS project should have a well-balanced developer community: (1) A relatively small core of developers who control the code base should be able to produce about 80% or more of the new functionality for an F/OSS system; (2) a group larger by an order of magnitude than the core should be available for bug fixing; and (3) yet another larger group by another order of magnitude should engage in testing and reporting problems.

Another thread of research on F/OSS focuses on its self-organizing community aspect, which distinguishes F/OSS from proprietary software. Moon and Sproull (2000) pointed out the importance of creating sustainable developer communities, with the support of both technical and social tools. Technical tools include mailing lists, newsgroups, and source code management systems that facilitate timely communications among the members and the coordination of distributed work. Social tools include differentiated roles and learning support. Those roles, with their corresponding obligations and responsibilities, should be made explicit and understood by all. Learning support, such as frequently asked questions (FAQs) and direct question-and-answer interactions within the community, enables newcomers to learn directly from more experienced and skilled long-term members, so that newcomers can progress into becoming skilled developers through informal training within the community.

Von Krogh, Spaeth, and Lakhani (2003) suggested that the success of an F/OSS project is related to the growth in the size of the F/OSS developer community. They analyzed the strategies and processes by which newcomers join an existing F/OSS community (the Freenet project) and make initial code contributions. They observed that for a newcomer to become an active developing member, the newcomer typically follows a "joining script" to go through different levels and types of activity. Newcomers tend to start with low-level activity that requires less efforts, then move gradually toward higher level activity as they gain more understanding of the system and community, until finally obtaining the right of committing code directly to the system.

# SCOPE OF THE CASE STUDY

Software Research Associates, Inc. (SRA) is a leading company in F/OSS development. SRA has been supporting the Free Software Foundation since 1987, and is currently involved in the development of a variety of F/OSS systems. We conducted our case study by interviewing the leaders and members of four F/OSS teams within SRA: the GNU team, the Linux support team, the PostgreSQL team, and the Jun team.

Although members of the four teams are employees of SRA, they have started their involvement with the related F/OSS projects as voluntary participants and are still participating in F/OSS communities-at-large as individuals. Contrary to the common belief that all F/OSS developers use their spare time only to code for F/OSS systems, company-sponsored participation has become a new trend as more companies start to support F/OSS (Hars & Ou, 2001; Hertel, Niedner, & Hermann, 2003). For these reasons, we believe our study still reflects the overall picture of F/OSS although its scope is limited to company-sponsored participants.

## The GNU Team

The GNU team at SRA supports the development of GNU software both for clients and for the community-at-large. The team leader started his involvement with GNU software in 1986 by submitting bug reports to GCC v1.3. Over the years, the team has ported several GNU systems, including GCC, GAS, and GDB, to three different platforms for its clients, and all the ports have been contributed back to the community. In addition to submitting numerous patches to different GNU systems, the team also helps its clients improve the quality of the patches developed by the clients and submit the patches back to GNU project leaders. This phenomenon is caused by the following. Viewing programs as "scientific knowledge to be shared among mankind," GNU project leaders execute tight and centralized control over the incorporation of patches into the official versions of systems to ensure high quality. This tight control mechanism sets very strict criteria for the quality of patches and creates a high barrier for newcomers to have their patches incorporated. Patches improved and submitted by the SRA team are more likely to be noticed and incorporated because the team members have collaborated closely with GNU project leaders for a long time.

## The Linux Support Team

The Linux support team provides user support for the Linux operating system, excluding the Linux kernel. We make this distinction because, similar to GNU, the development of the Linux kernel is under centralized control, whereas the remainder of Linux has been developed in the bazaar style (Raymond, 2001) with decentralized control. In contrast to GNU programs, multiple versions of programs for the same functionality exist in Linux, especially for device drivers (Godfrey & Tu, 2001), and many programs are not compatible with others.

A typical task of the SRA Linux support team is to help clients find appropriate distribution packages and to customize the software or develop patches according to the specific needs of clients. Surprisingly, the patches that the team develops for its clients are not contributed back to the community. The team leader explains that the clients do not care about version updates and prefer to stay with the current version of the system as long as the system is working, even if new versions are available. This is very different from the GNU team, in which it is critical that the patches developed and used at a client's site get incorporated into the core version.

## The PostgreSQL Team

The PostgreSQL team deals with the PostgreSQL database system. Because robustness is highly desirable in database systems, PostgreSQL is strictly controlled by the core development team (six members) and the major development team (14 members). Decisions about the development of PostgreSQL are made democratically by members of the two development teams. The leader of the SRA team belongs to the major development team.

The primary task of the SRA PostgreSQL team has been internationalization. The team has developed patches to deal with two-byte code languages, which have been incorporated into the core version, and the internationalized PostgreSQL is now the standard distribution. Another task for the team results from the fact that the software is a database system. For a bug report, it is often required to include a set of data to reproduce the bug because without the data it is very difficult for developers to debug. Such data, however, are often proprietary and cannot be made public. Therefore, many clients ask the SRA PostgreSQL team to debug for them, ensuring that their data are exposed only to the team rather than the whole community. The team members then develop patches and contribute them back to the PostgreSQL community.

## The Jun Team

The Jun team at SRA develops and distributes the Jun library, a Smalltalk and Java library for 3D objects and multimedia data handling. Different from the three teams discussed above, this team deals with the software that has been developed in-house. The source code of Jun, as well as its underlying object model, has been used by the Jun community. Jun has served as a reference model in the development of 3D objects and multimedia data handling (Aoki et al., 2001). Almost all of the Jun system is developed by a small group of programmers in SRA, and the development process is strictly controlled by the project leader. Although the community does not contribute much source code, it provides feedback, feature requests, and bug reports.

# ANALYSIS OF THE CASE STUDY

Unlike most previously related studies that focus on either the evolution of the system per se (Aoki et al., 2001; Godfrey & Tu, 2000) or the dynamics of the community (von Krogh et al., 2003), we take a broader perspective in our study: We examine both the evolution of the system and the evolution of the community, as well as their interrelationship.

## Roles and Structure of F/OSS Communities

The right to access and modify source code alone does not set F/OSS projects apart from proprietary ones because all developers in a project in most software companies would have the same access privilege. The fundamental difference is the *role transformation* of the people involved in a project. In proprietary software projects, developers and users are clearly defined and strictly separated. In F/OSS projects, there is no clear distinction between developers and users; all users are potential developers. Borrowing terms from programming languages, if we think of *developers* and *users* as types, and *persons* involved in a project as data objects, proprietary software projects are static-binding "languages" in which a *person* is bound to the type of *developer* or *user* statically, and F/OSS projects are dynamic-binding "languages" in which a *person* is bound to the type of *developer* or *user* dynamically, depending on his or her involvement with the project at a given time.

The distinct feature of *role transformation* in F/OSS projects leads to a different social structure. People involved in a particular F/OSS project create a community around the project, bonded by their shared interest in using and/or developing the system. Members of an F/OSS community assume certain roles by themselves according to their individual ability and personal interest, rather than being assigned roles by someone else. Benkler (2002) claims that this is the distinctive advantage of F/OSS, because it enables the matching of the best available person to a given job. Through the case study, we have found that, typically, a member may participate at a particular time in an F/OSS community in one of the following roles:

## Project Leader

The project leader is often the person who has initiated the project. The project leader oversees the direction of the whole project and makes most of the decisions about system development. Although all other members in a project are free to contribute and provide feedback, it is up to the project leader to decide which contribution should be included and which feedback should be addressed. Most GNU systems have a project leader.

## Core Member

Core members are responsible for guiding and coordinating, collectively, the development of a F/OSS project. Core members are those people who have been involved with the project for a long time and have made significant contributions to the development of the system. In those projects that have evolved into their second generation, a single project leader no longer exists, and the core members form a council to take the responsibility of overseeing the project. For example, PostgreSQL does not have a single project leader. Instead, it has six core members who collectively decide the direction of the system, and the inclusion of a new feature must be sponsored by one core member and approved by all other core members.

## Active Developer

Active developers regularly contribute new features and fix bugs; they are one of the major development forces of F/OSS systems and work very closely with the project leader or core members. The SRA GNU team members are active developers for GNU projects. PostgreSQL has 14 active developers, who make up the major development team in the community. Due to the limited amount of available time, the project leader and core members are not able to deal with all the contributions and feedback. Therefore, active developers, whose capability is well regarded and trusted by the project leader and core members and whose number is not very large, not only contribute their own code but also play, as we have seen in the SRA GNU team, an intermediary role by (1) improving the code contributed by less recognized developers, and (2) recommending the code to the project leader or core members.

## Peripheral Developer

Peripheral developers occasionally contribute new functionality or fix bugs. Their contribution is irregular, and the period of involvement is short and sporadic. The time that developers spend on a F/OSS project varies greatly, with most developers spending a rather short time (Koch & Schneider, 2002). The vast majority of developers make very small contributions. For example, in the GIMP project, 54% of developers contributed only once or twice (Ye & Kishida, 2003). Similarly, 91% of Linux developers contributed only one or two items (Dempsey, Weiss, Jones, & Greenberg, 2002). Most members in the four SRA teams are peripheral developers, as are the clients of the SRA GNU team who develop patches.

*Bug Reporter*

Bug reporters discover and report bugs. They do not fix the bugs themselves, nor do they necessarily read source code. They assume the same role as testers of the traditional software development model. The existence of many bug reporters assures the high quality of F/OSS systems because "given enough eyeballs, all bugs are shallow" (Raymond, 2001). Clients of the SRA PostgreSQL team are mainly bug reporters.

*Reader*

Readers are active users of the system. They not only use the system, but also try to understand how the system works by reading the source code. Given the high quality of most F/OSS systems, some readers read the systems to learn programming skills. Because GNU systems are meant to be "scientific knowledge to be shared" and are developed by very skilled programmers, their source code provides excellent educational resources for learning. As Michael Tiemann, founder of Cygnus, puts it: "It was this depth and richness that drove me to want to learn more, to read the GNU Emacs Manual and the GNU Emacs source code" (Tiemann, 1999: p. 72). At the same time, readers are also acting as peer reviewers or code inspectors who put implicit quality pressure on developers. Members of the Jun team pay special attention to keeping design simple, writing clear and high-quality code, following strict coding conventions, including documentation, and adding examples to show how it should be run because they are all aware that programmers worldwide will see their source code. Another group of readers exists who read a F/OSS system not for the purpose of improving the system but for understanding its underlying model and then using the model as a reference to implement similar systems. For example, the Jun system was used as a reference model by other developers who implemented a similar system in C++.

*Passive User*

Passive users use the system in the same way as the users of proprietary software. They are attracted to F/OSS mainly due to its high quality and its potential to be changed when needed.

Not all of these roles exist in all F/OSS communities, and the percentage of each type varies. Also, different F/OSS communities may use different names for the above roles. For example, some communities refer to core members as maintainers.

In addition to the above seven types of roles, another group of people is also an important factor to be considered in F/OSS development. We refer to these people as *stakeholders*. They are end-users who use computing services whose implementation is based on F/OSS systems. Although they are not members of a F/OSS community as they are not directly involved in using or developing the F/OSS system, they have stakes in the F/OSS system because they depend on it. As a database system,

PostgreSQL has a huge base of stakeholders. The existence of such stakeholders has implications on the decision-making process of F/OSS developers, which we will discuss further later.

Although a formally defined hierarchical structure does not exist in F/OSS communities, the structure of F/OSS communities is not completely flat. The influences that members have on the system and community are different, depending on the roles they play. Figure 1 depicts the general layered structure of F/OSS communities, in which roles closer to the center, or core, have greater influence. In other words, the activity of a project leader or a core member affects more members than that of an active developer, who in turn has a larger influence than a peripheral developer, and so on. Passive users have the least influence, but they still play important roles in the whole community. Although they do not directly contribute to the development of the system technically, their very presence contributes socially and psychologically by attracting and motivating other, more active members, to whom a large population of users is the utmost reward and compliment of their hard work (Raymond, 2001). Metaphorically speaking, those passive users play a role similar to that of the audience in a theatrical performance, which offers value, recognition, and applause to the efforts of actors.

Each F/OSS community may have a different percentage of each role. In general, most members are passive users. For example, about 99% of people who use Apache are passive users (Mockus et al., 2002). The percentage drops sharply from peripheral developers to core members (Mockus, Fielding, & Herbsleb, 2000; O'Reilly, 1999).

*Figure 1: General structure of a F/OSS community.*

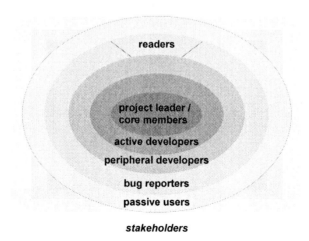

## Evolution of Systems

F/OSS systems evolve through the contributions of a large number of people. Different systems, however, have different evolution paths, depending on their objectives and policies regarding contributions. Any F/OSS developers can change the source code and share the change with other developers, but the way that the change is shared differs significantly.

Figure 2 summarizes how evolution takes place in the four projects we have studied. GNU software aims to have a single version. When people develop patches, they use dedicated mailing lists and newsgroups to share their patches. The patches are officially distributed and maintained only after they are incorporated into the core version by the project leader, who is not "obligated to include every change that someone asks [him or her] to include." The project leader checks not only the functionality enhancement of the patch but also its coding style and convention, documentation, and quality. Some patches are never incorporated. For this reason, many clients of the SRA GNU team, who are peripheral developers, ask the team members, who are active developers, to act as intermediaries to increase the possibility of patch incorporation.

In the Linux system, there is much less motivation and encouragement to contribute the developed patches back to the core version. Multiple implementations for the same functionality are allowed, especially in device drivers and in subsystems that support the features specific to particular CPUs (Godfrey & Tu, 2001). Many branches evolved from a single program may exist and compete with each other. Some branches die out eventually due to the lack of attention they receive from other developers and users. Consequently, the Linux community has a very large number of peripheral developers (e.g., 91.4% of contributors have contributed only one or two items (Dempsey et al., 2002)), and parallel evolution is normal in the Linux system.

*Figure 2: Evolutionary paths of the four projects.*

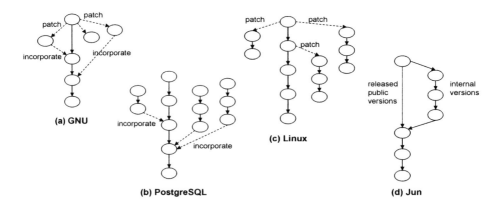

The major evolution of PostgreSQL takes place in the domain of dealing with new requirements. As new requirements emerge, interested users and developers debate about the importance and appropriateness of the new requirements. If the new requirements are deemed desirable after the debate, an active developer will organize a team to implement them. However, these new implementations will exist for a relatively long time as patches and are incorporated into the core version only after (1) they have been repeatedly tested, (2) their quality has been proven, and (3) they are approved by the core members through voting. During the debate about the requirements and the incorporation of the patches, the interests of a large number of stakeholders are constantly considered because the existing services provided by the systems that are based on PostgreSQL must not be disrupted.

Similar to GNU software, Jun evolves as a single-version tree. However, Jun differs from GNU software in that it has temporary branches that are created for internal use and test only and that cannot be accessed by the public. When the project leader decides that the system has been sufficiently tested internally, the tested version is released to public. Jun has evolved through 500 released versions over the last seven years. The evolution of Jun is driven by both the feedback from the community and the needs of several large projects that have used Jun. Unlike GNU in which patches that meet GNU quality control criteria and are approved by the project leader are directly incorporated into the official version, no patches submitted by community members are directly incorporated in Jun. The project leader often completely rewrites the code to assure the conceptual integrity of the system when he recognizes the ideas behind the submitted patches are worth incorporation. Jun has also undergone several refactoring processes (Fowler, Opdyke, Roberts, Beck, & Brant, 1999) that optimize the structure of the system. Therefore, the evolution of Jun is manifested as both the growth of system size and the restructuring of the system architecture that sometimes results in the reduction of system size (Aoki et al., 2001).

## Evolution of Communities

A F/OSS community is characterized by its members, the relations among its members, and the regularities of communication among its members. Therefore, the evolution of a F/OSS community is driven not only by the entry of new members and the exit of old members, but also by the role transformation of its members within the community. As community members change the roles they play in the community, they also change the social dynamics and reshape the structure of the community. The role that a F/OSS member plays in the community might constantly change, depending on how much the member wants to get involved and on how the member's involvement affects and is perceived by other members. The role is not preassigned; it is assumed by the member as he or she interacts with other members. An aspiring and determined member can become a core member through the following path or "joining script" as it is called by von Krogh, Spaeth, and Lakhani (2003).

New members are often attracted to a F/OSS community because the system can solve one of their own problems. New members start as passive users. While using the system, they may find a few bugs and report them back to the community, changing their role from passive users to bug reporters. The depth and richness of good F/OSS systems often drives motivated members to want to learn more, to read the system (Tiemann, 1999); in this case, new members migrate from being passive users to readers. As they gain more understanding of and experience with the system, they become capable of fixing bugs that either they encounter or are reported by others. They may also want to add a new twist to the system to make the system more powerful and more suitable for their own tasks. As their developed patches are made publicly available and benefited from by other community members, their roles as peripheral developers are recognized and established in the whole community. The more contributions they make, the higher recognition they earn, and finally, they will make into the highly selected "inner circle" of active developers and core members.

The above path describes an abstract and idealized model of the role transformation of aspiring members, which is common in all the four of the studied F/OSS projects. Not all members want to and will become active developers or core members. Some will always be passive users, and some will stop in the middle. In fact, most of the users served by the four teams at SRA remain passive users. Members of the SRA GNU team and the leader of the SRA PostgreSQL project have become active developers due to their long-term contributions to their respective communities. Members of the Linux support team at SRA remain peripheral developers because they have not contributed very much back to the community. Because all the development of Jun is conducted in SRA by the project leader and active developers, the evolution of the community is limited to the two outside layers as shown in Figure 1, i.e., from passive users to readers or bug reporters.

The evolution of a F/OSS community is thus determined by two factors: (1) the existence of motivated members who aspire to play roles with larger influence, and (2) the social mechanism of the community that encourages and enables such individual role transformation. This is consistent with the community-based learning theory called Legitimate Peripheral Participation (LPP) (Lave & Wenger, 1991). In LPP theory, a community of practice evolves by reproducing itself when new members (i.e., apprentices) gradually establish their identities as fully qualified members (i.e., masters) in the community. The process of identity establishment is also a process of learning—not as a result of being taught, but through legitimate peripheral participation in social, cultural, and technical practices within the community.

As new members join a F/OSS community, the right to access and change the source code of the system in collaboration with established members grants them the legitimate participation in the community. During the collaboration, new members are also given the legitimate access to the knowledge of existing members. At first, due to their limited capability and knowledge, new members can only peripherally participate by engaging in small yet authentic tasks, such as bug reporting and bug fixing. As they gain more experience and knowledge about the system and the community, new members become competent in undertaking more important tasks, thus making more contributions to the community. As their skills are gradually recognized by the community, based on their contributions, they are trusted to perform bigger and more challenging tasks, and move toward the "inner circle" of the community, establishing their identity as competent full participants, enjoying better reputation as skillful developers, and exerting larger influence.

The ontogenetic development of the identity of individual members changes their relations with the community as well as the relations among other members, resulting in the evolution, or phylogenic development, as well as the preservation and reproduction of the collective identity of the community. The collective identity of a F/OSS community that rests on the shared set of beliefs, knowledge, and practice is a critical precondition for the smooth, effective, and efficient collaboration during the development of the F/OSS system when developers are geographically and culturally distributed all over the world (Orlikowski, 2002). Because the collective identity of a community is constituted and reconstituted out of the social process of interaction among its members (Weick, 1995) and their common practice (Brown & Duguid, 2000), the collective identity of a F/OSS community can only be conserved and reproduced through the ontogenetic development of the identities of its new members, who become, through legitimate peripheral participation, masters who embody the mature practice and knowledge of the community.

## Co-Evolution of Systems and Communities

The evolution of F/OSS systems and the evolution of F/OSS communities are mutually constituent. In all the four of the studied projects, the evolution of the F/OSS community results from the contributions made by its aspiring and motivated members. Such contributions not only evolve the whole community by means of either transforming the roles of those relatively new contributors or reconstituting the central roles of those old and active contributors (because roles in a F/OSS community are not given and static; rather, they are achieved dynamically through recurrent practice and participation) (Orlikowski, 2002), but also are the sources for system evolution. The reverse is also true. Any modification, improvement, and extension made to a F/OSS system—whether it is a bug report, a bug fix, or a patch—not only evolves the system itself but also redefines or reconstitutes the roles of the contributing members and thus changes the social dynamics of the F/OSS community (Figure 3).

For a F/OSS system to have a sustainable development, the system and the community must co-evolve. A large base of voluntarily contributing members is one of the most important success factors of a F/OSS system. Without new members

*Figure 3: The co-evolution of F/OSS systems and F/OSS communities.*

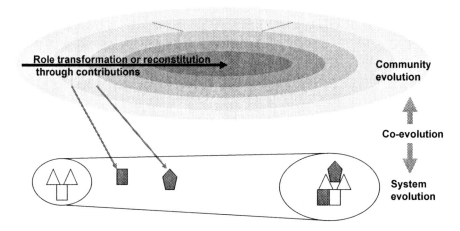

aspiring to become core members through continuous contributions to the system and the community, the development of the system will stop when some of the original core members decide to leave the community or simply stop contributing their time, knowledge, and services for some reason. Because members participate in F/OSS development voluntarily, such eventualities are always possible.

GIMP (Gnu Image Manipulation Program; see http://www.gimp.org/) is such an example. When the original two creators, Peter Mattis and Spencer Kimball, left college to take jobs, they cut their ties with GIMP because they thought they had done their services to the community and wanted to move on (HackVan, 1999). Because almost the entire system had been developed by these two developers, and at that time there was not a GIMP developer community to pick up immediately where the two had left off, the system stayed incomplete for more than a year. The development was resumed later when a community was finally formed (Ye & Kishida, 2003).

Because the evolution of F/OSS communities and the evolution of F/OSS systems are mutually dependent, it is essential to the long-term success of F/OSS development that enough attention should be paid to the creation and maintenance of a dynamic and self-reproducing F/OSS community. The project leader and core members of an existing F/OSS community should not only focus on the evolution of the system, but also strive to create an environment and culture that fosters the sense of belonging to the community and mechanisms that encourage new members to move toward the center of the community through continual participation.

# THREE TYPES OF F/OSS PROJECTS

Based on the analysis of the case study and the review of available research literature on F/OSS, we propose to classify F/OSS projects into three different types:

Exploration-Oriented, Utility-Oriented, and Service-Oriented. We do not mean that the three types cover all the F/OSS projects. Our attempt at defining three types of F/OSS is to create a general understanding that, although all of these systems are called free or open-source software, differences do exist in primary objective, control style, system evolution, community structure and evolution.

## Exploration-Oriented F/OSS

Exploration-oriented F/OSS projects, represented by GNU software and the Jun library, aim at pushing the frontier of software development collectively through the sharing of innovations embedded in freely shared F/OSS systems. This is very similar to the culture of a scientific research community in which scientific results are shared through conferences and journals for peer justification, mutual inspiration, and continued development (DiBona et al., 1999). Due to the epistemic nature of this F/OSS type, the quality requirements are often very high. Therefore, this type of software tends to be developed and maintained by expert programmers, such as project leaders, who keep a tight control over the system to maintain the conceptual and architectural integrity of the system so it reflects its original design goal.

This type of F/OSS systems evolves mainly at the hand of the project leader, and contributions made by the community exist as feedback. Direct code contributions to the system by community members are not frequent because those code contributions are incorporated only if they are consistent with the ideas of the project leader (Figure 2(a)). Most community members collaborate with the project leader as readers, bug reporters, or peripheral developers, who occasionally provide feedback and requests for new functionality. Because the project governance style of this F/OSS type is closer to the cathedral style than to the bazaar style (Raymond, 2001), it is more difficult for community members to move toward the center, and consequentially a very small number of active developers exists. From the perspective of the project leader, community members are more like assistant developers who help inspect, test, and find bugs in the system, rather than equal partners in system development. The community structure has a definite core, thin middle layers, and a large periphery. This is very similar to the "surgical team" suggested by Brooks (1995), with the project leader as the surgeon (or chief programmer), a few active developers as copilots, and other community members as testers.

Although community evolution in this type of F/OSS projects takes place mostly at the periphery through providing peer-to-peer assistance in understanding and using the software, it still provides important driving forces for system evolution by providing feedback for system improvement. System evolution enabled by contributions made by community members allows the contributors to develop a better understanding of the system and a better ability to offer more contributions in the future.

The success of such F/OSS projects depends greatly on the vision and leadership of the project leader. However, when the vision of the project leader conflicts with the needs of the majority of the community members, forking might happen. A new F/OSS project and community will be spun off the original one and embark on a similar but different development path. Two typical examples are the spinning off of EGCS from GCC, and XEmacs from Emacs.

## Utility-Oriented F/OSS

Utility-oriented F/OSS projects, represented by the Linux system (excluding the Linux kernel, which started as an exploration-oriented one and now is a service-oriented one), aim at filling a void in functionality. Most such F/OSS systems consist of many relatively independent programs (e.g., device drivers in Linux), and those programs are developed because the original developers could not find an existing program that completely met their needs. Rather than waiting for others to provide the needed functionality, they develop and share their own programs.

As we have observed from the SRA Linux support team, few F/OSS programs in utility-oriented projects are completely developed from scratch. Most developers search the Internet for a partial solution and then modify it to their own needs. Their primary concern is not to use the source code as a way of scientific exploration as the exploration-oriented F/OSS developers do, but to create a program that can meet their personal needs, or scratch their personal itch (Raymond, 2001). Because the development is driven by an individual need, developers are concerned with developing an operational system rather than delivering a refined solution as in the exploration-oriented type.

As the program is released for sharing, other users who have a similar problem will pick it up and either use it as is or modify it further. As we have mentioned before in the discussion of the SRA Linux support team, the original developers are not very much concerned if they receive feedback or improvement from potential users, as long as the current program works to their satisfaction. This type of F/OSS software development is a typical bazaar style; no centralized control exists. Unlike the exploration-oriented F/OSS in which forking is rather rare and the evolution of the system takes place in the form of improving the original system, utility-oriented F/OSS has a lot of forks, evolving in the form of developing new programs by reusing and modifying existing programs rather than replacing the old programs with the new ones (Figure 2(c)). This leads to the existence of multiple, often incompatible programs. Programs that implement a similar functionality compete with each other and evolve simultaneously, but the implementation that wins the most support in the community will finally excel and eliminate other competing versions. This evolution pattern can be called tournament style.

One program of the utility-oriented F/OSS may not have an independent community associated with it. Instead, it often exists within a larger F/OSS community of the system of which the program is a part. For example, many Linux device driver developers are a part of the larger Linux community. From the perspective of the larger F/OSS community, those developers of F/OSS programs are specialized peripheral developers. Because most such developers only want to develop a program for their own particular need, they remain peripheral developers, such as the members of the SRA Linux support team. However, those peripheral developers might also move on to become active developers or core members if they extend their involvement beyond their specialized areas and make more general contributions to the whole system after having demonstrated their mastery in coding through their initial code contributions in the specialized areas (von Krogh et al., 2003).

The system evolution of the utility-oriented F/OSS is driven by the diversified needs of, and contributions made by, its community members. Due to the high specialization of most utility-oriented F/OSS programs, they have a lower entry barrier that makes it easier for newcomers to join the community as peripheral developers. As the utility-oriented program will eventually interact with other components of the larger system, those peripheral developers are also given the opportunity to spread their efforts into the whole system and establish themselves as active developers or core members in the larger community.

## Service-Oriented F/OSS

Service-oriented F/OSS projects, represented by PostgreSQL and Apache, aim at providing stable and robust services to all the stakeholders of F/OSS systems. In a service-oriented F/OSS project, because the population of stakeholders is much larger than that of the F/OSS community, any changes made to the system have to be carefully considered so they do not disrupt its provided services, on which many end-users rely. Therefore, service-oriented F/OSS projects are usually very conservative against rapid evolutionary changes. It takes four to 11 months for a major version of PostgreSQL to be released.

In accordance with its conservative nature, the control style of service-oriented F/OSS is neither cathedral-like nor bazaar-like. Although the cathedral style has a tight control over the system, it is often controlled by one project leader whose creative idea may not reflect the best interests of all the stakeholders. On the other hand, the bazaar style encourages too many rapid changes to provide stable services. As we can see in the PostgreSQL community, service-oriented F/OSS is often collectively controlled by a group of core members, and there is no single project leader. Any changes are subject to debate by the group and only those changes that won the support of the majority of the group are incorporated. We call this kind of control the council style.

Although the control over the development of the F/OSS system is still centralized in the council style, it is not controlled by any individual person. The council is the assembly of core members, who earn their rights by long-time devotion and contributions to the F/OSS community (Mockus et al., 2002). Furthermore, the membership of the council is not fixed. Most F/OSS communities of this type have a mechanism of accepting new council members whose contributions and competence are well recognized and who are trusted by community members.

Most members of service-oriented F/OSS communities exist as passive users, while some of them may become bug reporters and peripheral developers as they report or submit bug fixes back to the community. Active developers emerge when some big changes are needed, such as the requirements of dealing with the Japanese language in PostgreSQL. Active developers often work with other peripheral developers to develop a patch for the new requirements, and the patch is finally incorporated into the official version of the system if it has been widely tested and approved by core members (Figure 2(b)). The system evolution of service-oriented F/OSS is contingent on the increase of the size of its user community. As more users start to use the system, more bugs are reported and more new requirements are generated. The reported bugs result in the incremental quality improvement of the system as peripheral developers submit bug fixes. New requirements often result in major development efforts, generating a new crop of active developers in the community who can evolve the system further.

# PRACTICAL IMPLICATIONS

Realizing that co-evolution of systems and communities is a key to the sustainable development of F/OSS projects and recognizing that different types of F/OSS projects exist have practical implications in managing and developing F/OSS projects, as discussed here.

## Creating Opportunities for Legitimate Peripheral Participation

F/OSS communities evolve through the legitimate peripheral participation of new members. The openness of the produced system, the development process, and the communication among community members is essential to provide new members with legitimate participation in practice and legitimate access to learning resources—products and processes—available in the community. Although all F/OSS communities are open to certain forms of participation and access, the different control structure inherent in each F/OSS community due to considerations of system quality creates different degrees of openness that allow legitimate participation and access of community members.

Table 1 shows the possible combinations of openness in two dimensions: product (row) and process (column). In the product dimension, open release means that only formally released versions are accessible to all community members, and open development means that all interim versions are accessible. In the process dimension, closed process means that the discussion of system development is conducted mostly within the "inner circle," often through a strictly controlled mailing list that is not accessible to other members; transparent process means that although only the "inner circle" is involved in the development process, their discussion is readable by other members; open process means that development decisions are made in public

*Table 1: Openness of OSS communities.*

|  | Open release | Open development |
|---|---|---|
| **Closed process** | GNU; Jun | Apache |
| **Transparent process** | Tcl/Tk | PostgreSQL |
| **Open process** |  | GIMP; Linux (excluding kernel) |

space, allowing the participation and access of all interested parties. Encouragement of broad participation requires the highest degree of openness in both dimensions because it offers more learning resources and opportunities for participation. However, it may also reduce the project leader's control over the system. This conflict needs to be carefully balanced by those who want to make their systems open source. To maintain control over the quality, exploration-oriented and service-oriented F/OSS may not adopt an open process but should make the process at least transparent to offer more learning opportunities, while utility-oriented F/OSS can have the maximum openness in both dimensions.

The possibility for newcomers to participate peripherally is another key aspect in LPP. To attract more users to become developers, the system architecture must be designed in a modularized way to create many relatively independent tasks with progressive difficulty so that newcomers can start to participate peripherally and move on gradually to take charge of more difficult tasks. The way a system is partitioned has consequences for both the efficiency of parallel development—a prerequisite for F/OSS—and the possibility of peripheral participation. The success of Linux is due in large part to its well-designed modularity (Torvalds, 1999).

Another approach to afford peripheral participation is perhaps to intentionally under-design the system by leaving some non-critical parts unimplemented to facilitate easy participation. The TODO list of most F/OSS systems creates guidance for participation. Rather than just listing TODO items, grouping them according to their estimated difficulty might provide a better roadmap for newcomers to start participation at the periphery.

## Advice for F/OSS Practitioners

Developers at the center of F/OSS communities should strive to create an environment and culture that encourage and enable newcomers to move toward the center of the community. It is very important for the community to be responsive to the questions and contributions of newcomers in order to sustain their interest and encourage their further participation. Established members should remember that they are also the learning resources for newcomers. Core members and active developers of PostgreSQL have devoted a lot of attention to educating newcomers by writing a flowchart of its modules to outline the purpose of each. One possible mechanism is to have skilled members take turns in the mailing list to answer questions of newcomers and to help new contributors perfect their code contributions.

People who want to start a F/OSS project need to consider how many opportunities for participation it offers and how easy the project is for others to participate legitimately and peripherally. A system with a large size and cumbersome architecture, such as the early version of Mozilla, makes it difficult to attract new F/OSS participants (Baker, 2000).

Different types of F/OSS projects require different types of efforts to sustain the continuous evolution of systems and communities. By identifying their projects with the particular type, F/OSS developers can take appropriate measures to guide the management and operation of the projects.

The project leader of an exploration-oriented F/OSS should pay extra attention to the quality and readability of the source code by enforcing strict coding, formatting, and documenting conventions so that it can fully fulfill its goal of disseminating good programming skills and knowledge. To avoid unnecessary fragmentation of the community resources caused by forking, project leaders need to adapt and respond to the needs of the community members.

Project leaders of utility-oriented F/OSS do not need to worry too much about forking. Instead, they need to develop a social mechanism that coordinates and encourages peer support among the community members and to facilitate the easy choice of different implementations of the same functionality.

Project leaders or core members of service-oriented F/OSS should always keep in mind that many stakeholders rely on the stable services the systems provide. The simple strategy of "release early, release often" (Raymond, 2001) is not the best one for this type of projects, because less knowledgeable, passive users might pick up those unstable versions to provide services for their end-users. To balance the conflicting needs of both providing stable and robust systems and leveraging the whole community as testers and code inspectors, developers might consider a two-stage version release strategy, i.e., releasing beta versions explicitly for testing purposes while maintaining robust versions simultaneously for passive users.

## Changing the Type of F/OSS Projects to Accommodate New Needs

As the F/OSS system and community evolve over time or the socio-technical environment for the F/OSS changes, the type of the F/OSS project might need to change accordingly. Exploration-oriented and utility-oriented types are good for the initiation of a F/OSS project, while a service-orientation is more suitable for mature F/OSS projects.

The success of an exploration-oriented F/OSS attracts many followers, who, as users, will demand stability at some point because they have invested much effort into the project and have used it to develop systems for end-users. For the benefit of the F/OSS system itself and the community, it is better for such a project to mutate into a service-oriented type; otherwise, the F/OSS community may split or the F/OSS system may simply be abandoned by most community members, making the F/OSS project a victim of its own success.

One example that has successfully completed the mutation from an exploration-oriented F/OSS to a service-oriented one is the Tcl project (http://www.scriptics.com/). Tcl was initially created in 1988 by John Ousterhout who wanted to explore a different style of system programming through the creation of a scripting language for "gluing" existing applications (Ousterhout, 1998). More than a decade later, Tcl is now used by thousands of companies and over 500,000 users, often for mission-critical applications. In the earlier years, Tcl was developed as an exploration-oriented type, with Ousterhout deciding what extensions should be included based on his own interest and feedback from the community. Since August 2000, the development of Tcl is at the helm of the Tcl Core Team (TCT); the team is made up of 14 members who were elected by the Tcl community in recognition of their long-time devotion and support of Tcl.

Utility-oriented F/OSS projects also need to change into service-oriented ones at some point. Developers of competing implementations for similar functionality can join forces to create a team to develop a system collaboratively that can accommodate the different needs of each developer. The Apache project is a typical example. The project, since it assumed the name of Apache, has been a typical service-oriented F/OSS project under the control of the Apache Group. However, the Apache Group was formed because its members felt the need to combine the competing and overlapping extensions and bug fixes they had developed individually and simultaneously for the NCSA HTTPd system in a utility-oriented style (Fielding, 1999).

Exploration-oriented and utility-oriented F/OSS projects experience rapid evolution, mostly in a super-linear fashion (Aoki et al., 2001; Godfrey & Tu, 2000). As the projects mature into service-oriented ones, the speed of evolution slows down to a stable growth. A new round of rapid evolution may start again if the stable F/OSS systems inspire new ideas or new requirements, giving birth to new exploration-oriented or utility-oriented F/OSS projects.

## CONCLUSIONS

We reported a case study of four F/OSS projects that we have conducted to understand the evolution of F/OSS. We analyzed the evolution of each F/OSS system and the evolution of its corresponding community. Our analysis has identified both the similarities and differences among the four projects in terms of their objective, collaboration model, control style, system evolution, and community structure and evolution. The similarities among those projects have led us to propose the theory that F/OSS systems and communities co-evolve, with evolution in one inevitably causing evolution in the other. The differences have prompted a theoretical classification of F/OSS projects into three different types: exploration-oriented, utility-oriented, and service-oriented. Each type requires different strategies to manage the sustainable evolution of system and community. Practical implications of recognizing the similarities and differences among the different types of F/OSS projects are also discussed.

Our theories, developed through qualitative rather than quantitative analysis, are still at a primitive stage. The goal of this chapter is to provide an initial theoretical framework for discovering the best strategies to guide the sustainable development of F/OSS projects by investigating the mutual constituency of system evolution and community evolution. We are currently conducting quantitative studies of other F/OSS projects to substantiate our theories (Ye & Kishida, 2003), and we hope that we have stimulated the interests of some readers to join us in this direction.

# REFERENCES

Aoki, A., Hayashi, K., Kishida, K., Nakakoji, K., Nishinaka, Y., Reeves, B., Takashima, A., & Yamamoto, Y. (2001). A case study of the evolution of Jun: An object-oriented open-source 3D multimedia library. *Proceedings of 23rd International Conference on Software Engineering (ICSE'01)*, Toronto, Canada, pp. 524-533.

Baker, M. (2000). *The Mozilla Project and Mozilla.org*. Retrieved February 2, 2002 from: http://www.mozilla.org/editorials/mozilla-overview.html

Benkler, Y. (2002). Coase's Penguin, or Linux and the nature of the firm. *Yale Law Journal, 112*(3), 369-446.

Brooks, F. P. J. (1995). *The mythical man-month: Essays on software engineering, 20th Anniversary edition*. Reading, MA: Addison-Wesley.

Brown, J. S. & Duguid, P. (2000). *The social life of information*. Boston, MA: Harvard Business School Press.

Dempsey, B. J., Weiss, D., Jones, P., & Greenberg, J. (2002). Who is an open source software developer? *Communications of the ACM, 45*(2), 67-72.

DiBona, C., Ockman, S., & Stone, M. (Eds.). (1999). *Open sources: Voices from the open source revolution*. Sebastopol, CA: O'Reilly & Associates.

Fielding, R. T. (1999). Shared leadership in the Apache project. *Communications of the ACM, 42*(4), 42-43.

Fowler, M., Opdyke, W., Roberts, D., Beck, K., & Brant, J. (1999). *Refactoring: Improving the design of existing code*. Reading, MA: Addison-Wesley.

FSF. *GNU Philosophy*. Retrieved February 2, 2002 from: http://www.gnu.org/philosophy/philosophy.html.

Godfrey, M., & Tu, Q. (2000). Evolution in Open Source Software: A Case Study. *Proceedings of 2000 International Conference on Software Maintenance*, San Jose, CA, pp. 131-142.

Godfrey, M. & Tu, Q. (2001). Growth, evolution, and structural change in open source software. *Proceedings of the 4th International Workshop on Principles of Software Evolution (IWPSE'01)*, Vienna, Austria, pp.103-106.

HackVan, S. (1999). *Where did Spencer Kimball and Peter Mattis go?* Retrieved February 2, 2002, from: http://devlinux.com/.

Hars, A. & Ou, S. (2001). Working for free? - Motivations of participating in open source projects. *Proceedings of the 34th Hawaii International Conference on System Sciences*, Maui, HI, January 3-6, p. 7014.

Hertel, G., Niedner, S., & Hermann, S. (2003). Motivation of software developers in open Source projects: An Internet-based survey of contributors to the Linux Kernel. *Research Policy, 32*(7), 1159-1177.

Koch, S. & Schneider, G. (2002). Effort, co-operation and co-ordination in an open source software project: GNOME. *Informaton Systems Journal, 12*(1), 27-42.

Lave, J. & Wenger, E. (1991). *Situated learning: Legitimate peripheral participation*. Cambridge, UK: Cambridge University Press.

Mockus, A., Fielding, R., & Herbsleb, J. (2000). A case study of open source software development: The Apache server. *Proceedings of 2000 International Conference on Software Engineering (ICSE2000)*, Limerick, Ireland, pp. 263-272.

Mockus, A., Fielding, R. T., & Herbsleb, J. D. (2002). Two case studies of open source software development: Apache and Mozilla. *ACM Transactions on Software Engineering and Methodology, 11*(3), 309-346.

Moon, J. Y. & Sproull, L. (2000). Essence of distributed work: The case of the Linux kernel. *First Monday, 5*(11), November, http://firstmonday.org/issues/issue5_11/moon/index.html.

O'Reilly, T. (1999). Lessons from open source software development. *Communications of the ACM, 42*(4), 33-37.

Orlikowski, W. J. (2002). Knowing in practice: Enacting a collective capability in distributed organizing. *Organization Science, 13*(3), 249-273.

Ousterhout, J. (1998). Scripting: Higher level programming for the 21st Century. *IEEE Computer, 31*(3), 23-30.

Raymond, E. S. (2001). *The cathedral and the bazaar: Musings on Linux and open source by an accidental revolutionary*. Sebastopol, CA: O'Reilly.

Tiemann, M. (1999). Future of Cygnus Solutions. In C. DiBona, S. Ockman & M. Stone (Eds.), *Open sources: Voices from the open source revolution*, pp. 71-89. Sebastopol, CA: O'Reilly.

Torvalds, L. (1999). The Linux edge. *Communications of the ACM, 42*(4), 38-39.

Turski, W. M. (1996). Reference model for smooth growth of software systems. *IEEE Transactions on Software Engineering, 22*(8), 599-600.

von Krogh, G., Spaeth, S., & Lakhani, K. R. (2003). Community, joining, and specialization in open source software innovation: A case study. *Research Policy, 32*(7), 1217-1241.

Weick, K. (1995). *Sensemaking in organizations*. Thousand Oaks, CA: Sage.

Ye, Y. & Kishida, K. (2003). Toward an understanding of the motivation of open source software developers. *Proceedings of 2003 International Conference on Software Engineering (ICSE'03)*, Portland, OR, pp. 419-429.

# SECTION II:

# F/OSS Development and Software Engineering Practices – "Extensive Analysis"

## Chapter IV

# The Role of Modularity in Free/Open Source Software Development

Alessandro Narduzzo, Università di Bologna, Italy

Alessandro Rossi, Università di Trento, Italy

## ABSTRACT

*Software design and development in Free/Open Source projects are analyzed through the lens of the theory of modularity applied to complex systems. We show that both the architecture of the artifacts (software) and the organization of the projects benefited from the paradigm of modularity in an original and effective manner. In particular, our analysis on empirical evidence suggests that three main shortcuts to modular design have been introduced and effectively applied. First, some successful projects inherited previously existing modular architecture, rather than designing new modular systems from scratch. Second, popular modular systems, like GNU/Linux kernel, evolved from an initial integrated structure through a process of evolutionary adaptation. Third, the development of modular software took advantage of the violation of one fundamental rule of modularity, that is, information hiding. Through these three routines, the projects can exploit the benefits of modularity, such as concurrent engineering, division of labor, decentralized and parallel development; at the same time, these routines lessen some of the problems posed by the design of modular architectures, namely imperfect decompositions of interdependent components. Implications and extensions of Free/Open Source projects experience are discussed in the conclusions.*

# INTRODUCTION

In this chapter we argue that modularity, a well-known paradigm for the design and the production of complex artifacts (Schilling, 2000), is a key element in explaining the development and the success of many Free/Open Source (F/OSS) projects, since it offers a comprehensive explanation of many key issues, such as how division of labor takes place between developers, how coordination is achieved, how code testing and integration is deployed and how innovation occurs.

Our reconsideration of the accounts of many F/OSS projects highlights how they benefited from the typical advantages of implementing modular architectures (e.g., fast speed of development, recombination of modules, innovation through projects competition, reuse of previously developed code) (Feller & Fitzgerald, 2000; Hatch, 2001; Jackson, 1998), while, at the same time, many critical pitfalls typically related to managing modularity (the architectural design of modules and interfaces) were avoided. Three interrelated strategies, or design shortcuts, appear to have been particularly effective in this respect.

First of all, the architectural guidelines of many complex systems were clearly inherited from previously existing modular software projects (see, for instance, the GNU project and the FreeBSD project, which closely resemble the UNIX architecture). By imitating a well-established architecture, developers were able to avoid the problems related to the design of modular architectures from scratch, namely, devising a decomposition of the whole system in independent sub-parts or modules (as we will see further on, this is not a trivial task).

Secondly, when devising modular architectures that are considerably innovative, it is not possible to rely on blueprints of existing software; in such cases, another design shortcut is to think of modularity not in terms of a static and ex-ante design principle but, rather, as a dynamic activity of problem solving that starts from fairly interconnected architectures that are repeatedly fine-tuned and reworked, leading over time to more modular outcomes ("evolving modularization"). In this respect, we found it useful to analyze the evolution of the GNU/Linux kernel. Conversely, pursuing full modularity from the beginning may be very risky, and may eventually lead to serious difficulties (as in the case of the development of the HURD microkernel).

Finally, F/OSS development style seems to suggest a third effective design shortcut. While information hiding has been traditionally viewed as the key principle guiding both the design and the implementation of modular software artifacts, F/OSS seems to substantially disregard this principle at the implementation level. For instance, empirical evidence shows that F/OSS developers systematically improve parts of the project on which they are working by tinkering with the code of multiple modules, taking advantage of the source availability and of the absence of code ownership.

Our examination is based on published and unpublished data: interviews and papers written by key actors, analyses developed by scholars, and quite a large mass of original documents made available through Internet websites.

The chapter is organized as follows: The next section surveys some of the most relevant topics of modularity in management and organization science. Then, it turns to software development, characterizing modularity as one of the fundamental topic in the software engineering debate. We then interpret F/OSS development through the lens of the theory of complex modular system. We summarize the accounts of specific projects and we advance some stylized facts of F/OSS development. The final section sketches some reflections on how F/OSS methodologies may fully benefit from a mindful modular design and suggests how evidence collected by many F/OSS projects may help in refining both existing theories of modularity and their practical application to domains different from software production.

# MODULARITY

Modularity has been receiving an increasing amount of attention in a variety of fields, from neuroscience and artificial intelligence to architecture, urban design, and management (Baldwin &Clark, 2000). Nowadays, a modular approach is applied to complex projects in R&D, industrial manufacturing, and software engineering, and modularity has been assumed as a key-concept in the design and production of a great number of artifacts, both physical (e.g., buildings, cars, furniture) and immaterial ones (e.g., software) (Schilling, 2000).

This interdisciplinary interest is largely due to the fact that modularity is regarded as a general property of complex systems, pertaining to the degree of decomposability of a system in loosely coupled sub-parts made by tightly coupled components. Literature on modularity emphasizes the importance of structures and relationships, and the outlined models all rely on an underlying system theory that provides a comprehensive framework for understanding and pertinently describes the specific object of study (artifacts, objects, machines, tasks, molecules, spaces, projects). A modular system is thus represented as a complex of components or sub-systems, where designers try to minimize and standardize the interdependencies among modules.

Herbert Simon's (1957, 1981) influence in the way modularity has been conceived is particularly evident. First of all, modularity is often introduced within a problem-solving framework, and modular design is regarded as a solution to cope with uncertainty and variability. Second, as in Simon's analysis of the artificial, modularity in complex systems regards both goals and hierarchies. Third, modular solutions are based on problem decomposition; fourth, since complex systems are not quite entirely decomposable, modular design eventually needs to deal with residual interdependencies (Simon, 1981).

## Modularity in Management and Organization Science

In management and organization science literature, modularity has been introduced as an innovative paradigm in firm manufacturing (Schilling, 2000; Ulrich,

1995), organization design (Baldwin & Clark, 2000), and in the theory of the firm (Langlois, 2002). Modularity provides relevant advantages that have been neatly identified in the literature. Modularity allows for product variety that is obtained by recombination (mix and match) of components (Langlois & Robertson, 1992). Modularity is viewed as a base for differentiation strategies; firms may enrich their products catalog and adapt to customers' needs with limited additional costs (Camuffo, 2002). Moreover, modularity also has a great impact on production processes as it positively affects flexibility, division of labor, and specialization (Devetag & Zaninotto, 2001).

According to Baldwin and Clark (2000), modularity in production systems is obtained by following some general rules, originally drawn from computer science and software development, concerning two different categories of information: visible and hidden information. Modular systems design needs to specify only the visible rules, namely, the information about: *(a)* the definition of the architecture, *(b)* interface specifications, and *(c)* module integration tests. The inner description of each module (how it works) is concealed from outside—it does not need to be defined ex-ante or communicated during the process, since module interactions exclusively follow the rules specified by the interfaces parameters.

Unfortunately, the design of modular complex systems is not as smooth as it has been described; most of the times, after the integration of the independently developed modules, inconsistencies surface and the system does not work properly. The most common reason for this failure is that the decomposition of a complex system is not a trivial business at all. Most of the times, the activities of decomposition are suboptimal and result, at best, in a quasi-decomposable architecture where some degree of interdependency between modules is still at work. As we will see in the next subsection, residual and unforeseen interdependencies seem to be particularly relevant in the production of software artifacts.

## Modularity in Software Development
*The "Power of Modularity" in Software Engineering*

The notion of modularity is central in the design and production of software artifacts, especially for large and complex projects. Since the early days of software engineering, the issue of designing, developing, testing, and releasing a large software project brought into discussion the trade-off between simplicity and speed of development.

Frederick Brooks (1975), the author of one of the most influential software project management handbooks, clearly recognized that small sharp teams performed better than large ones, but they were not sufficiently staffed to deliver large software projects under schedule pressure. Conversely, while larger teams potentially increased the pace of the development process, they also resulted in an overwhelming need for coordination of individual efforts and in diminishing marginal returns of manpower on productivity (also known, in the extreme case of negative marginal returns, as the "Brooks' Law").

Brooks' recipe for coping with the design and the production of complex software was to vertically divide labor in order to separate high-level activities (such as the design of a software artifact) from lower ones (such as the implementation of code) as much as possible. Then, the implementation of each part of the project is assigned to small and focused teams (the so called "surgical team").

In terms of a modern theory of modularity, the basic assumption inside Brooks' seminal work is that large software projects are integral and non-decomposable systems, where interactions among parts are nontrivial and generate high communication and coordination needs. What is clearly overlooked from Brooks' perspective is that interdependencies may not only be considered as given constraints. As a matter of fact, the introduction of a fully modular approach in modern software engineering methodologies has been fostered by the recognition that the degree of interdependencies may be greatly reduced if a complex software project can be decomposed in independent subparts, that is, dividing the whole project into smaller components that are loosely coupled and highly independent on each other (Langlois, 2002; von Hippel, 1990).

Conceiving the design of a complex software artifact as a modular system means to apply the basic principle of "information hiding," originally developed by Parnas (1972), that prescribes treating software modules as opaque entities whose relevant information is only available to its inner programmer, while not being accessible to external programmers. Here, the only information revealed is embedded in the interfaces, while the information regarding the design and how the module works is not communicated.

Nowadays, the widespread adoption of object-oriented languages and the diffusion of component-based development as well other popular trends in software engineering seem to have affirmed this information hiding principle and the paradigm of modularity as common software practices aimed at speeding up the development process.

### *Modularity as a Complex Design Activity: Managing Unforeseen Interdependencies in Software Modules*

A software product architecture may be defined as a mapping of required functions of the product in functional components. The system as a whole is decomposed in a set of functional modules whose interactions provide the overall functionality of the system (Sanchez & Mahoney, 1996; Ulrich, 1995). When it comes to the topic of component interactions, software seems to be a particular artifact with respect to physical artifacts. Both software engineering literature (Brooks, 1975; Pressman, 2000; Schach, 2002) and empirical case studies of software product development (Glass, 1997; Solheim & Rowland, 1993) suggest that integrating software components may be harder than assembling hardware artifacts.

Brooks' famous essay on the difficulties of software engineering techniques in granting improvements in productivity, reliability, and simplicity in developing software programs describes the fundamental properties of software entities that may account for the difficulties in separating interdependencies and decompose large software projects (Brooks, 1986). Software entities differ from physical artifacts in that their highly nonlinear complexity leads to the impossibility of enumerating (not to mention understanding) all the possible states of a program. As the size of a software project increases, it becomes more and more difficult to decompose interdependencies and to design an architecture that preserves the initial conceptual integrity of the software project by a combination of loosely coupled functional software components.

As a consequence, the process of modular software design tends to be a faulty one, where testing, debugging, and integration phases may be much more relevant in terms of resources needed when compared to the production of physical artifacts. This is largely due to two intertwined aspects: *(a)* designers are boundedly rational decision makers (Simon, 1957), and *(b)* the nature of interdependencies between modules is mainly multidimensional and invisible. As a result, the act of decomposing a large software project into components is an activity that results, at best, in a suboptimal outcome; some sources of interdependencies are well determined and are taken into account in the design of components and interfaces, while others are not.

# MODULARITY IN F/OSS DEVELOPMENT
## Imitating a Previously Existing Design

Modular design of complex systems is a demanding job since all the modules interfaces have to be defined ex ante. How can designers cope with such degrees of complexities? One of the lessons coming from the stories of some well-known F/OSS projects is to take advantage of existing templates, rather than to develop a brand new project from scratch. The Free Software Foundation (FSF) GNU project and the FreeBSD operating system project are two relevant instances of this approach to the modular design of complex artifacts, as their kinship with UNIX operating system is openly recognized.

UNIX operating system was a milestone in computer software history, and it is usually described as a highly modular, scalable and portable platform (Gancarz, 1994; Ritchie & Thompson, 1974). The UNIX architecture is a complex and massively decomposed architecture of independent modules, characterized by high specialization of programs ("programs that do one thing and do it well"), working together by means of structures, "pipes"[1], and sharing text streams as a fundamental interface of communication (also known as the "UNIX philosophy" as formulated by Douglas McIlroy, the inventor of pipes (Salus, 1994)). UNIX was the first modern operating system not developed using a hardware-dependent assembly language. The kernel was written in C, ensuring portability to various hardware platforms (Johnson & Ritchie, 1978; Miller, 1978).

UNIX highly modular architecture had strong consequences, both at the level of developers' coding activities and at the level of users' experience. Developers were able, thanks to its modular design, to carry out development of specific parts of the system in autonomy and without any need to coordinate their efforts with other sub-projects. Modularity allowed for both parallel development and contribution of new components; furthermore, the overall design of the system was significantly improved by the development of innovative modules and competition between similar projects (Baldwin & Clark, 2000). At the end-user level, modularity invited mere users to employ mix and match strategies (recombination of different modules), allowing them to generate a wide variety of different implementations of the operating system where a large part of the modules pertaining to the user space were highly customizable and were chosen according to specific tastes or needs.

The GNU project, started in 1984 by Richard Stallman, represented at its beginning a titanic effort to offer a free alternative to currently existing commercial and proprietary operating systems. In this respect, Stallman's design strategy consisted of cloning an already existing project—a stable and mature architecture that had been originally conceived around 15 years before. As suggested by Rosenberg (2000), "Stallman says that he chose UNIX as his model because that way he would not have to make any design decisions."

FreeBSD is another important operating system that deliberately imitated the architecture of the UNIX operative system. Again, by adopting an existing architecture, the community spent its attention on incremental development rather than on design discussions (Jørgensen, 2001). To conceive a new operating system characterized by a modular architecture is a challenging cognitive activity of modules and interface definition. First, the designer needs to conceive a system of modules by decomposing the whole system in quasi-independent components. Second, failures in the decomposition phase result in extra costs for fine-tuning and fixing activities aimed at solving unexpected and unforeseen interdependencies. In this respect, the FSF and the FreeBSD community were able to consciously handle what, through the lens of the theory of modularity, is a fundamental trade-off between threats at the design level and opportunities at the implementation level. As a result, the decision of establishing the GNU and the FreeBSD projects upon a stable, mature, and carefully modularized architecture was the key element to benefit from the typical advantages of modularity (concurrent engineering, division of labor, decentralized development, innovation via module-based evolutionary dynamics, and much more), while at the same time avoiding the classic pitfalls and drawbacks of modularity, concerning the risks of imperfect decomposition in the design of an innovative modular architecture as the backbone for the project.[2]

## Horizontal Division of Labor, Task Interdependencies and Brooks' Law

One of the most criticized principles of the otherwise seminal and evocative essay, "The Cathedral and the Bazaar" (Raymond, 1999), is the one prefiguring the demise of Brooks' Law within F/OSS development. This view is supported by a *reductio ad absurdum* argument, claiming that if Brooks' Law were at work, it would not be possible to observe such a thing as Linux development. Conversely, the observation of the Linux case study suggests to the author that the effects of Brooks' Law may be overcome by other forces, such as the project leader's capabilities in attracting, motivating, and coordinating a team of skilled and talented developers in a distributed process strongly facilitated by Internet connectivity as a shared medium of communication. This argument, i.e., that Brooks' Law does not apply to Internet-based distributed development, has been widely criticized by many authors (see for instance, Bezroukov, 1999; Jones, 2000).

Modularity allows us to refine and clarify these criticisms suggesting that a large number of participants in a project may be not a sufficient condition to generate dysfunctional effects, such as diminishing or negative marginal return of manpower to productivity. The key aspect in this regard is represented by the degree of task interdependency between the various members belonging to the project. Thus, the high productivity experienced in the GNU/Linux development is interpreted as largely due to the massively modularized structure of the project, enabling the existence of highly independent sub-projects joined by a limited number of developers, resembling in essence the theory of Brooks' surgical team (small, skilled, and focused) (Brooks, 1975); the role of the Internet in this interpretation is of mere medium of exchange allowing distant communication.

Actually, our latest claim seems to be straightforward if we look at a typical sub-project within the GNU/Linux architecture. Furthermore, if we look more generally at the world of F/OSS projects, there is growing empirical evidence showing that the number of participants involved in a project is on average very small (Capiluppi, Lago, & Morisio, 2003; Krishnamurthy, 2002). Despite this, in some specific cases (such as in the development of the kernel for the GNU operating system, that has been undertaken thanks to the coordinated effort of hundreds of contributors), we need to clarify our point and to address the relationships between Brooks' Law and division of labor in the case of vertical division of labor.

## Vertical Division of Labor and Organization and Architectural Ladders

Many authors have criticized Raymond's cathedral versus bazaar metaphor (see, for instance, Kuwabara, 2000). In our view, there is a serious misinterpretation of this metaphor when it comes to the topic of the architectural characteristics of GNU/Linux.

The misinterpretation of the above quote runs, simplifying slightly, as follows: GNU/Linux comes out of the blue from a chaotic mess of contributions and organizes itself as a coherent system in an apparently self-regulating way, showing a mysteriously spontaneous order. This emergent view of the genesis of GNU/Linux is misleading in that it suggests the existence of a deregulated and emergent flat architecture. In contrast, the modular architecture of GNU/Linux is characterized by being quite hierarchical, rather than flat.

Basically, it boils down to the distinct possibility of distinguishing at least two different and hierarchically ordered ladders in GNU/Linux: a higher level, the kernel space, and a lower one, the user space. As it happens, the celebrated babbling bazaar, representing the decentralized and anarchic distributed process, takes place at the user level and it is fostered by the highly modular architecture, as previously described. Conversely, at the higher inner level of the operating system, the development process seems to be rather different. Linux inner core started to be developed as a one-person hack, and only at a subsequent stage of the process were contribution from other developers introduced. Moreover, while contributions to the kernel represent an open process, the integration of code within the kernel has been a process firmly regulated by the same Torvalds, at the beginning, and later supported by a small group of "trusted lieutenants" (Franck & Jungwirth, 2002; Dafermos, 2001). In order to preserve integrity and coherence within the most important and complex part of the system, at the kernel space ladder all initial relevant design decisions were largely made by Torvalds and by an inner team of developers. The same holds for most of the subsequent activities of kernel development. While one has to acknowledge the role of code contribution from the bottom (the hacker community), it is also indisputable that its incorporation in the project has been fueled by a highly structured and hierarchical process of review and selection (albeit not based on formal authority but rather on competence and reputation).

Sanchez and Mahoney (1996) diffusely discuss a basic feature of modular product architectures, namely, the isomorphic relationship between product architecture and organization traits. This seems to be indeed the case for GNU/Linux, which emerged as a stable system not by a succession of miracles, but rather by exploiting modularity at the user space level, encouraging decentralization, and carefully crafting and controlling the overall consistency of the design at the kernel space, imposing a cathedral-like hierarchy in code evaluation and integration.

We have emphasized that the GNU operating system is a massive modular architecture, mostly inherited from a previous design and characterized by a hierarchical two-ladder architecture that hardly resembles the flatness of a bazaar at all. To further refine our analysis we need to admit that, albeit largely based on the UNIX architecture, there does exist something truly innovative and original in the GNU operating system, i.e., its kernel. In the following subsection we turn to the different approaches to modularity and interdependencies decomposition followed by two different competing projects: the Linux project and the HURD project.

# Ex-Ante Modularity Versus Evolving Modularization: The Development of a Kernel for the GNU Operating System

If we underestimate the problems posed by module identification and decomposition of new architectures, we hardly understand why modular design of complex products is so difficult and unpredictable. Another way to grasp this issue is to consider that many modular products were originally developed from interconnected solutions. While this is not a general rule, it was definitely true for Linus Torvalds' kernel; the GNU operating system is known for being a modular complex artifact and its successful development, accomplished by a distributed community of hackers, largely benefited from that. Thus, it may be surprising that its core component— the so-called kernel—was initially conceived as a highly integrated product and only eventually acquired a modular structure.

At the time Linus Torvalds started to work on his kernel, a long debate was mounting around the advantages offered by an alternative architecture called microkernel, that was designed to work with all possible and different processors.[3] Torvalds decided to develop his kernel in less general terms, thinking that microkernels at the beginning of the 1990s were still experimental, too complex, and exhibited a much worse performance (Torvalds, 1999). By the way, when Torvalds started to work on his kernel, the Free Software community and the GNU partisans were already involved in the development of a microkernel (called HURD), even though the task seemed to be far away from its conclusion.

Therefore, the very first version of Linus' kernel had a monolithic structure and was also hardware specific, since it was conceived for working on Intel 80386 processors only. The first effort to port the Linux kernel to another processor (Motorola 68K) demonstrated all the drawbacks of having a hardware-specific architecture, since the developers of 68K Linux had to write another hardware-specific kernel from scratch. When Torvalds started to think about porting Linux to the Alpha platform, he realized that the original design was no longer effective, and in 1993 he started to rewrite the kernel code completely. He decided to keep a monolithic architecture, but he introduced some degree of modularity in the system design in order to simplify the portability task and to encourage parallel development in some less critical parts of the system.

Therefore, the general kernel model made use of modules, and it was conceived bearing in mind those elements common to all typical modular architectures (even though it was not as rigorous and general as microkernels are). Following this scheme, Torvalds could deal with them separately and confine all the hardware-specific pieces of code in modules out of the core kernel (de Goyeneche & Apolinario Fernández de Sousa, 1999). These modules could be later updated or changed by Torvalds himself and the other Linux developers with no effect on the kernel core.[4] Device driver structure is a good example of the *third way* followed by Torvalds.

In later discussions, Torvalds explained the reasons for his choice: a fully modular architecture, like the one adopted for HURD, would have posed problems to a degree of complexity that it could have compromised the accomplishment of the project. To avoid such risks and keep the degree of complexity of the project as low as possible, Torvalds decided to design a monolith, and he actually wrote all the architectural specs himself,[5] avoiding all the problems related to collective projects (e.g., division of labor, coordination, communication). On the other hand, the HURD micro-kernel, a project in direct competition with the Linux kernel, has paid for the choice of pursuing a fully modular approach from the beginning in terms of the continuous delays that have plagued its development. Nowadays, it is still under active development and still lacks the stability and performance assured by the Linux kernel.

While the modular structure adopted by Torvalds for the Linux kernel happened to be successful, it does not prevent the system from the emergence of unforeseen interdependencies within the modules that may arise with the future development of hardware and software. While HURD established itself as an attempt to develop a fully general and modular system, Linux kernel took advantage of some architectural shortcuts. As it is, the problem related to emergent interdependencies that were not expected at the beginning may become a problem for the future enduring success of Linux, even though this can be regarded as a future cost for the straightforwardness of its design.

The emergent interdependencies sometimes are solved by tinkering, reworking, and re-designing (Ratto, 2003); in other cases, unanticipated interdependencies may end up in forks or complete failures of the projects.

The stories of the development of the Linux kernel and of the HURD kernel suggest the following reflections: the design of modular architectures from scratch may reveal itself to be an extremely complex task; therefore, designers may prefer interconnected solutions that are easier to devise and handle, even when task partitioning and division of labor considerations might suggest modular ones. Moreover, a modular architecture is more vulnerable to design faults, especially when the task is complex and the amount of resources are limited; in particular, an ineffective definition of modules that are not coupled loosely enough produces an increasing amount of interdependency. Thus, individuals, rather than groups of developers may more efficiently accomplish the early stages of new projects. Some successful F/OSS stories experienced this fate, as they have been started as one-man projects aimed to solve specific problems and eventually evolved into structured projects involving a large number of people.[6] Finally, Torvalds' kernel story enriches the perspective offered by Conway's law about the isomorphic structure of product and process (Conway, 1968). Modularity, in fact, seems to be pursued not as a dogmatic feature of the product, but it arises as a general design rule, and it is boosted only when it provides some direct advantage. Therefore, the evolution of the Linux kernel towards modular design suggests that it is possible to combine both modular components and integrated parts under the same architecture. Later on, the designers may introduce a higher degree of modularity by adapting the originally interconnected architecture. In other words, modularity arises more as a process of evolutionary design (modularization) than as an ultimate ex ante property of an artifact.

## Beyond the Principles of Modularity

In the previous subsections, we have argued that the paradigm of modularity has a great explanatory power in characterizing the F/OSS development style and the success of many software projects;: well-decomposed architectures seem to reconcile considerations about division of labor and size of a project with concerns of high speed of development. Nevertheless, as mentioned earlier, for complex software artifacts it may be almost impossible to separate ex ante all interdependencies, and unforeseen coupling between components at later stages (like, for instance, integrating new and existing modules) may strongly affect the final outcome of the process. We argue that F/OSS development style has originally adapted the principles of modularity in order to lower the impact of this "dark side" of modularity.

It is worth mentioning that many scholars have radically criticized the modular approach to the design of software artifacts since its introduction. As noted by Brooks (1995): "Harlan Mills has argued pervasively that 'programming should be a public process,' that exposing all the work to everybody's gaze helps quality control, both by peer pressure to do things well and by peers actually spotting flaws and bugs."

Brooks (1975) argued that information should be completely available in order for failures in the design of software to become evident and be corrected. Conversely, in accordance with the principles of modularity, these processes of peer review, control, and contribution to others' source code are strongly limited by information-hiding constraints, since modules are not available to other developers. Despite these criticisms, information hiding has nowadays become almost ubiquitous in software engineering. Even Brooks (1995), in the 20th year anniversary edition of his *The Mythical Man-Month*, admits the following: "Parnas was right, and I was wrong on information hiding."

We claim that the fundamental innovation of F/OSS practices lies in how the basic postulate of information hiding is adapted to overcome these pitfalls, suggesting a step further in the software engineering debate on the pros and cons of modularity. While information hiding is clearly at the core of designers' activities when initially decomposing a software project in modules, the same principle is later disregarded at the implementation level, in day-to-day coding, testing, and integration activities. As a matter of fact, in the F/OSS community, hackers actually are overexposed to, rather then shielded from, a huge amount of code.

The free availability of the source and the absence of code ownership make programming a truly public process, since good coding solutions are shared and adapted to solve similar problems (Pavlicek, 2000), and ex post interdependency conflicts are handled by employing a wider set of fine-tuning strategies. A well-known feature of F/OSS methodologies is parallelized and distributed code debugging, where bugs are highlighted and corrected by others' "eyeballs" (Iannacci, 2003; Raymond, 1999). Kuan (2000), for instance, shows that F/OSS has a higher rate of quality improvement than closed-source software. Similar results were obtained by Succi, Paulson, and Eberlein (2001). Jørgensen (2001) reports that half of the respondents to his research survey claimed to have received a bug report from someone else within the previous month and nearly half of them credited an external contributor with fixing a bug in their code. Likewise, at the code review level, similar parallel and distributed processes of peer review highlight design incoherencies introduced by others.

In other circumstances the "no hiding" principle allows developers to undertake much more sophisticated software engineering activities, such as redefining module and interface specifications in response to the emergence of new interdependencies between separate modules. This is often the case in the introduction of radically new or substantially complex features in stable projects. For instance, the introduction of cryptography in the *Freenet* project (von Krogh et al., 2003) affected many different modules and demanded the whole redefinition of the architecture. The availability of other module source code is what allowed the developers to disentangle the complex web of interdependencies introduced by adding a public key to cryptography. Similarly, Jørgensen (2001) underlines how the free flow of information about the whole project helped FreeBSD developers to introduce a radically innovative feature to support multiprocessing (Symmetric Multiprocessing) within a mature software architecture.

The no-hiding policy bears one additional consequence: it makes it possible for individual hackers or entire groups to write patches or variations of the original code that are not completely compatible with previous work carried out in the same software project or with respect to other related pieces of software. While incompatibilities are usually unintentional and marginal and may be fixed by subsequent coding activities, sometimes these modifications are large and/or intentional and may result in forking, i.e., the introduction of an independent and partially incompatible version of the original software.

As a matter of fact, within the software industry, advocates of corporate closed-source software development have argued that, due to the lack of code ownership, F/OSS seems to be particularly prone to developing "multiple incompatible versions of programs, [plagued by] weakened interoperability, [and] product instability" (Mundie, 2001). With respect to software development activities, this may lead to duplication of effort and may result in the inefficient allocation of scarce resources at the level of the whole F/OSS community. Nevertheless, other studies have suggested that forking in F/OSS may be much less frequent than one might expect at a first glance, and may eventually lead to positive, rather than catastrophic, outcomes. Many F/OSS projects have a governance structure (ranging from the project leader benevolent dictatorship to the formation of complex coalitions) that prevents attempts to fork (Kogut & Metiu, 2001). Moreover, the widespread diffusion of the GNU General Public License (GPL) seems to mitigate the incentives to fork an existing F/OSS project since, in essence, it prevents the appropriability of innovations. In fact, while anyone may fork any software project at any time, his or her subsequent work would be available to the whole community as well due to the "infectious" nature of GPL. Thus, others may take advantage of the improvements of the fork. In this perspective, forking rarely happens and even when it occurs, this often translates in being beneficial to both competing projects, since the GPL allows each one to study the other and implement the most innovative features (e.g., this seems to have been the case in the rivalry between the Emacs and the XEmacs projects (Moen, 2003)). As a result, forking seems to take place largely in cases of ultimate and irreconcilable differences in views and priorities in the development of a software project, and forks take off and succeed only if they are able to occupy different ecological niches (see, for instance, the existence of various GNU/Linux distributions), thus offering specialized solutions for a differentiated audience (van Wendel de Joode, de Bruijn, & van Eeten, 2003). Finally, it has been noted that it is not uncommon for forks to merge back with the original project as benefits, and drawbacks of "running alone" may change over time (as in the case, for instance, of the *egcs* project, re-merged by the FSF with the original *gcc* project in 1999 (Moen, 2003)).

## DISCUSSION

In the end, modularity may be conceived as simple as it is, as long as we do not open the "black box" and keep track of the organizational processes behind the structure. Most quoted contributions in management studies (Baldwin & Clark, 1997, 2000; Sanchez & Mahoney, 1996; Ulrich, 1995) unfold a neat and smooth theory of modularity, introduced as a cornerstone for artifact design.[7] According to this Olympic version, modularity is defined as a "particular design structure, in which parameters and tasks are interdependent within units (modules) and independent across them" (Baldwin & Clark, 2000, p. 88). Unfortunately, this perspective underestimates the fact that the decomposition of complex systems generally results in a quasi-decomposition and not in a full decomposition, as some interdependencies may not be predicted or are left out on purpose, simply because they are regarded as marginal ones.

Our reconsideration of the development of some F/OSS projects shows how the modularity principles may in practice differ from what the theory prescribes. GNU/Linux and, more generally, F/OSS represent an instance of *unorthodox modularity*; the information-hiding principle is significantly disregarded as the artifact evolves mainly through a repertoire of practices (e.g., peer coding and debugging, frank discussions, and open decisions) where developers and users work apart, tinkering and patching the original modular product.

In our view, reading the GNU/Linux case from a modularity perspective provides a complementary understanding of the F/OSS phenomenon and, at the same time, offers some insights about the way we conceive a theory of modularity for complex systems.

With respect to the first issue, taking advantage of existing architectures like UNIX and related standards (e.g., POSIX) has been a successful strategy, as the community of developers avoided designing a modular structure from scratch. The comparison between the HURD project and Torvalds' monolithic kernel shows that developing decomposable architectures for complex products exposes the designers to the risk of unforeseen interdependencies that may ultimately endanger the whole project. Besides, as F/OSS projects are developed by distributed organizations and the community members communicate only remotely, coordination and collective decision making seem to be two fundamental issues in F/OSS development.

In other words, our study of F/OSS projects viewed through the lens of the theory of modularity outlines three main strategies that characterize the design and the development of complex systems: 1) inheriting existing modular architecture, 2) evolving towards increasing degrees of modularity, and 3) violating the information-hiding principle. This repertoire of practices, or shortcuts as we called them in our introduction, emerge as effective and robust routines that seem to fit very well with the actors involved (i.e., distributed communities of developers) and the problem-solving activity they embrace.

GNU/Linux case, on the other hand, suggests some general reflections on modularity and modularization. F/OSS developers exploit all the advantages of a modular architecture, as the massive parallel activity within modules/programs shows; on the other hand, the modularization does not stop with the architecture design. The unforeseen interdependencies that come to the surface as the operative system evolves, revealing some inconsistencies, are met by violations of the information-hiding principle.

In questioning how this experience may be extended to other contexts where modularity has already started to represent a promising approach, there are at least two fundamental conditions that need to be clearly spelled out. First, F/OSS distinctive trait is represented by the open access to knowledge (source code and documentation) stored in the modules. In the F/OSS world, imitation and copying are encouraged and protected by a reverse form of copyright (copyleft). According to the Economics of Innovation standard models, copyleft should inhibit any investment in innovations, since anybody may take advantage of any innovation and there are no incentives for the innovators. F/OSS apparently contravenes this rule, and motivational analyses based on various perspectives (i.e., psychological, cultural, sociological, etc.) seem to be urgently needed to support an economic explanation of this phenomenon. From our viewpoint, the apparent paradox of compelling innovations in a copyleft regime is due to the second fundamental condition that characterizes the F/OSS movement, that is, a deep overlap between producers and users. At the beginning, at least, most users were developers or had some skills that allowed them to perform successful adaptations. Again, most of the traditional ways of conceiving innovation and product development in other domains keep producers and users separated, even though today customers are more and more often directly involved in the definition of their own product.

As long as developers and users communities deeply overlap, copyleft regime does not inhibit innovation but, rather, ensures its open and free diffusion. On the other hand, when the communities start to be more and more different from each other, when developers are viewed as producers and users as customers, the natural system of reciprocal benefits becomes less and less salient. Therefore, looking for a possible generalization of F/OSS experience should push us towards other economic contexts where developers and users are able to establish strong relationships; in this respect, settings where customers actively participate in the development of new products (see, for instance, von Hippel, 1998) seem to represent a promising milieu for empirical investigation.

# REFERENCES

Baldwin, C. Y. & Clark, K. B. (1997). Managing in the age of modularity. *Harvard Business Review*, 75(5):84–93.

Baldwin, C. Y. & Clark, K. B. (2000). *Design rules. Vol. I: The power of modularity.* Cambridge, MA: MIT Press.

Bezroukov, N. (1999). A second look at the cathedral and bazaar. *First Monday*, 4(12).

Brooks, F. P. (1975). *The Mythical Man-Month. Essays on Software Engineering.* Reading, MA: Addison-Wesley.

Brooks, F. P. (1986). No silver bullet. In H.J. Kugler, H. J. (Ed.), *Information Processing 1986, Proceedings of the IFIP Tenth World Computing Conference*, pp. 1069-1076. Amsterdam: Elsevier Science.

Brooks, F. P. (1995). *The Mythical Man-Month. Essays on Software Engineering,* 20th Anniversary ed. Reading, MA: Addison-Wesley.

Camuffo, A. (2002). Rolling out a "world car": Globalization, outsourcing and modularity. 2nd EURAM Conference, Stockholm, Sweden.

Capiluppi, A., Lago, P., & Morisio, M. (2003). Characterizing the OSS process: A horizontal study. 7th European Conference on Software Maintenance and Reengineering, Benevento.

Conway, M. (1968). How do committees invent. *Datamation*, 14(10):28-31.

Dafermos, G. (2001). Management and virtual decentralized networks: The Linux project. *First Monday*, 6(11).

de Goyeneche, J. & Apolinario Fernández de Sousa, E. (1999). Loadable kernel modules. *IEEE Software*, 16(1):65-71.

Devetag, M. & Zaninotto, E. (2001). The imperfect hiding: Some introductory concepts and preliminary issues on modularity. DISA Working Paper, Università degli Studi di Trento.

DiBona, C., Ockman, S., & Stone, M. (1999). *Open sources: Voices from the open source revolution*. Sebastopol, CA: O'Reilly & Associates.

Feller, J. & Fitzgerald, B. (2000). A framework analysis of the open source software development paradigm. In *Proceedings of the 21st International Conference on Information Systems*, pp. 58-69. Atlanta, GA: Association for Information Systems.

Franck, E. & Jungwirth, C. (2002). Reconciling investors and donators. The governance structure of open source. Lehrstuhl für Unternehmensführung und–politik Universität Zürich.

Gancarz, M. (1994). *The UNIX philosophy*. Newton, MA: Digital Press.

Glass, R. (1997). *Software runaways. Lessons learned from massive software project failures*. Upper Saddle River, NJ.: Prentice Hall.

Hatch, N. (2001). Modular stepping stones along the firm's technology path. Nelson and Winter Conference, Aalborg, Denmark.

Iannacci, F. (2003). The Linux managing model. In *Proceedings of the 3rd International Conference on Open Source, ICOS 2003*.

Jackson, I. (1998). Why is software freedom useful and what does it mean? SANE'98 November 18-20.

Johnson, S. & Ritchie, D. (1978). Portability of C programs and the UNIX system. *The Bell System Technical Journal*, 57(6):2021-2048.

Jones, P. (2000). Brooks' law and open aource: The more the merrier? Retrieved January 2, 2003, from http://www-106.ibm.com/developerworks/library/merrier.html

Jørgensen, N. (2001). Putting it all in the trunk: Incremental software development in the Free BSD open source project. *Information Systems Journal*, 11(4):321-336.

Kogut, B. & Metiu, A. (2001). Open-source software development and distributed innovation. *Oxford Review of Economic Policy*, 17(2):248–264.

Krishnamurthy, S. (2002). Cave or community? An empirical examination of 100 mature open source projects. *First Monday*, 7(6).

Kuan, J. (2000). Open source software as consumer integration into production. Retrieved July 12, 2003, from http://opensource.mit.edu.

Kuwabara, K. (2000). Linux: A bazaar at the edge of chaos. *First Monday*, 5(3).

Langlois, R. N. (2002). Modularity in technology and organization. *Journal of Economic Behavior & Organization*, 49(1):19-37.

Langlois, R. N. & Robertson, P. (1992). Networks and innovation in a modular system: Lessons from the microcomputer and stereo component industries. *Research Policy*, 21(4):297–313.

Miller, R. (1978). UNIX – A portable operating system? *ACM Operating Systems Review*, 12(3):32–37.

Moen, R. (2003). Fear of forking essay. Retrieved July 1, 2003, from http://linux-mafia.com/Erick/essays/forking.html.

Mundie, C. (2001). The commercial software model. Retrieved July 1, 2003, from http://www.microsoft.com/presspass/exec/craig/05-03sharedsource.asp.

Parnas, D. L. (1972). On the criteria for decomposing systems into modules. *Communication of the ACM*, 15(12):1053-1058.

Pavlicek, R. C. (2000). *Embracing insanity: Open source software development*. Indianapolis, IN: Sams Publishing.

Pressman, R. (2000). *Software engineering: A practitioner's approach, 5th ed*. Boston, MA: McGraw-Hill.

Ratto, M. (2003). Re-working by the Linux kernel developers. Department of Communication, University of California, San Diego, CA.

Raymond, E. S. (1999). *The cathedral and the bazaar: Musings on Linux and open source by an accidental revolutionary*. Sebastopol, CA: O'Reilly & Associates.

Ritchie, D. & Thompson, K. (1974). The UNIX time-sharing system. *Communications of the ACM*, 17(7):365-375.

Rosenberg, D. K. (2000). *Open source. The unauthorized white papers*. Foster City, CA: IDG Book Worldwide.

Salus, H. P. (1994). *A quarter century of UNIX*. Reading, MA: Addison Wesley.

Sanchez, R. & Mahoney, J. T. (1996). Modularity, flexibility, and knowledge management in product and organizational design. *Strategic Management Journal*, 17(Winter special issue): 63-76.

Schach, S. (2002). *Object- oriented and classical software engineering, 5th ed*. Boston, MA: McGraw-Hill.

Schilling, M. A. (2000). Toward a general modular systems theory and its application to interfirm product modularity. *Academy of Management Review*, 25(2):312-334.

Simon, H. (1957). *Models of man*. New York: Wiley.

Simon, H. A. (1981). *The sciences of the artificial, 2nd ed*. Cambridge, MA: MIT Press.

Solheim, J. & Rowland, J. (1993). An empirical study of testing and integration strategies using artificial software systems. *IEEE Transactions on Software Engineering*, 19(10):941-949.

Succi, G., Paulson, J., & Eberlein, A. (2001). Preliminary results from an empirical study of open-source and commercial software products. International Conference on Software Engineering, Toronto, Ontario, Canada.

Torvalds, L. (1999). The Linux edge. In C. DiBona, S. Ockman, & M. Stone (Eds.), *Open sources: Voices from the open-source revolution*. Sebastopol, CA: O'Reilly & Associates.

Ulrich, K. (1995). The role of product architecture in the manufacturing firm. *Research Policy*, 24:419-440.

van Wendel de Joode, R., de Bruijn, J. A., & van Eeten, M. J. G. (2003). Protecting the virtual commons. Self-organizing open source and free software communities and innovative intellectual property regimes. *Information Technology and Law Series, No. 3*, The Hague: T. M. C. Asser Press.

von Hippel, E. (1990). Task partitioning: An innovation process variable. *Research Policy*, 19(5):407-418.

von Hippel, E. (1998). Economics of product development by users: The impact of "sticky" local information. *Management Science*, 44(5):629-644.

von Krogh, G., Spaeth, S., & Lakhani, K. (2003). Community, joining, and specialization in open source software innovation: A case study. *Research Policy*, 32(7):1217-1241.

# ENDNOTES

[1] By pipe technology it is possible to connect the output of one program to the input of another one, and thereby execute complex tasks by sequences of elementary programs linked together.

[2] See also, later in the chapter, how, in the case of the GNU project, the failure to correctly modularize the architecture resulted in serious troubles for the developers of the HURD micro-kernel.

[3] As Torvalds put it, "When I began to write the Linux kernel, there was an accepted school of thought about how to write a portable system. The conventional wisdom was that you had to use a microkernel-style architecture. (Torvalds, 1999). See also the well-known "Linux is obsolete" flamewar in the comp.os.minix newsgroup (reported in Appendix A of DiBona et al., 1999), where Linus Torvalds, Andrew Tanenbaum, and other relevant hackers passionately debated OS design issues and the strength and weakness of micro versus monolithic kernels.

[4] Version 2.1.110, released in July 1998, counts around 1,5 million lines of code: 29% makes up the kernel and the file systems, 54% the platform-independent drivers, and the remaining 17% is architecture-specific code.

[5] Releasing Version 0.11 in December 1991, he credited three other people.

[6] Apart from the Linus/Linux case, see also Sendmail, initially developed by Eric Allman to route email to other users within UC Berkeley, Perl by LarryWall to solve some annoying problems in system administration, the World Wide Web by Tim Berners-Lee as a group environment for academic information sharing among high-energy physicists, and so on.

[7] For an insightful assessment of this topic see also Langlois (2002) and Devetag & Zaninotto (2001).

Chapter V

# A Quantitative Study of the Adoption of Design Patterns by Open Source Software Developers

Michael Hahsler
Vienna University of Economics and Business Administration, Austria

## ABSTRACT

*Several successful projects (Linux, Free-BSD, BIND, Apache, etc.) showed that the collaborative and self-organizing process of developing open source software produces reliable, high quality software. Without doubt, the open source software development process differs in many ways from the traditional development process in a commercial environment. An interesting research question is how these differences influence the adoption of traditional software engineering practices. In this chapter we investigate how design patterns, a widely accepted software engineering practice, are adopted by open source developers for documenting changes. We analyze the development process of almost 1,000 open source software projects using version control information and explore differences in pattern adoption using characteristics of projects and developers. By analyzing these differences, we provide evidence that design patterns are an important practice in open source projects and that there exist significant differences between developers who use design patterns and who do not.*

# INTRODUCTION

The growing need for reliable software has made software engineering an important industry in the last few decades. The steady progress produced an enormous number of different approaches, concepts, and techniques: structured analysis and design, the object-oriented paradigm, agile software development, component-based systems, frameworks, and design patterns, just to mention a few. These techniques were developed with traditional software development in a commercial environment in mind. Recently, a new organizational form of collaborative software development, the open source movement (Raymond, 1999), gained popularity. Open source software (OSS) development differs from traditional forms in many respects. For example, the source code is publicly shared and therefore rigorously peer reviewed, the development teams are often geographically dispersed, and massive system-level testing by large user communities is conducted instead of extensive unit tests (Dempsey, Weiss, Jones & Greenberg, 2002; Jorgensen, 2001; Perpich, Perry, Porter, Votta, & Wade, 1997; Vixie, 1999). Quantitative research is the key to understanding how these differences influence the adoption of existing software engineering practices or how software engineering practices are adapted to the needs of open source development.

In this chapter we study how design patterns are used in open source software development teams. We are interested in the question "Are design patterns are useful for open source development and, if so, are there factors that influence their adoption?" To gain an insight into the application of design patterns, we analyze historic data of the development process of OSS projects.

This chapter is organized as follows: As the starting point for the chapter we review literature about design patterns to identify how patterns are used for traditional software development. Next, we describe the research method employed by the study. We present the used data set and its main characteristics followed by the analysis of the data set and the discussion of the results. We conclude the chapter with the main findings and point out directions for further research.

# BACKGROUND AND RELATED LITERATURE

Design patterns describe non-obvious solutions in a standard written form for recurring software design problems in a certain context. They are normally developed by experts from their experiences with many existing systems and represent good and flexible solutions.

Since the introduction of the first software pattern catalog containing 23 design patterns by Gamma, Helm, Johnson, and Vlissides (1995), design patterns were rapidly accepted by the software engineering community, and their use is now strongly facilitated by the Unified Modeling Language (UML), the standardized notation for object-oriented analysis and design. The number of publications about design patterns has soared, and even several conference series on the topic were initiated.

In the United States, the conference series is called Pattern Languages of Programs (PLoP), and in other parts of the world conference series like EuroPLoP, KoalaPLoP, and ChiliPLoP were started. These conferences, as well as most publications, focus on the development of new and improved design patterns, but the research on the actual adoption of design patterns by commercial and especially by open source software developers is still underdeveloped.

In their book about design patterns, Gamma et al. (1995) state that they expect design patterns will provide (a) a common vocabulary, (b) a documentation and learning aid, (c) an adjunct to existing methods, and (d) a target for refactoring. To underpin these expectations there were some early publications by practitioners that describe experiences with design patterns in an industrial setting as well as for training (Beck et al., 1996; Goldfedder & Rising, 1996; Helm, 1995). However, these publications represent personal experience reports with no empirical data to support their claims. In the joint paper "Industrial Experience with Design Patterns" (Beck et al., 1996) co-authored by Kent Beck (First Class Software), James O. Coplien (AT&T), Ron Crocker (Motorola Inc.), Lutz Dominick and Frances Paulitsch (Siemens AG), Gerard Meszaros (Bell Northern Research), and John Vlissides (IBM Research), these seven experts describe the efforts they and their companies put into design patterns and the resulting experiences. The paper contains a table of the most important observations sorted by the number of experts who mentioned them. This can be interpreted as the results of interviewing experts. The following observations were mentioned by all experts (Beck et al., 1996):

1. Patterns are a good communications medium.
2. Patterns are extracted from working designs.
3. Patterns capture design essentials.

The first observation is the most prominent benefit of design patterns. In Gamma et al. (1995), two of the expected benefits are that design patterns provide "a common design vocabulary" and "a documentation and learning aid" that also focus on the communication process. Prechelt, Unger-Lamprecht, Philippsen, and Tichy (2002) studied this aspect with two controlled experiments using undergraduate and graduate students to perform maintenance tasks for small programs. The experiments showed that explicit documentation of a used pattern has a positive influence on the speed and the error rate of the maintenance task.

Efficient communication is certainly also a key issue for OSS development. Often, there is no explicit design document for new OSS projects, and the design emerges later on from the implementation (Vixie, 1999). This means that collaborating developers need to communicate the design in part directly by code. But since infering design from code artifacts is known to be hard, this would make collaboration almost impossible unless there are some widely accepted design practices that are known within the software engineering community. Design patterns try to capture such design practices and give them a unique name to communicate them efficiently, even as short comments within the code.

Seen, Taylor, and Martin (2000) look at the adoption of design patterns in a commercial environment from a perspective of the diffusion of innovation theory. Although they rate the overall adoption rate for design patterns as moderate, they identify an area where design patterns score very high. The area is concerned with properties that influence the individual motivation for adoption, which are especially interesting for open source developers. Specifically, patterns require no infrastructure investment; they can be adopted bottom-up; and visible pattern adoption advertises competence. All three properties are certainly more important in an open source environment than in a traditional company where the necessary infrastructure is provided and the management controls the development process.

## RESEARCH METHOD

To investigate the adoption of design patterns by open source software developers we analyze the development process of OSS projects by using publicly available version control data accessible via the Source Forge Web site (http://www.sourceforge.net). This approach is inexpensive and non-intrusive (Atkins, Ball, Graves, & Mockus, 1999; Cook & Votta, 1998) and was already successfully used to analyze the development of the Apache Web Server project (Mockus, Fielding, & Herbsleb, 2000) and of the GNOME project (Koch & Schneider, 2002), both large-scale open source projects.

Source Forge currently hosts over 59,000 open source projects and has over 597,000 registered users (as of April 2003). It provides the projects with a version control facility as well as a presentation platform and communication channels for developers and users. For Source Forge, each developer has a unique pseudonym, a user name, that can be used throughout all projects in which he or she participates. Each project has a home page with general information about the project, like the project name, a short description of the project, the developers in charge of the project (administrators), the development status of the project (alpha, beta, production, etc.), the intended audience (e.g., developers or end users), the programming languages used, and more general information.

For this study we analyze projects that use the object-oriented programming language Java and employ the version control tool Concurrent Versions Control (CVS) (Fogel, 1999). The CVS tool supports parallel development by several developers, comments for modifications of the code, control of releases, reversing modifications, generating history logs for the projects and for each individual file, and much more. A project is defined as a collection of individual files that are stored together with version control information in a CVS repository. New files can be added to the project, and existing files can be modified by the developers of the project. To modify files using version control, the developer has to obtain a version of the files from the repository, change the files locally, and commit the modifications to the repository (execute a *check-in* for the files). During the check-in (often called modification request), the developer is encouraged to add a short log message that explains the purpose of the modifications and therefore makes it easier to understand the changes in the code later on.

CVS records the modifications in each file in lines of code (LOCs) added and LOCs deleted by the developer. The definition of LOCs used by CVS is the number of physical lines. There is no distinction between program statements, comments, or other arbitrary text. We adopt this definition for our study. Furthermore, CVS does not explicitly record changes in a line; instead it records a changed line as a line deleted and a new line added. Therefore, the growth of LOCs for a check-in (the delta) is the difference between the LOCs added and the LOCs deleted.

To analyze the application of design patterns we have to identify the patterns in the projects. Design patterns are design artifacts that result in special constructions in the final code, e.g., several objects that interact in a certain way. It is very difficult to infer the application of design patterns automatically from code (see, e.g., Antoniol, Fiutem, & Cristoforetti, 1998, for an automatic approach). However, CVS provides us with additional information, i.e., the log messages for changes, which Mockus and Votta (2000) already successfully used to classify maintenance activities. For this study we analyze the log messages to identify the application of design patterns. Of course, design patterns can be applied without mentioning them in the log message. Therefore, we can only identify the application of design patterns when they are used for documentation of changes and to support communication between developers. This is one of the major contributions of patterns stressed throughout the literature, namely, that the names of design patterns become part of a common language that developers use to communicate design more efficiently (Buschmann, Meunier, Rohnert, Sommerlad, & Stal, 1996; Gamma et al., 1995; Vlissides, 1998). However, it is important to note that with this approach we only analyze the communication aspect of patterns. To analyze the influence of patterns on software quality or the usage of patterns in general, other approaches are necessary that might include analyzing bug tracking databases, interviews with developers, extensive manual code reviews, and controlled experiments.

For this study we only use the original set of the 23 design patterns introduced by Gamma et al. (1995). Although many other design patterns were introduced in the literature (e.g., in Coplien & Schmidt, 1995; Martin, Riehle, & Buschmann, 1998; and Vlissides, Coplien, & Kerth, 1996), these 23 patterns are still the most popular and best known patterns. For the analysis we first extracted the CVS log messages for each project using Java. We parsed the messages using regular expressions, identified changes to files that constitute together a check-in, and stored the information in a relational database. All analyses in the subsequent sections are performed using standard SQL select statements on the data base and a standard statistical package.

# THE DATA SET

The used data set includes 988 open source projects from Source Forge using Java as the main programming language. The projects were downloaded between August and September of 2001 and were selected by the following criteria: only projects that enabled CVS and that have had already more than 1,000 LOCs Java code. In total, the selected projects contain almost 120,000 files (with the extension *.java*) with more than 19.5 million LOCs and 1,487 different developers having worked on them. Figure 1 depicts the distribution of the size of the projects in LOCs. The project sizes in number of files follows a similar distribution. Most of the projects are rather small, with a mean around 20,000 LOCs or 120 files, but there is a significant amount of much bigger projects. The biggest project has almost 1 million LOCs and almost 6,000 files. The sizes of the developer teams for the projects show a similar distribution, with many projects with only a single developer (see Figure 2). The mean of the team size is 1.84, and the biggest team consists of 73 developers.

Figure 3 shows the number of projects by the development status used by Source Forge. The status ranges from 1 to 6 (1 planning, 2 pre-alpha, 3 alpha, 4 beta, 5 production/stable and 6 mature) and gives an idea about the project in its

*Figure 1: Number of projects by size in LOC.*

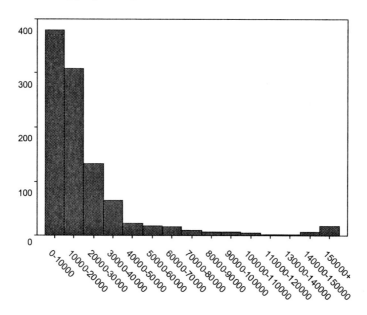

*Figure 2: Number of projects by team size.*

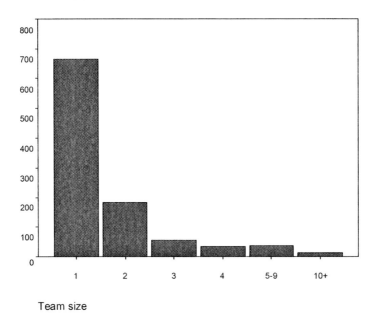

*Figure 3: Number of projects by status.*

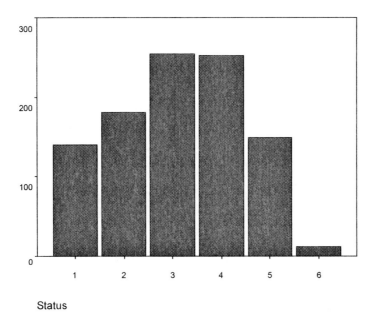

development life cycle. Most of the selected projects with more than 1,000 LOCs have a status of 3 and 4. Fewer projects with status 1 and 2 already reached the minimum LOCs. For the status 5 and 6, there are fewer Java projects in Source Forge, indicating that most of the analyzed projects can be considered still in the design and implementation phases of the software life cycle. In Table 1 we summarize the statistics of the analyzed projects.

For the analyzed changes made to the projects, we excluded the initial check-in of new files since new files added to the Source Forge CVS repository often already have a considerable size and it is impossible to say how many developers, and who, already worked on this file. We only know the developer who checked-in the file. This would distort the results in an unpredictable way. We observed about 97,300 check-ins where 6.65 million lines of code were changed or added in almost 329,000 files. A check-in affected, on average, 3.38 files and increased the code base of the project by almost 20 lines. Table 2 summarizes the descriptive statistics of the analyzed changes.

*Table 1: Descriptive statistics of the main project variables.*

|          | Status | Size in LOCs | Size in Files | Team Size |
|----------|--------|--------------|---------------|-----------|
| N        | 988    | 988          | 988           | 988       |
| Minimum  | 1      | 1001         | 1             | 1         |
| Maximum  | 6      | 961961       | 5951          | 73        |
| Median   | 3.00   | 7279.50      | 50.00         | 1.00      |
| Mean     | 3.13   | 19908.79     | 121.14        | 1.84      |
| Variance | 1.703  | 2952336139   | 104534.022    | 9.468     |

*Table 2: Descriptive statistics of the analyzed changes (check-ins).*

|          | Number of files | LOCs added/changed | LOCs deleted | Increase in LOCs |
|----------|-----------------|--------------------|--------------|------------------|
| N        | 97292           | 97292              | 97292        | 97292            |
| Minimum  | 1               | 0                  | 0            | -5375            |
| Maximum  | 644             | 27189              | 26998        | 18018            |
| Median   | 1.00            | 7.00               | 3.00         | .00              |
| Mean     | 3.38            | 68.40              | 48.92        | 19.48            |
| Variance | 77.328          | 172224.284         | 143837.124   | 25235.507        |

# RESULTS

In this section we present the results of our explorative analysis of the adoption of design patterns by open source developers. The section is divided into three parts. First, the results of the pattern identification are presented. Second, we try to find out if characteristics of the project (e.g., size) make the usage of design patterns more likely. And finally we determine if there are specific characteristics of developers who use design patterns (e.g., extent of participation).

## The Identified Design Patterns

To identify patterns in the log messages we first searched all messages for the names of the 23 design patterns introduced by Gamma et al. (1995). In the log messages of 1,475 of the 97,292 observed check-ins we found the name of at least one pattern. The names found most often were "command" (142 counts) and "state" (80).

Of course we expect a high number of false positives since several pattern names (e.g., Prototype) are also terms frequently used in software engineering with a meaning other than the design pattern. To quantify the rate of false positives, we manually inspected the log message and the code of a random sample of 100 out of the 1,475 check-ins where pattern names were found. For 69 changes it was clear from the log messages whether a design pattern was meant or not, and for the remaining 31 changes we had to inspect the source code. In the inspected sample we found a false positive rate of 69%. For example, "command" was almost always used in connection with changing the command-line interface of the software, which is not related to the application of a design pattern. To reduce this problem we required the pattern names that were also common terms for software development or that had a meaning in the projects' problem domains (command, interpreter, prototype, proxy, state, and strategy) to be accompanied by the term "pattern," which reduced the number of check-ins with patterns to 343.

*Table 3: Accuracy of pattern identification from log messages.*

|  |  |  | Contains a pattern | | Total |
|---|---|---|---|---|---|
|  |  |  | no | yes |  |
| Classified as pattern | no | Count | 65 | 7 | 72 |
|  |  | % | 90.3% | 9.7% | 100.0% |
|  | yes | Count | 4 | 24 | 28 |
|  |  | % | 14.3% | 85.7% | 100.0% |
| Total |  | Count | 69 | 31 | 100 |
|  |  | % | 69.0% | 31.0% | 100.0% |

Table 3 depicts the accuracy of this approach. 85.7% of the check-ins identified as using a pattern really use the pattern, reducing the rate of false positives to 14.3%. Overall 89 out of the 100 inspected check-ins were classified correctly. Note that this approach only identifies the application of design patterns where their full original names are used in the log messages. Therefore, the actual usage of design patterns is considerably higher and includes the following cases:

1. Alternative names are used (e.g., Virtual Constructor instead of Factory Method; some alternative names can be found in Gamma et al., 1995).
2. Design patterns are not mentioned in the log message but are obvious from comments in the code or the names of classes, methods and files.
3. Design patterns are used without giving any clue.

Although it is possible to incorporate some of these cases into an identification heuristic, we use the simpler and therefore more robust approach described above.

In Figure 4 we show the number of projects in which we identified each individual pattern. The pattern mentioned most by far in the analyzed projects is the Singleton pattern. The reason for this is that the Singleton pattern is very simple and is easy to implement in Java. This leads us to the conclusion that for Java the application of the pattern Singleton seems to provide important design advantages

*Figure 4: Number of projects using individual patterns.*

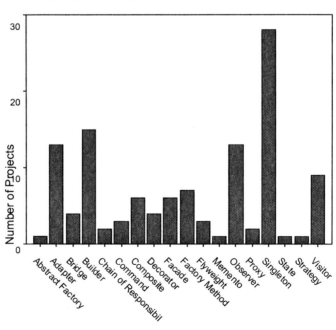

at little implementation effort. During manual inspection of the log messages, we observed that out of eight inspected changes concerning the Singleton pattern, four changes consisted of removing existing Singletons. This perhaps indicates overuse of the pattern caused by its simplicity. Further analysis is needed to investigate the usage of different patterns and pattern types, but this is outside the scope of this chapter.

## Analysis of Differences between Projects

For Source Forge all development efforts are organized in projects as the basic unit of coordination. A project has one or several administrators who coordinate the development of the project and organize the cooperation between the developers. In this section we investigate whether characteristics of projects (e.g., development status, project size, team size, number of check-ins) are related to the application of design patterns for documenting changes.

Most of the projects in the data set do not apply patterns for documenting changes in the code. Only for a small fraction of projects—86 out of 988 projects—did we find one or more patterns. To exclude very small projects that are still in their planning phase and therefore do not have enough code that can contain design patterns, we only select the projects for which we observed at least 1,000 LOCs being added or changed. This leaves 519 projects for analysis. The general trend of the distributions of project sizes and team sizes of this sample is similar to the distributions of the whole data set. Only for the development status do more projects with a status smaller than three not meet this criterion, resulting in a shift towards projects with status three and up. We used the number of developers using patterns and the number of distinct patterns used in a project as the indicators of pattern usage in a project. Table 4 contains the descriptive statistics for the selected projects.

In Figure 5 we show the number of projects using 0 to 5 different design patterns in the project. For 83 projects (16% of the 519 projects), at least one design pattern was used.

*Table 4: Descriptive statistics for the selected projects.*

|  | Status | Size in LOCs | Size in files | Number of check-ins | Team size | Developer using patterns | Number of distinct patterns |
|---|---|---|---|---|---|---|---|
| N | 519 | 519 | 519 | 519 | 519 | 519 | 519 |
| Minimum | 1 | 1038 | 1 | 4 | 1 | 0 | 0 |
| Maximum | 6 | 961961 | 4265 | 4146 | 73 | 10 | 5 |
| Median | 3.00 | 12126.00 | 79.00 | 68.00 | 2.00 | .00 | .00 |
| Mean | 3.30 | 27172.79 | 164.34 | 164.47 | 2.47 | .20 | .22 |
| Variance | 1.604 | 3.9E+09 | 113925 | 1197551.852 | 17.029 | .401 | .359 |

*Figure 5: Projects by number of different design patterns used.*

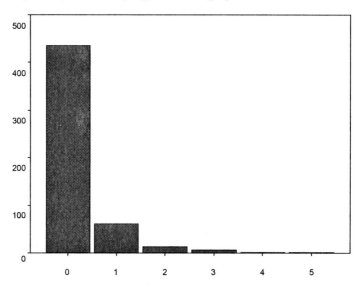

Number of distinct patterns

To explore the relationship between the main project variables (development status, size of the project, the number of check-ins, and the team size) and the use of design patterns for documenting changes, we calculated the correlation coefficients. Since the distributions are highly skewed, we use non-parametric Spearman's rank order correlation. All reported correlation coefficients are significant with $p < .01$.

The two measures of project size, LOCs and the number of files, are highly correlated with a coefficient of .90. Both measures are correlated with the number of check-ins, with a coefficient around .60. Team size is correlated with the number of check-ins (.48) and with the project size (around .30). These correlations reflect the intuitive relationship, with larger projects having more check-ins and also tending to be developed by bigger teams.

Both indicators of pattern usage (number of different patterns and developers using patterns) seem highly correlated with each other in the data set (a significant coefficient of .99) and are significantly correlated with the number of check-ins (.33), the size of the project (around .19), and the team size (around .16). However, all these correlations are artifacts occurring only because we have a very small proportion of projects with patterns (creating the high correlation coefficient between the two indicators) and because the probability of detecting a pattern increases with the number of check-ins we analyze and this number is in turn correlated to the other variables. To improve this analysis, object-oriented software metrics like the Chidamber and Kemerer's Metrics Suite (1994) can be used. These metrics can reflect differences in the complexity of classes and their interactions better than simple

*Table 5: Cross-tabulation of team size and the number of developers who use patterns.*

|  |  | Team size | | | | | | Total |
|---|---|---|---|---|---|---|---|---|
|  |  | 1 | 2 | 3 | 4 | 5-9 | 10+ |  |
| Developer using patterns | 0 | 228 | 107 | 43 | 26 | 29 | 5 | 438 |
|  | 1 | 29 | 21 | 8 | 6 | 6 | 2 | 72 |
|  | 2 |  | 2 |  | 1 | 2 | 3 | 8 |
|  | 3 |  |  |  |  |  | 1 | 1 |
|  | 4 |  |  |  |  |  | 1 | 1 |
|  | 10 |  |  |  |  |  | 1 | 1 |
| Total |  | 255 | 139 | 51 | 33 | 37 | 13 | 518 |
| Observed projects with patterns |  | 11.4% | 17.7% | 15.7% | 21.2% | 21.6% | 61.5% | 16.0% |
| Estimated projects with patterns[a] |  | 11.4% | 21.5% | 30.4% | 38.3% | 50.8% | 82.2% | 18.4% |

[a] *Using team size = 1 to estimate probability of a developer using patterns*

LOCs. However, it was shown by Masuda, Sakamoto, and Ushijima (1999) and by Reissing (2001) that the introduction of design patterns can increase the complexity measured by object-oriented metrics by using additional classes, more inheritance, and increased coupling between the classes of which the pattern is composed. Since this would create an additional artificial relationship between complexity and the usage of design patterns, we restrict our analysis here to the simple LOCs-oriented analysis made possible by the output of the CVS tool.

To analyze the application of patterns in more detail, we cross-tabulate team size by developers using patterns (see Table 5). A simple assumption would be that the application of patterns is independent from the projects and that there exists a fixed subset of developers who know and use patterns. Under this assumption we can estimate the proportion of developers using patterns from the projects with a team site of 1 in Table 5. We found patterns in 11.4% of the projects with one developer. If the usage of patterns is independent of the projects, this percentage can be used as an estimate of the total proportion of developers using patterns. With this estimate we can calculate the probability that we will see at least one developer using a pattern in teams of sizes 2 and up. This simple model largely overestimates the observed proportion of projects with developers who document changes with patterns (compare in Table 5). The observed proportion of projects with developers who use patterns stays almost constant around 20% for team sizes 2 to 9. Only for team sizes 10 and larger does the proportion jump to over 60%, which is still considerably below the estimate. Therefore, the assumption that the developers who use patterns are evenly distributed over all projects does not hold.

Next, we only look at the projects where patterns are used (rows with developers using patterns greater than 0 in Table 5). Interestingly, we found in almost all these projects only one developer who used patterns for documentation, regardless of team size. Again, only at a team size of 10 and larger did we find more developers using patterns. The projects with team sizes 1 through 9 all had, on average, similar sizes and number of check-ins. For the projects with team sizes 10 and higher, the average of all variables (except the project status) was by the factor 2 to 4 higher, which seems to make them different. However, the fact that for almost all projects only one developer uses design patterns indicates that there must be factors that influence the usage of design patterns that are not dependent on the team size and other characteristics of the project but on the developer and his or her position inside the project.

An interesting fact is that the development status of the project has only very small significant correlation (between .13 and .24) with the other variables and no significant correlation with the indicators of pattern usage. This means we have very different projects in terms of size in the data set (ranging from small tools to large applications) and that we cannot find evidence that design patterns are used more often in later stages of the life cycle to refactor code (replace code and design with more flexible design provided by a design pattern) as suggested by Gamma et al. (1995). To analyze this in more detail we split the projects into two groups. The first group contains projects that are still in an early phase (277 projects with status 1-3) and the second contains projects that are more mature (242 projects with status 4-6). If design patterns are used frequently for refactoring, the more mature projects should contain significantly more different design patterns or a higher proportion of check-ins containing patterns. We used the non-parametric Mann-Whitney U test to compare the two groups of projects. We found that project size, team size, and number of check-ins is significantly higher for the more mature projects ($p < .0033$, significant at the .01 level using Bonferoni correction for three independent tests). But there is no significant difference between the groups for the indicators of pattern usage ($p > .9$). This results either from the fact that design patterns are not used for refactoring and documenting these changes in our data set or that if design patterns are used at all, they are already applied in the initial stages of development to create new design. For OSS development, the second explanation seems appealing since in the early stages of the project developers need to create and communicate the design of a system using code (Vixie, 1999), and supporting communication is a major advantage attributed by experts to design patterns.

## Analysis of Differences between Developers

In this section we use the developer as the main unit of analysis. We analyze the usage of patterns for each developer and compare that with observed characteristics of the developer to investigate whether a relationship exists. As the developer's characteristics, we use the number of projects in which a developer participates,

*Figure 6: Number of projects to which developer contributes.*

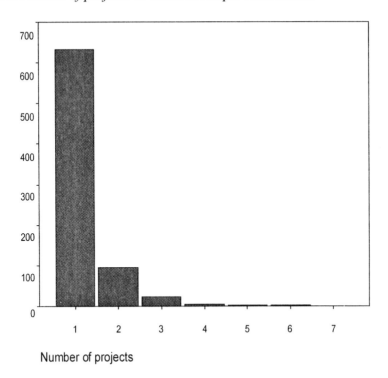

the LOCs he or she modified in the analyzed period, and the increase of LOCs and the check-ins the developer produced. As the indicator of pattern usage, we use the number of different patterns used by a developer. To exclude developers who did not apply patterns only because of their small contribution of code to the analyzed projects, we restrict our analysis to the 761 of the 1,487 developers who added or changed more than 1,000 LOCs.

First, we look at the number of projects in which a developer participates (Figure 6). More than 80% of all developers only participate in one project. Although participation in several projects could indicate more experience, no significant correlation was found between the number of projects in which a developer participates and the indicator of design pattern usage.

The increase in LOCs produced by developers was also used by Mockus, Fielding and Herbsleb (2000) and Koch and Schneider (2002) to analyze two big, open-source projects. A main finding of these studies was that there exists a relatively small, very active group of core developers who typically produce more than 80% of the total code. This group is responsible for implementing most of the new functionality, which also involves making design decisions. The histogram of the increase in LOCs by developer for the analyzed projects is depicted in Figure 7. Only 278 of the 1,487 developers are responsible for 80% of the total increase in

*Figure 7: Histogram of the increase of LOCs by developer.*

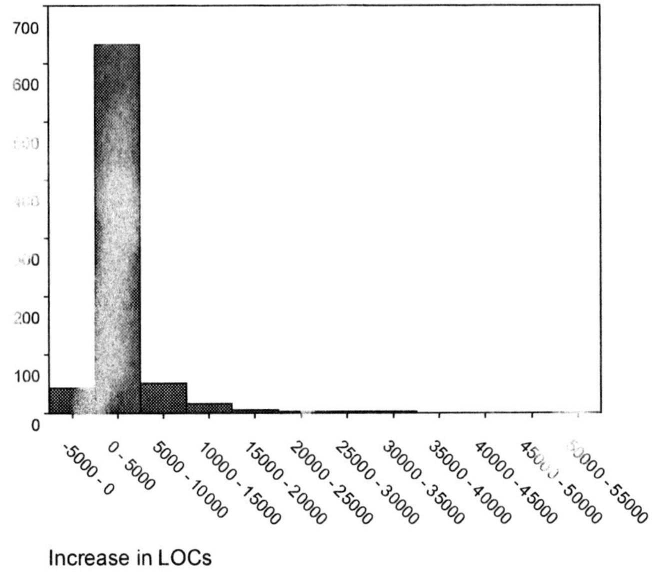

LOCs for all projects. With 18.7% of the total developers, the proportion is similar to the proportion found for the GNOME Project analyzed in Koch and Schneider (2002). An observation that agrees with the Pareto Principle (also known as the 80/20 rule), which was found to hold in many different settings and disciplines. The 278 most dedicated developers can be seen as the core developers in the development community of the analyzed projects where, depending on the project's size, none (for very small projects), one, or several core developers work together with less active developers on a project.

We use rank order correlations to check if the usage of design patterns is related to indicators of activity of different developers. Significant correlation coefficients were found between the number of different patterns a developer uses and the LOCs added or changed (.19), the increase in LOCs (.20) and the number of check-ins conducted (.27). All three indicators are significantly intercorrelated (with correlation coefficients between .57 and .63). These correlations could mean that developers who work more intensely on a project (more check-ins) and therefore increase the code size are more likely to document their changes using design patterns. But this is a very tentative interpretation given the low correlation coefficient, the influence check-ins have on the opportunity to use a design pattern, and the probability of detecting a design pattern with the used heuristic.

To analyze this further, we divided the population of developers (the developers who added/changed more than 1,000 LOCs) into a group of developers for which we never found patterns in the analyzed projects and a group of developers

*Table 6: Differences between developers who use patterns and who do not.*

| | Uses patterns | | Number of projects | LOCs added/changed | Increase in LOCs | Number of check-ins |
|---|---|---|---|---|---|---|
| no | N | | 667 | 667 | 667 | 667 |
| | Minimum | | 1 | 1001 | -4157 | 1 |
| | Maximum | | 7 | 173869 | 52797 | 2430 |
| | Median | | 1.00 | 3585.00 | 1068.00 | 49.00 |
| | Mean | | 1.22 | 7988.10 | 2117.14 | 97.26 |
| | Variance | | .382 | 213576579.378 | 15432652 | 36096.841 |
| yes | N | | 94 | 94 | 94 | 94 |
| | Minimum | | 1 | 1251 | -3485 | 5 |
| | Maximum | | 6 | 60955 | 32970 | 1665 |
| | Median | | 1.00 | 6866.00 | 2858.00 | 123.50 |
| | Mean | | 1.37 | 12022.61 | 4371.15 | 210.72 |
| | Variance | | .731 | 177046413.317 | 31091095 | 67273.148 |

*Table 7: Cross-tabulation for normal developers and core developers who are responsible for more than 80% of the project's code base with pattern usage.*

| | | | Uses patterns | | Total |
|---|---|---|---|---|---|
| | | | no | yes | |
| Core developer | no | Count | 447 | 36 | 483 |
| | | % | 92.5% | 7.5% | 100.0% |
| | yes | Count | 220 | 58 | 278 |
| | | % | 79.1% | 20.9% | 100.0% |
| Total | | Count | 667 | 94 | 761 |
| | | % | 87.6% | 12.4% | 100.0% |

for which we did. Table 6 contains the statistics for both groups. Except for the number of projects, the medians and means of all variables are about twice as high for the group that uses patterns. The Mann-Whitney U Test confirms that the location of the distributions of LOCs added or changed, the increase in LOCs, and the number of check-ins differ significantly (applying Bonferroni correction for four independent tests which reduces the maximum accepted $p$-value to .0025 at the .01 level). This suggests that there is a highly significant difference between developers who use patterns for documentation and developers who do not. Developers who use patterns tend to create more new code (increase of LOCs) and therefore are also involved in creating new design.

In Table 7 we show the cross-tabulation for developers (core developers identified above in this section and other developers who added or changed at least 1,000

LOCs) and pattern usage. On average, 12.4% of developers use pattern names in the log messages. The table shows a big difference (a significant relationship, $p < .01$ using the chi-square test) in pattern usage between core developers (20.9%) and other developers (7.5%). The observation of more core developers commenting their changes with patterns could have the following reasons:

1. Core developers simply check-in more changes and therefore have more opportunities to use patterns and document these changes with the names of the used design patterns.
2. Design patterns represent an efficient way to apply best practices in the form of flexible and robust design and to communicate these design decision (Beck et al., 1996). This is needed more often by experienced developers who create most of the new design.
3. The open-source community has a reputation-based culture (Feller & Fitzgerald, 2000; Raymond, 1999). Visible pattern adoption by core developers advertises competence (Seen, Taylor, & Martin, 2000) and can therefore be used to strengthen their position in the project team and in the open-source community.

It seems reasonable to believe that all three reasons contribute to adopting design patterns to document changes in source code. However, it is not possible to quantify the extent to which each reason influences the observed pattern usage based on historic version control data alone. Another study that includes interviews of developers who use patterns is necessary.

# SUMMARY OF RESULTS AND CONCLUSION

Open source software development represents a new and efficient way to produce high quality software. Many traditional software engineering practices are adopted by open source developers. However, different software engineering practices support the collaborative and implementation-driven development style of OSS better than others. In this chapter, we analyzed the adoption of design patterns by open source developers. Design patterns are interesting for OSS development since their adoption does not require an infrastructure investment and they can be used by a developer without interfering with the development style of others. For almost 1,000 open source projects using Java, we checked if the 23 original design patterns introduced by Gamma, Helm, Johnson, and Vlissides (1995) were used to document changes in the code. There are three main results of analyzing the factors that influence pattern usage:

1. The characteristics of projects have little influence on the developers' adoption of design patterns. For most team sizes, the percentage of projects where

patterns are used for documentation is around 20% in the data set. Only the projects with the largest team size (10+) show a significantly higher proportion. These projects differ from the rest by a much higher level of activity (number of check-ins, increase and changes of LOCs are, on average, two to four times higher than the other projects). Also, we discovered that for most projects where design patterns were used, only one developer used them (again except the few most active projects).

2. No relationship was found between the usage of design patterns in a project and the project's phase in the software life-cycle. Therefore, in the analyzed OSS projects, we found no evidence that design patterns are widely used for refactoring in later stages of the software development. A reason for this finding could be that the analyzed projects are still too early in their life-cycle to make major restructuring necessary. Another explanation could be that open source development generally favors a more flexible design by already using patterns for developing new code and frequent modifications and expansion of the code. This constant change would reduce the need for an explicit refactoring in later phases of the life-cycle.

3. There exist differences between developers who use patterns and developers who do not. There are significant differences in the level of activity of developers in the analyzed projects. These differences indicate that the small number of developers that create most of the code (for OSS projects, often called core developer) are more likely to use design patterns. About 20% of these developers used the 23 design patterns to document changes.

This first study of the adoption of design patterns by open source software developers has many limitations: e.g., information on the quality of the produced code is not included; there is no differentiation between types of changes (bug fixes, new features, etc.); the complexity of the projects is not analyzed using object-oriented metrics; only a very limited number of known design patterns and only one programming language are used; and no further information from the developers (actual effort used for changes, reasons for using patterns) is incorporated. Each of these limitations provides a direction for further research. However, even with these limitation, the results of this study show that design patterns are adopted for documenting changes and thus for communication in practice by many of the most active open source developers.

# REFERENCES

Antoniol, G., Fiutem, R., & Cristoforetti, L. (1998). Design pattern recovery in object-oriented software. *Proceedings of the 6th Workshop on Program Comprehension (WPC)*, Ischia, Italy, June 24-26, 153-160.

Atkins, D., Ball T., Graves, T., & Mockus, A. (1999). Using version control data to evaluate the impact of software tools. *Proceedings of the 21st International Conference on Software Engineering*, Los Angeles, CA, May 16-22, 324-333.

Beck, K., Coplien, J.O., Crocker, R., Dominick, L., Meszaros, G., Paulisch F., & Vlissides, J. (1996). Industrial experience with design patterns. *Proceedings of the 18th International Conference on Software Engineering*, Berlin, Germany, March 25-30, 103-114.

Buschmann, F., Meunier, R., Rohnert, H., Sommerlad, P., & Stal, M. (1996). *Pattern-oriented software architecture: A system of patterns*. Chichester, England: John Wiley & Sons Ltd.

Chidamber, S.R. & Kemerer, C.F. (1994). A metrics suite for object oriented design. *IEEE Transactions on Software Engineering, 20(6),* 476-493.

Cook, J.E. & Votta, L.G. (1998). Cost-effective analysis of in-place software processes. *IEEE Transactions on Software Engineering, 24(8),* 650-663.

Coplien, C.O. & Schmidt, D.C. (1995). *Pattern languages of program design*. Reading, MA: Addison-Wesley.

Dempsey, B.J., Weiss, D., Jones, P., & Greenberg, J. (2002). Who is an open source software developer? Profiling a community of Linux developers. *Communications of the ACM, 45(2),* 67-72.

Feller, J. & Fitzgerald, B. (2000). A framework analysis of the open source software development paradigm. *Proceedings of the 21st Annual International Conference on Information Systems*, Brisbane, Queensland, Australia, December 10-13, 58-69.

Fogel, K. (1999). *Open source development with CVS*. Scottsdale, AZ: CoriolosOpen Press.

Gamma, E., Helm, R., Johnson, R., & Vlissides, J. (1995). *Design patterns: Elements of reusable object-oriented software*. New York: Addison-Wesley.

Goldfedder, B. & Rising, L. (1996). A training experience with patterns. *Communications of the ACM, 39(10),* 60-64.

Helm, R. (1995). Patterns in practice. *Proceedings of the 10th Conference on Object-Oriented Programming, Systems, Languages, and Applications*, Brisbane, Queensland, Australia, December 10-13, 337-341.

Jorgensen, N. (2001). Putting it all in the trunk: Incremental software engineering in the FreeBSD open source project. *Information Systems Journal, 11(4),* 321-336.

Koch, S. & Schneider, G. (2002). Effort, co-operation and co-ordination in an open source software project: GNOME. *Information Systems Journal, 12(1),* 27-42.

Martin, R.C., Riehle, D., & Buschmann, F. (1998). Pattern languages of program design 3. Reading, MA: Addison Wesley.

Masuda, G., Sakamoto, N., & Ushijima, K. (1999). Evaluation and analysis of applying design patterns. *Proceedings of the International Workshop on the Principles of Software Evolution (IWPSE99),* Fukuoka City, Japan, July 16-17.

Mockus, A. & Votta, L.G. (2000). Identifying reasons for software changes using historic databases. *International Conference on Software Maintenance*, San Jose, CA, October 11-14, 120-130.

Mockus, A., Fielding, R., & Herbsleb, J. (2000). A case study of open source software development: the Apache server. *Proceedings of the 22nd International Conference on Software Engineering,* Limerick, Ireland, June 4-11, 263-272.

Perpich, J.M., Perry, D.E., Porter, A.A., Votta, L.G., & Wade, M.W. (1997). Anywhere, anytime code inspections: Using the Web to remove inspection bottlenecks in large-scale software development. *Proceedings of the 19th International Conference on Software Engineering*, Boston, MA, May 17-23, 14-21.

Prechelt, L., Unger-Lamprecht, B., Philippsen, M., & Tichy, W.F. (2002). Two controlled experiments assessing the usefulness of design pattern information in program maintenance. *IEEE Transactions on Software Engineering, 28(6),* 595-606.

Raymond, E.S. (1999). The cathedral and the bazaar: Musings on Linux and open source by an accidental revolutionary. Sebastopol, CA: O'Reilly and Associates.

Reissing, R. (2001). The impact of pattern use on design quality. Position paper, Beyond design: Patterns (mis)used, Workshop at the *ACM Conference on Object-Oriented Programming, Systems, Languages, and Applications (OOPSLA 2001),* Tampa Bay, FL, October 15.

Seen, M., Taylor, P., & Martin, D. (2000). Applying a crystal ball to design pattern adoption. *33rd Technology of Object-Oriented Languages and Systems (TOOLS-33 2000),* St. Malo, France, June 5-8, 443-454.

Vixie, P. (1999). Software engineering. In C. DiBona, S. Ockman & M. Stone (Eds.), *Open sources: Voices from the open source revolution.* Cambridge, MA: O'Reilly and Associates.

Vlissides, J. (1998). Pattern hatching: Design patterns applied. *Software Patterns Series.* New York: Addison-Wesley.

Vlissides, J., Coplien, J. O., & Kerth, N. L. (1996). *Pattern languages of program design 2.* Reading, MA: Addison-Wesley.

# SECTION III:

# F/OSS Projects as Social Constructs

## Chapter VI

# Coordination and Social Structures in an Open Source Project:
## VideoLAN

Thomas Basset, Centre de Sociologie des Organisations, France
and Ecole Normale Superieure de Chachan, France

### ABSTRACT

*This chapter tackles the issue of the distribution of work in an open source project through the influence of social relationships among developers. The author demonstrates that the concentration of code in the VideoLAN project —already pointed out in other projects—does not only depend on technical expertise but is strongly influenced by the nature of social relationships among developers. Face-to-face relationships have a great importance, as does friendship which can favor the circulation of advice. In addition to technical expertise, a second kind of expertise —the ability to be aware of who is working on what—determines the hierarchy within this entity that looks like a collegial organization. The author hopes that this work will help to reduce the hiatus between technical and social considerations on open source software.*

Copyright © 2005, Idea Group Inc. Copying or distributing in print or electronic forms without written permission of Idea Group Inc. is prohibited.

# OPEN SOURCE AND FREE SOFTWARE AS A SUBJECT OF STUDY FOR SOCIAL SCIENCES

## Free Software as a Public Good

As shown by the huge number of papers written on the subject during the past two years, sociologists and economists take a great interest in the development of free software. The main reason for such an activity may be that free software shares many characteristics with well-known objects in social sciences, e.g., public goods. Whereas open source is often only a technical measure, by making the source code of a program available, the choice of the license has great importance as it may exclude some uses of the source code. Free software appears as a surprising object, as its users are very careful about this public good status and have invented and try to promote[2] the GNU General Public License. This license allows anyone to use, modify, and distribute copies or improved versions of the software, but also requires any software that uses any part of its code to be published under the same conditions. By a contagious effect, the software remains "free" and conserves its public good characteristics.

## Participation and Coordination

An intense research activity has taken place during the last few years on the part of both economists and sociologists. If the production of products that look like public goods is quite surprising for economists, it is even more so when it is done by people who communicate and coordinate themselves over a new medium. It has therefore raised the interest of many sociologists. According to Healy and Schussman (2003), studies have focused on two main issues: on the one hand, participation, i.e., trying to understand why people spend time on the elaboration of a product that is available for free, and, on the other hand, coordination, as free software may have appeared to be quite an unstructured phenomenon.

The answers to the first question may be broadly brought together in three categories that correspond to three kinds of actors that have taken interest in the subject. A first interpretation of the free software phenomenon has been given by its members themselves. Part of them consider that their work is based on an alternative way of production to the capitalist system: the participation of some people in free software projects may be explained by a *don contre-don* model (Mauss, 1923). For example, some contributors give code in exchange for user feedback, and because of the existence of such a generalized exchange, free-rider attitudes are rare. But such an explanation has always been difficult to elaborate with classical economic tools. A second set of answers has been developed to explain the participation in the free software movement by the existence of individual incentives (e.g., Lee, Moisa, & Weiss, 2003; Lerner & Tirole, 2000; Dalle & David, 2001). The participation in a free software project provides personal advantages that outweigh its costs, since it could be an opportunity to acquire higher skills or to gain a reputation as a good developer, manager of project, etc.[3] A third stream of studies has developed by taking into account earlier research that has been done on social movements (e.g., the civil rights movements or the peace movement) where the gains from an action can not be individually captured, and by insisting on the relationship a developer has with the small team with which he or she works[4] (Hertel, Niedner, & Herrmann, 2003).

The second issue that has been tackled — mostly by sociologists — is the way people coordinate within organizations that are virtually open to anyone who wants to take a look into them and where access costs are low. Eric Raymond's "The Cathedral and the Bazaar" (1999) was a fundamental article written by a developer. It compares two ways of coordination: cathedrals (or proprietary) software and open source software that look like a bazaar where there is little hierarchy and where development appears chaotic. In cathedrals, plans are drawn and have to be closely followed, whereas in a bazaar-type environment, anyone can join and leave the team after making a contribution in an unplanned kind of coordination. According to Raymond, this free entrance characteristic gives a great technical advantage to open source software over proprietary software — bugs are more rapidly detected because more people with different skills tend to take a look at the code. This idea of a good allocation of resources because of a low cost of gathering information has been theorized by Benkler (2002), who uses the concept of transaction costs[5] (Coase, 1988; Williamson, 1985). Because it has little hierarchy and participants to the project are linked within a peer-to-peer network, an open source project lowers transaction costs. Therefore, resources could be allocated in a much better way than in a market or in a firm. According to Benkler (2002), "peer production has a systematic advantage over markets and firms in matching the best available human capital to the best available information inputs to create the most desired information product."

One may consider that the studies that focus on the distribution of work inside the open source projects are based on such a hypothesis. They rely on the analysis of the Concurrent Version System (CVS)[6], a tool that allows participants to work together and researchers to know who has done what within each file. By focusing on the CVS, the only criterion taken into account is the source code that is exchanged. The main result obtained by these studies is that the majority of the code is written by a small number of people (Koch & Schneider, 2002; Mockus, Fielding, & Herbsleb, 2000, 2002), which goes against the myth of a myriad of people working together on a software. However, such a result has not been that much discussed. It may just reflect that people have only offered a punctual contribution and that, once it has been accepted, they no longer participate. This is coherent with the fact that developers tend to often join and leave a project (Koch & Schneider, 2002). But another interpretation of these facts is possible. Since many people are able to participate, it may prove that the costs of information acquisition are low, but the very important concentration of code in the hands of a few developers could also be a sign of a huge cost for acquiring information that prevents developers from gathering enough information to join the core team.

# LITERATURE LACUNA AND RESEARCH PROJECT

These two approaches to the study of the open source[7] way of development face an issue that has already been pointed out by Granovetter (1985) in economic exchanges. They are often treated in an over- or under-socialized way. The study of a CVS tends to be quite the same as an analysis of an economic exchange, without taking into account the importance of the community. In the same way, the analysis in terms of incentives — however interesting it may be — goes against the myth of a spontaneous and totally original movement tending to be under-socialized as it focuses on personal interests. Reconnaissance in a small community is more satisfactory, as it points out the importance that is given to belonging to a community, as does the *don contre-don* model. But in these cases, exchanges are over socialized; community is considered as a whole, almost separated from the writing of the code. Exchanges are considered as natural and their impact on the distribution of work is not questioned further.

Consequently, if both these approaches have made interesting interpretations about *why* work was done, current literature lacks studies on *how* work is done. The under-socialized studies that focus on the analysis of the CVS provide results about how work is distributed but do not explain how such a state has been elaborated, as coordination between the developers does not appear in a dynamic way within a CVS. They give an interesting view of the state of work at a given time, how work is *distributed* in terms of contributions, but do not say much about how work is *done*, what kind of processes have produced such a state of the work.

In order to complete these approaches and escape from this situation, exchanges of code, joining a community and belonging to it should be considered as the results of embedded interactions. The questions of participation and coordination would therefore be joined, as von Krogh, Spaeth and Lakhani (2003) have already begun to argue. I suggest here that we should try to conciliate both approaches by studying how some micro social structures see to it that cooperation within an open source project is not only led by technical expertise (that is, by the amount of code contributed) or by communitarian considerations, but that both influence each other. Taking into account social aspects of the open source phenomenon may help explain how — and not only why — work is done in a dynamic prospect by considering the influence of social relationships. This views social aspects not as incentives but as variables that may be used as resources.

## Open Source Software as Collegial Organizations

The fact that the eventual impact of social interactions on coordination has been little studied is all the more surprising as studies have already been devoted to structures that are close to open source projects. I make the hypothesis that open source software may be considered as collegial organizations. Waters (1989) identifies six criteria that could characterize a collegial organization. They fit with common characteristics of many open source projects. At first, what characterizes a collegial structure is that theoretical (as opposed to practical) knowledge is at the core of the work. The theoretical knowledge we are talking about is the ability to write software, which is the expertise familiar to many studies devoted to open source. The second criterion concerns the "career" of the members of the collegial organization. It is composed of two steps: a period of learning is necessary before a new member is considered equal to the other members. This fits with the idea often developed within open source teams that anybody is able to join the team but that he or she has to prove his skills, for example, by submitting valuable patches (enhancements) in the mailing-lists. The third criterion is that performance of each member is difficult to compare with that of others. In an open source project, how could one determine who is the performing the most? Is it the one who has written the main part of the code, the one who is in charge of the website[8], the one who has written the documentation that helps people to use the software and allow developers to know what needs to be done and how it could be? Even if one only considers the code, the amount of code written by each team member is a very doubtful measure of performance, and the use of CVS statistics for such a task should be done very carefully, as all parts of the code do not all have the same importance. Because the performance of each team member is very difficult to compare, most members of a project are on equal footing with each other from a formal point of view. The fourth criterion is that a collegial organization controls itself, as opposed to a subcontractor whose performance is evaluated by the firm that has bought its services. The collegial organization therefore gives itself ways to evaluate its own production (fifth criterion), this evaluation being mostly done by exchanging opinions among peers. Finally, the last criterion is the existence of collective forums where decisions are made. In most open source projects, the mailing-list (and in a way IRC channels) are a place to ask the members of the project their opinion on some technical choices. The very existence of flame wars proves that decisions are often very debated before and even after being made.

As members of a collegial organization are permanent, the comparison between collegial organizations and open source projects is limited to the core developers who almost permanently work on the software. Considering open source projects as collegial organizations could shed a new light on how they work and help to go further than the simple statement that the core teams are very narrow. Lazega's work on corporate law partnerships (1995, 2001) could be a source of inspiration. Corporate law partnerships are firms that group highly skilled lawyers who work together on judicial cases, the structure of the team being decided according to the cases and the knowledge that is supposed to be needed. As in open source projects, what is at the core of these teams is each member's knowledge. But if one only considers technical skills, one may miss many aspects that explain how people work together. Lazega points out that professional cooperation between people is deeply influenced by the fact that they have social interactions outside of their work. For example, friendship may be used to control one peer who is straying from the norm, this resource being used in very different ways according to the status of the person to be controlled and the person who tries to exercise the social control (Lazega, 1995). Far from being guided only by technical considerations, coordination within a collegial organization depends on more "social" variables. The understanding of the workings of open source projects may benefit greatly from taking into account the role of such social embeddedness, rather than only considering social incentives.

The fact that interactions have rarely been studied at a mezzo level, as being embedded in a small network of relationships, may result from the particular nature of the phenomenon that is studied. The open source movement requires great technical skills in programming that are essential in order to understand exactly what is at stake in interactions. The interactions between developers are by themselves difficult to observe, as some of them are private (e.g., mail between the core developers that are not made public) and occur between people who are geographically located throughout the world. The cost of a traditional study with face-to-face interviews would be astronomical and mail or Internet Relay Chat (IRC) interviews would introduce other biases. Moreover, although some of the interactions are publicly available and archived, they are very numerous. For a sociologist, the cost of entry into an open source project is indeed very high, and very little ethnographic observation has been done. However, some projects are more accessible than others. What follows is a case study of an open source project that was by chance accessible to various sociological methods such as ethnographic observation, network analysis, and CVS analysis.

# A CASE STUDY: VIDEOLAN

## Presentation

The VideoLAN[9] project is quite a particular open source one. From a technical point of view, it is a video-streaming solution over high-bandwidth networks. It is made of two major pieces of software: the VideoLAN Server (VLS) that sends the video over the network and the VideoLAN Client (VLC) that receives and displays the video. As with Xine, Mplayer or Ogle (with which it shares some pieces of code), the VLC can work as stand-alone software to play videos on DVDs. Though it is not as well known as Xine or Mplayer, in the small world of open source video players, VideoLAN has a good reputation, being considered the best video player by a majority of MacOS X users, who prefer it to Apple's DVD player.

But VideoLAN was not only developed over the Internet by people who did not see or even know each other before they began to work together. VideoLAN has strong "real" roots, and the ties between some of its members are therefore very strong, deeply rooted in traditional social links. VideoLAN was developed for years at the Ecole Centrale Paris (ECP)[10] before being released under the GPL and made publicly available in 2001. Though many external developers worked on the project, a huge number of those working on VideoLAN are students from the ECP. These students live on a campus, most of them on the same floor in a small building devoted to them. As students in a college who live together, strong bonds of friendship often develop between them. Moreover, VideoLAN still remains in parallel a "club" of the school, i.e., a team that the students may choose to join in order to validate a part of their studies.

## Questions

This provides VideoLAN with two major advantages for a sociologist. It is a place where ethnographic observation can take place, which allows gathering much more information than with interviews over the Internet. Moreover, the existence of a network of strong relationships allows the study of the role of embeddedness on the coordination within the project[11]. Three points need a particular attention. First, it should be interesting to see whether the concentration of expertise in the hands of a few people is a global characteristic of open source projects or whether it was only the case for some software, such as Apache and Mozilla (Mockus, Fielding & Herbsleb, 2000, 2002) or GNOME (Koch & Schneider, 2002). Second, as the students from the ECP constitute a strong community, it should be interesting to see whether there is a gate-keeping phenomenon for the external contributors who would like to work on the project. If not, one may think that the first participation in an open source project relies primarily on technical expertise and that being part of a network of embedded relationships has little impact on the issue of joining an open source project. Last, the impact of the social variables on the place taken within the team and the work done should be tackled. In other words, it is an attempt to determine to what extent the distribution of work is structured by technical expertise and by the nature of the relationships between the developers.

## Metrics Used

In order to tackle these issues, three methods have been used for the gathering of data: analysis of the CVS, ethnographic observation, and network analysis.

The CVS was analyzed using StatCVS[12] 0.1.3. The analysis was run on the CVS version of VideoLAN dated March 29, 2003. StatCVS does not allow the distinction between added and deleted lines, as has been done in previous studies (e.g., Koch & Schneider, 2002; Mockus, Fielding & Herbsleb, 2000, 2002) but returns only the modified lines. As it gives significant results that are consistent with ethnographic observations, this measure was considered sufficient.

The ethnographic observation consisted of a personal involvement in the project, in reading the mailing-lists, and being on the IRC channel and at the meeting of the ECP staff that takes place every week. The observation began in October 2002 and ended in March 2003. At the end of it, I was almost considered as a member of the tribe, knowing each developer, even the external ones. My position was the one that a member of the ECP would have held in the project, except for the fact that I never lived where most of VideoLAN developers from the ECP are grouped.

In order not to rely only on subjective observations (which were moreover highly asymmetric between the developers linked to the ECP and those who were not), a network analysis has been done. A list of 53 active developers out of 184 people identified as having taken part in the development of VideoLAN[13] was elaborated with the two main coders of the VLC. An email was sent to these 53 people, inviting them to complete web questionnaires on a private server. Forty-seven people registered on the server. Because of uncompleted answers, only 34 people furnished usable data. This number is quite low and does not allow serious statistical treatment. However, these people represent 85% of modified lines for the VLC, 100% for the website[14] and only 37% for the VLS. Data were analyzed thanks to Pajek[15] 0.90.

## Concentration of the Code and Weight of Technical Expertise

One of the first and most spectacular results concerning the distribution of work within an open source team is the fact that only a very limited number of people contribute to a huge part of the code that is written. In the case of Apache, the top 15 developers contributed 88% of lines added (Mockus, Fielding & Hersbsleb, 2000). The concentration of contributions is not always that high, as found for the GNOME project where the top 15 developers contributed 48%, 52 being necessary to reach the 88% (Koch & Schneider, 2002). A similar result has been found for the Mozilla project where for each of the seven main parts of the software, between 22 and 35 developers were necessary to reach the 88% of lines of code added (Mockus, Fielding & Hersbsleb, 2002). However, if one considers that at the time of the different studies 220,000 lines had been added to Apache, 2,037,000 to Mozilla and 6,300,000 to GNOME, these differences were probably due to the size of the project (the wider the project, the wider the group of core developers), because it is almost impossible for a small group of people to write millions of lines of code.

An analysis was run on VideoLAN in order to know whether this result was common to any open source project or not, even if it has a very particular social structure. The VLC is a piece of software that is constituted of 355,014 lines of code. 860,686 lines have been modified between 1999 and March 29, 2003. Forty-four people are quoted in the AUTHORS file and 58 in the THANKS file, i.e., at least 102 people have been involved in VLC development. Within the VLC, the core team, defined by Mockus, Fielding, and Herbsleb (2000) as the people that have together contributed 83% of the modified lines or 88% of added lines, is made up of 10 people among the 46 who did commit in the CVS at least once. The concentration is even stronger if one takes a look at the very top developers: the most active developer has contributed 46% of modified lines[16], and the top two most active developers 61% of the code. Almost 75% of the code was written by four people. This is consistent

*Figure 1: Distribution of work between developers in VLC.*

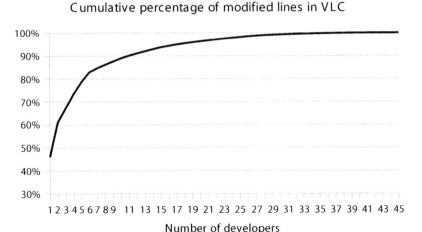

with the statement of Mockus, Fielding, and Herbsleb (2002) who considered that a consequent part of the code was written only by a few people.

The distribution within the VLS is quite different. Only 15 different people did commit some work in the CVS; 16 are quoted in the AUTHORS file and contributed to many more modified lines of code; and six people were needed to reach 85% of modified lines.

If the distribution of work is more "egalitarian," we know by ethnographic observation that expertise concerning this software is still very concentrated. The VLS is a relatively small piece of software (73,223 lines on March 29, 2003) but, according to the members of the VideoLAN project, it is quite hard to understand how it works. This is all the more true as the two main developers have not worked on it for at least six months. In fact, only four people actively work on the VLS (three on its very core and one on peripheral features) and are therefore the only ones who have the necessary technical skills to understand its workings. Technical skills are very concentrated among some happy few for both the VLC and the VLS.

This result is another indication that technical expertise is very concentrated in open source projects. The role of these experts is therefore probably very strong in the project. That expertise may be one of the major ways to be recognized and to gain an important place in the project. However, the issue of how such a distribution is elaborated still remains. Are so few people writing so much of the code because they are highly skilled developers and therefore take a central place within the team or aren't there any social variables that may play a role by giving these developers more opportunities to gather information and hence acquire more expertise on the software in order to take a central place in the project? Technical expertise should be the only explanatory variable of such a distribution of work if one makes the hypothesis that the costs required to gather information are very low and equal

*Figure 2: Distribution of work between developers in VLS.*

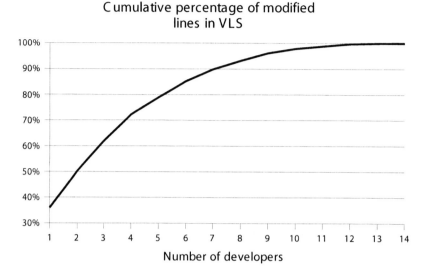

between every developer. This is a strong premise that needs closer examination. As the VideoLAN project has a structure in which a strong network of friendship exists, it offers the opportunity to examine whether technical expertise leads the way or whether the coordination within the team may be influenced by other variables that may have an influence on costs for gathering information.

## Ability to Join the Team

If costs for gathering information are not the same for each developer and depend on the intensity of relationships between developers, technical expertise may play a less important role in the participation to a project than has been stated.

For the members of the ECP, the goal of the release under the GPL license was to attract external contributors. However, they were conscious that it might be a difficult task because of the existence of a strong community with almost daily face-to-face interactions, whereas external contributors would only be accessible via the Internet through which it is (according to them) much harder to exchange information than face-to-face. Furthermore, within the ECP, information is more easily available thanks to the meeting that takes place every Sunday evening and that allows people to share their knowledge about the state of the development. Such information is also available to external contributors, but they have to ask people on an individual basis through the IRC channel, which constantly necessitates a new interaction, as Internet usage of all involved rarely overlaps. One may therefore think that the costs necessary to get the information are much higher between people of the ECP and external contributors than among people of the ECP. It should therefore be harder for exterior participants to join VideoLAN because it would be much more difficult for them to know what needs to be done and how than the people from the ECP.

A first look at the distribution of work tends to point out that it is much easier to participate to the VideoLAN project when one is linked with the ECP. Among the top developers, a huge majority are students or past students at the ECP; among the 10 most active developers of the VLC, only two of them do not belong to the ECP—one contributed 6.31% of modified lines (third place) and the other 1.47% (eighth place). Only two external contributors have worked on the VLS, their contribution being of 6.2% of modified lines. There seems to be a huge entry barrier for people who are not related to the ECP.

However, this does not take into account the time that people have been involved in the project. As VideoLAN has been released under the GPL in 2001, it is not surprising that external contributors should not yet have written as many lines of code as people who worked on VideoLAN at the ECP before this date. The number of modified lines per months of participation to the project reflects in a much better way whether a developer had enough information to work and was able to participate to the development. Out of the 10 developers who wrote the most important part of the VLC since the beginning of the project, seven are still active (i.e., they sent work to the CVS at least once during the past five months) as of March the 29, 2003, most of them sending commits every week. For this reason, on April 1, 2003, the tenth most active developer who still works on the VLC has the 13$^{th}$ rank in terms of modified lines since the beginning of the development of the VLC. The comparison

*Figure 3: Distribution of work per month in VLC.*

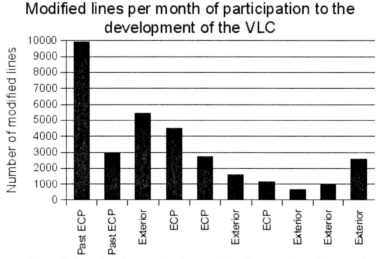

of the number of lines written per month of participation to the development of the VLC shows no clear link between the intensity of the development and whether the developer is a member or a past member of the ECP.

Among these 10 most active developers who are still working on the VLC, two were past students of the ECP and are often present at the Sunday evening meeting, three are currently students at the ECP, and five have no link to the ECP. Whereas three of the five people who have modified the greatest number of lines are related to the ECP, there is no clear link between whether a developer is a student or a past student from the ECP and the amount of modified lines per month. Information does seem to be accessible to everybody and the fact that a group with strong social relationships exists is not an obstacle for new contributors to join the group. From the large point of view of the amount of work done, being part of a group that is located in the same place does not seem to have any influence on one's ability to become an expert by writing a huge amount of code.

One could share the assumption that for the experts of the core team, information is available at a very low cost. This is possible because of the existence of the IRC channel #videolan and the mailing-lists. In the network analysis, people were asked with which other members of the team they had private interaction (that is, a conversation where they answer to each other, not only giving their thoughts to everyone in a public discussion) about any aspect related to the VideoLAN project during the last month and on which medium. Both the diameters of the networks concerning IRC and the mailing-lists of the project were of 3. Five people out of 34 were not connected to the IRC network, and two of them were not connected to the mailing list. As long as one is able to write good code, the fact of not belonging to the ECP does not seem to have an impact on the ability to work on the VideoLAN project, thanks to the various existing avenues of communication between developers. External developers have always gained their places in the project by submitting some patches in the mailing-lists. Technical expertise is evaluated by the team according to the importance of the contribution (e.g., a developer has been immediately integrated after he had posted a conversion of the VLC to Windows) or the fact that several small but good works have been done over time and give CVS access when a developer is considered as competent. Embeddedness has indeed little impact on the ability to join the project for the people who have already good technical skills.

## How Do People Coordinate?

The process of joining VideoLAN is often the same: a developer identifies a lacking feature that he is able to code, works on it, and submits his work to the VideoLAN team. If it is considered good enough, he is given CVS rights. The Windows version and a new design of the Mac OS X interface have been two contributions that have allowed their authors to integrate the team. These patches mainly need technical expertise. But once inside the team, where choices on the very architecture of the software are made, is technical expertise sufficient to explain the participation on the project? One may think that it is not and that the issue of coordination needs other variables in order to be correctly studied.

Two facts tend to prove that the distribution of the work, which is a result of the coordination between developers, depends on a second kind of expertise called *panoptic* expertise. The first fact is the way the documentation has been written. Though it has been started by an external developer, the user documentation of Video LAN is mostly written by three first-year students from the ECP under the direction of a third-year student. According to this third-year student, they were indeed astonished that an external developer may have been able to write documentation, considering that they were in a much more comfortable situation to do so because "[they] had easier access to anything." The distribution of the work done on the website which gives the users information concerning the state of the project has the same kind of distribution as over 75% has been done by two people related to the ECP, one of them having committed only 0.35% of the modified lines of code for VLC and VLS, whereas the other represents the major contribution (46%) to the VLC. Most of the documentation and the website, which are the two main ways for a user to gather information, have been done by people related to the ECP who have very little technical skill. This is surprising, because if we consider that costs necessary to gather information are low, anybody—including people who are not linked to the ECP—should be able to gather information and take part in the writing of the documentation. And that proves that technical expertise may be supplemented by another expertise, a "panoptic" expertise that is an awareness of the state of the work.

The second fact that favors consideration of an expertise other than technical as an important part of coordination among the different developers has been furnished by the ethnographic observation. If, during the meetings that take place at the ECP and the exchange on the IRC channel, the persons who have contributed most of the code are carefully listened to, other people with a weaker technical legitimacy are also listened to very carefully, even on technical choices. The reason why such attention is given to them is that they often are very knowledgeable about the state of the developments, sometimes even more so than the developers themselves who do not always share information concerning what they are doing. Whereas the way the documentation has been written was an argument in favor of considering that a second kind of expertise existed, the argumentation used in the decision making proves that such an expertise is legitimate.

But the assessment that coordination is influenced by two kinds of expertise, one mostly technical and the other that is an ability to have a wider view and access to information, does not help very much in explaining the way coordination is driven. It only points out that technical expertise may not be considered as the only way of explaining how people work together. It leads to consider, on the one hand, that these kinds of expertise that are not equally distributed within a team may result in various hierarchical structures and, on the other hand, that the variability of the relative legitimacy of these two kinds of expertise has an influence on the kind of management that is adopted. This is to some extent the point made by Healy and Schussman (2003), who consider that the success of an open source project depends on three factors, two of them being the hierarchy and the kind of management. I would like to go one step further and not take these two kinds of expertise for granted but, rather, try to understand how people acquire them and what influence they may have on the distribution of work within the VideoLAN team.

## How Does Technical Advice Circulate? Authorities and Hubs

In order to tackle this point, the circulation of technical advice within the team has been studied. People were asked to indicate the persons from whom they have already sought advice when they faced an issue concerning any part of the Video LAN code. This study was guided by the hypothesis that people who are the most (directly or indirectly) sought out for advice are the technical experts, recognized as such by the members of the team, but that the circulation of advice within the team is not only determined by technical expertise but also by the "panoptic" expertise. The comparison between the structure of the network and the distribution of technical expertise, determined by the study of the CVS, would help to understand how both kinds of expertise are distributed among the team, how they are mobilized, and how they are acquired.

In order to protect their anonymity, people have been divided into six populations according to their technical expertise, depending on the amount of code written and the ownership of a unique skill. In Category 1 are the four top developers of the VLC, who have modified between 54,000 and 400,000 lines of code (out of 810,000). The specialists of the VLS, who are almost the only ones to work on it, are regrouped in Category 2. The people who recently gave significant amounts of code (from 8,000 to 35,000 modified lines of code) to the VLC have been marked as Category 3. People who have written the documentation but have committed almost no source code to any program are grouped in Category 4. A fifth group has been constituted with the people who have given significant amounts of code (around 7,000 modified lines of code) but who have stopped their development for months and with the people who have made only limited contributions (from 2,000 to 4,000 modified lines of code) though they are still considered as active. A last group was created, grouping the people who are considered to be active members of the project but do not have CVS access or did not use it, all of them needing to relay their action through a member of the team because they are first-year students or very specialized people who make limited and punctual contributions, such as the making of packages for Linux distributions—Red Hat or Mandrake—when a new release is coming.

## Partial Discrepancy between Technical Expertise and the Circulation of Advice

The results of the study show a clear discrepancy between the technical expertise and the network of advice. The people who have the highest authority[17] score (between 0.26 and 0.47) are the four people in Category 1 because of their technical skill. The higher the technical skill, the higher the authority in the technical advice network. This is also consistent with the fact that among the ten people whose advice is most sought by others, there is one of the three experts working on the VLS and the person who is in charge of most of the documentation and the website. However, even if technical skill explains a large part of the circulation of advice among the team as it induces who are the authorities, it is strongly influenced by other factors. Within the group of the three most skilled people, the person not belonging to the ECP has an authority score of 0.26, whereas the other people who are members or past members of the ECP have authority scores from 0.43 to 0.47. In the same way, people from the ECP have been quoted between 21 and 32 times as having been asked for advice, whereas this person has only been quoted 10 times[18]. Despite the fact that this expert has taken more and more weight in the technical choices made by the VideoLAN team and is now one of the three developers able to work on the VLC kernel, the circulation of advice to the top depends on the links to the ECP. One may suppose that the cost necessary to gather information is lower among developers of the ECP than if an external developer is involved, and that it therefore has influence on all authority scores. The same interpretation could be made if one considers a second group made up of people having authority scores from 0.10 to 0.23. In this group, the two persons working on the VLS who are related to the ECP have been sought out for advice much more than the one who comes from outside the ECP. Even if the technical expertise allows all three of them to be considered as experts of the VLS, the costs to contact the person are less important when one knows each other. In the same way, if one excludes the "core" team, the people who have written the main part of the contributions over the last months are people who do not belong to the ECP and they are not considered as advice authorities in the network. One may therefore again suppose that the costs to gather information are not equally low among the members of the team but that people from the ECP tend to seek advice from each other.

## Importance of Face-to-Face Relationships

The structure of the advice network is therefore much more intricate than if it was only based on technical expertise. It is quite clear with regard to the external contributors who may have joined the project mainly because of technical interest—they ask advice from the members of the core groups working on VLS or VLC and from the secondary contributors in terms of code. However, in most cases, they have only asked advice from only one or two people, whereas most of these external contributors have worked on several parts of the code. If costs to reach the people who have the technical expertise were null or equal among all developers, one may have found a much wider spectrum of relationships. Concerning the students from the ECP, the question is more complex, as they see each other every day. Because of the technical gap that may exist between two members of the project, it may be difficult for the new and less skilled people to contact more experienced developers, as such a contact may reveal the technical gap between them. It is particularly true in the case of the students from the ECP, but one may think that this fact is only outlined by the specific relationships among them and exists at a broader scale with less intensity. In both cases, a developer does not always dare start a new interaction in order to gather information. The costs to gather information should therefore be reintroduced through the notion of "face" as it has been developed by Goffman (1956); i.e., in order not to lose face, the developers use elaborated strategies, trying to gather information through hubs, which explains why the advice network is so much disconnected from the technical skills. A member of Group 3 reveals that he prefers to leave his room and ask advice from a friend who is one of the core developers rather than trying to join any of the core developers over the IRC. Despite VideoLAN being a collegial organization with equals, the circulation of advice is not free from social constraints but depends on individual strategies.

## A Network Structured around Hubs

This notion of face helps to understand how panoptic expertise is acquired and mobilized. The advice network is indeed very much structured around hubs that have been given this role not because of their technical skills but because they both are accessible to advice-seekers and themselves have access to technical experts. They are therefore able to gather information at an cost acceptable to them. And people are able to ask advice from them. The four people who have the highest hub scores[25] have modified very few lines of code (from 600 to 15,000) as compared to other members of the project. But for the students from the ECP, these four have the great advantage of having known the project for a long time and therefore have easy access to the main coders. Three of them are third-year students who are able to join the "core" developers without too much cost, as long as they have demonstrated a great involvement in the project and have been at the ECP with the two main developers. One of these hubs is the main writer of the documentation, and another is one of the experts of the VLS. Newcomers use these persons who have panoptic expertise in order to get technical information indirectly. Within the VideoLAN team, the first-year students have easier contacts with second-year students because they are integrated in clubs driven by their predecessors, where older people are less active. On a wider scale, the forum related to an open source project may be the place to find some of the people with high panoptic expertise, as the core developers do not always have much time to spend on guiding newcomers.

But the way panoptic expertise is being mobilized by people who need it does not explain how it is acquired. Within the VideoLAN project, the panoptic expertise is acquired through a very specific process. As indicated by ethnographic observation and confirmed by network analysis, friendship is a resource that may be used to develop panoptic expertise. It is particularly true concerning the students from the ECP: the friendship network[19] is made of groups that roughly tie together people from the same year of study. Some individual people are the links between various groups, one first-year student being, for example, a friend of one of the second-year students who has a wide friendship network, thus allowing the first-year student to slowly integrate the group and have indirect contact with third-year students who themselves know external developers. This explains the high betweenness centrality scores[20] concerning friendship from people 2 and 18 who are at the intersection of two groups of friends. Such a structure is extended to the relationships with external contributors. The core developers (e.g., 17 and 33) tend to have developed friendship with the people with whom they work, which may explain why the core developers who are linked with the ECP also have high betweenness centrality scores. Because some people are indispensable in helping two persons to communicate, their panoptic expertise grows. Newcomers ask them for advice, and developers are therefore aware of their projects. And they may have to ask core developers about some functions within the kernel and are therefore aware of the state of its development. Panoptic experts benefit both from ascending and descending information. Because of their position based on their panoptic expertise, they have to face many more technical problems than if they were isolated and did not have to gather information for other people. Their technical expertise consequently grows, and they have the ability to convert some part of their panoptic expertise into technical expertise. But as their technical expertise grows, the cost to contact developers of the kernel declines and the cost to be contacted by newcomers grows, and thus they tend to become authorities rather than hubs. Therefore, a new cycle has to begin in order to keep the link between newcomers and experimented coders. At the ECP, this process is helped by the arrival of a new *promotion* (entering class) every year. The two kinds of expertise, though they do not strictly go together, influence each other and are therefore sometimes difficult to distinguish.

## Expertise and Its Influence on the Regulation System

The fact that panoptic expertise is partially based on friendship — and not only on a technical expertise that could allow information gathering without losing face — has important consequences on the regulation of the relationships within the team. Friendship plays a double role in the VideoLAN project. It allows newcomers to the ECP to integrate the team after a period of acclimatization, without losing face in front of experts who often impress them. But these bonds of friendship are not only visible from inside the ECP team; they are often publicly demonstrated, especially on the IRC channel which is a permanent circus where people exchange news about their lives, in addition to answering questions of users or developers who would like to do some work. According to the external developers, such an atmosphere is quite rare on an IRC channel devoted to software. Therefore, the friendship that unites the members of the ECP has spillover effects, as external contributors tend to join the conversation and speak about their private lives. By doing so, they develop a link with the team that is much stronger than if it was only technical. Even if some of the external developers join the team because they would like to implement a new feature into the software, the atmosphere of the team may be a criterion for the choice. According to most of the external developers, the feeling that the VideoLAN project is supported by a community that has more than technical ties is very important. One may think that such importance is given to these non-technical considerations partly because it helps to prevent the risk of a fork, a phenomenon that is quite common within open source software, even very important ones (e.g., PHPNuke — PostNuke and their followers, the various BSD systems, or Xfree86 — Xwin). VideoLAN is therefore a project that produces both code and social link, this one being voluntarily maintained by external developers who come to France and see the students from the ECP, or people who have stopped working on the project but who are still present on the IRC channel[21].

## CONCLUSION

Even in a particular project such as VideoLAN, in which a significant number of people are able to have physical contact and share bonds of friendship, the importance of technical expertise remains absolutely preponderant. In a coherent way with previous studies (Koch & Schneider, 2002; Mockus, Fielding & Herbsleb, 2000, 2002), technical expertise is highly concentrated in the hands of few contributors, who are disconnected from their geographical situations and who coordinate themselves over non-physical ties — the IRC channel being the main medium of communication. However, one should moderate the assumption that this abstraction from physical links means that open source software succeeds because costs to gather information required to join the development (which include cost to contact a person, if possible, an expert) are very low and therefore information is available to anyone who wants to participate. Relationships that are not only based on technical considerations may have an effect on the distribution of work within the team. The fact that these relationships (sometimes with strong forces such as friendship) link embedded people should be seriously considered when one studies an open source project, as it introduces differentiated costs of communication that influence the way information is distributed. It is an illusion to consider that all information is easily available (even when it is publicly available) and that the network of active developers is a flat one. In this particular study, students from the ECP have the opportunity to seek out technical experts in order to gather pertinent information through a medium such as the IRC that allows people to contact anyone at the same cost. But they do not, rather preferring to use their social network in order not to lose face. Such considerations thus introduce new costs that have to be considered in order to join the project and that define what is possible. Because of differences in costs of communication, one should consider a panoptic expertise that plays a significant role and has non-trivial links with technical expertise. Concerning the external contributors, their cost to join the project is inferior, as they do not already belong to the social network constituted by the VideoLAN members. They may therefore directly work on their field of expertise.

As the ways to determine the position taken in the project are not the same, the distribution of work inherits from this difference. The coordination and documentation tasks are endorsed by students from the ECP who have easy access to information, whereas people who have technical expertise focus on the development of the software.[22] The agent who joins a project takes advantage of the resources (technical expertise or position within the network of relationships) that he is able to use and that gave him access to some places. As stated by David Zeitlyn (2003), "relationships are made through action, mutually directed and reciprocal" (p.8), and through action people acquire and confirm a symbolic capital, as it has been defined by Bourdieu, that allows the development of panoptic expertise. They are then able to redefine what is possible. The place that is occupied in the social network of developers and the resources that have been used to access it should therefore be taken into account in a model that tries to determine the development of an open source software, particularly when it focuses on the choice of developing a new module or working on the kernel of the software (Dalle & David, 2003).

These results shed a light on VideoLAN analyzed as a collegial structure. The fact that developers are theoretically equal and that information is made publicly available does not imply that everybody communicates with everybody, the structure of the circulation of advice being explained by technical expertise. Face-to-face interactions have a great importance in the structuration of this network of advice and their analysis can help us to understand how work is done within VideoLAN.

Of course, the VideoLAN project remains a very unique open source project because of its links with the ECP. The issue of the generalization of the conclusions that have been drawn should not be ignored and this study calls for comparative works. However, quite a few facts indicate that taking into account the embeddedness of relationships between the *core*[23] developers may help us to understand open source development better. Some projects are based on the relationship between a limited number of people. Several projects have been started by people who were present at the same place, such as the Apache web server or Emacs at MIT, the DVD player Ogle at the University of Copenhagen, or Galeon at the University of Milano. The main project that reflects the importance of interpersonal links is probably the Debian distribution in which co-optation is the key to participation. One may think without too much risk that personal interactions within this project play an important role and constitute an exciting field of study.[24] But Debian is quite a unique project. If one looks at the libraries that are used by many video players in order to decode specific audio and video formats (which are not marginal, such as the MPEG2 and A52/AC3 decoders used for the video and sound of the DVD), one can see that the same people are working together on it. It may be the result of a technical specialization, the worldwide experts on the subject working together, but it also may be the result of a lowering of communication costs and the people who already worked together on a project tending to work on other ones (close in technical terms) with the same people. This is consistent with some results from the FLOSS report (FLOSS, 2002): 50.1% of the developers had less than six regular contacts with other members of the open source/free software community and 64.2% of the developers were involved in less than three projects. If the cost to join a project is high, it is not only because of the technical expertise to acquire and information to gather, but it may also be because one has to develop relationships with new people. These relationships are not only based on technical consideration but also on social ones, and the difference in communication costs induces differences in the network that link the members of a team together.

In the long run, one may finally wonder whether embeddedness would not take a growing place in the open source development as firms invest in it and therefore hire people. One may see the emergence of structures that share many common characteristics with the VideoLAN project. For example, Netscape and IBM have involved several people in Mozilla and the Linux kernel, and it is quite possible that the development of these projects may be influenced by the existence of teams that work together in offices and in which information may circulate more easily than among isolated people. However, I consider that embeddedness already has a great importance in existing projects; if the involvement of firms in the open source software field follows the movement of the last few years, one may see projects with a structure close to VideoLAN appear and reinforce these aspects.

# REFERENCES

Benkler, Y. (2002). Coase's penguin, or, Linux and the nature of the firm. Yale Law Journal. 112(3). Retrieved in January 2003 from: http://www.yale.edu/yalelj/112/BenklerWEB.pdf.

Coase, R. (1988). *The firm, the market and the law*. Chicago, IL: University of Chicago Press.

Dalle, J.-M. & David, P. (2003). The allocation of software development resources in 'open source' production. Retrieved in April 2003 from: http://opensource.mit.edu/papers/dalledavid.pdf.

Dalle, J.-M. & Jullien, N. (2001). Turning fads into institutions. Retrieved in September 2002 from: http://opensource.mit.edu/papers/Libre-Software.pdf.

Floss Report. Free/Libre and open source software: Survey and Study (2002). University of Maastricht and Bercelon Research GmbH. Retrieved in September 2002 from: http://www.infonomics.nl/FLOSS/report/.

Friedberg, E. (1993). *Le pouvoir et la règle*. Paris, France: Le Seuil.

Goffman, E. (1956). Presentation of self in everyday life. Edinburgh, Scotland: University of Edinburgh, Social Sciences Research Center.

Granovetter, M. (1985). Economic action and social structure: The problem of embeddedness. *American Journal of Sociology.* 91(3), 481-510.

Healy, K. & Schussman, A. (2003). The ecology of open source software development. Retrieved in February 2003 from: http://opensource.mit.edu/papers/healyschussman.pdf.

Hertel, G. (2002). Management virtueller teams auf der dasis sozialpsychologischer modelle. In E.H. Witte. (Ed.), *Sozialpsychologie wirtschaftlicher Prozesse.* Lengerish, Germany: Pabst Publishers, 172-202.

Hertel, G., Niedner, S., & Herrmann, S. (2003). Motivation of software developers in open source projects: An Internet-based survey of contributors to the Linux Kernel. Retrieved in March 2003 from: http://opensource.mit.edu/papers/rp-hertelniednerhermann.pdf.

Hirschman, A. (1970). *Exit, voice, and loyalty: Responses to decline in firms, organizations and states.* Cambridge, MA: Harvard University Press.

Koch, S. & Schneider, G. (2002). Effort, cooperation and coordination in an open source software project: GNOME. *Information Systems Journal.* 12(1), 27-42.

Lazega, E. (1995). Capital Social et Contrainte Latérale. *Revue Française de Sociologie.* 36(4), 759-777.

Lazega, E. (2001). *The collegial phenomenon: The social mechanisms of cooperation among peers in a corporate law partnership.* Oxford, UK: Oxford University Press.

Lee, S., Moisa, N., & Weiss, M. (2003). Open source as a signaling device – An economic analysis. Retrieved in April 2003 from: http://opensource.mit.edu/papers/leemoisaweiss.pdf.

Lerner, J. & Tirole, J. (2000). The simple economics of open source. Retrieved in September 2002 from: http://opensource.mit.edu/papers/Josh%20Lerner%20and%20Jean%20Triole%20-%20The%20Simple%20Economics%20of%20Open%20Source.pdf.

Lerner, J. & Tirole, J. (2002). The scope of open source licensing. Retrieved in September 2002 from: http://opensource.mit.edu/papers/lernertirole2.pdf.

Mauss, M.. (1923). Essai sur le don. Forme et raison de l'echange dans les Sociétés Archaïques. *L'Année Sociologique.* 2(1), 30-186.

Mockus, A., Fielding, R., & Herbsleb, J. (2000). A case study of open source software development: The Apache server. In *Proceedings of the 22th International Conference on Software Engineering.* pp.263-272. Retrieved in September 2002 from: http://opensource.mit.edu/papers/mockusapache.pdf.

Mockus, A., Fielding, R., & Herbsleb, J. (2002). Two case studies of open source software development: Apache and Mozilla. *ACM Transactions on Software Engineering and Methodology* (TOSEM*).* 11(3), 1-38.

Raymond, E. (1999). *The cathedral and the bazaar: Musings on Linux and open source by an accidental revolutionary.* Sebastopol, CA: O'Reilly and Associates.

von Krogh, G., Spaeth, S., & Lakhani, K. (2003). Community, joining and specialization in open source software innovation: A case study. Retrieved in March 2003 from: http://opensource.mit.edu/papers/rp-vonkroghspaethlakhani.pdf

Wasserman, S. & Faust, K. (1995). *Social network analysis: Methods and applications.* Cambridge, UK: Cambridge University Press.

Waters, M. (1989). Collegiality, bureaucratization and professionalization: A Weberian analysis. *American Journal of Sociology.* 94(5), 945-972.

Williamson, O. (1985). *The economic institution of capitalism.* New York: Free Press.

Zeitlyn, D. (2003). Gift economies in the development of open source software: Anthropological reflections. Retrieved in March 2003 from: http://opensource.mit.edu/papers/rp-zeitlyn.pdf.

# ENDNOTES

[1] I would like to thank Erhard Friedberg and Antoine Roullet for their stimulating comments, as well as Martha Zuber, Thomas Constantinesco and Paula Chesley for their contributions. However, I am the only one who should be considered responsible for what follows.

[2] The promotion of the GNU GPL has been quite successful, as a huge majority of software developed on SourceForge.net is released under its terms (Lerner & Tirole, 2002, p.38). This majority does not reflect the exact part of the GPL among open source software, as many great pieces of software have created their own licenses (e.g., Mozilla, Apache), but it shows how popular and well-known it is.

[3] The influence of such analysis may be found in the questionnaire whose results have been published in part II of the Free/Libre Open Source Survey (FLOSS 2002).

[4] This is based on the VIST (for valence, instrumentality, self-efficacy and trust) model developed by Hertel (2002) that explains individual motivations to work in a virtual team. Valence stands for the adequation between the personal goals and the team goals. Instrumentality is the perceived importance or indispensability of one's contribution to the work of the team. Self-efficacy is the fact that the individual perceives himself as being able to accomplish is part of the team task. And trust relies on the consideration that the other members of the team will not use team efforts for their personal interest.

[5] I disagree with the way Benkler (2002) uses this notion: transaction costs are used in a much more precise way by Coase and Williamson and reflect the costs that would be required to elaborate a contract between two persons. It includes costs for discovering adequate prices and costs for establishing a

contract for each transaction (i.e., negotiation costs). I will only focus in this article on costs that are required to gather the information necessary to take part in the development of the project.

6   A CVS is a piece of software that is designed to help people to work together. As a file system it stores data files but it stores every difference of a file since its creation. Write permission is limited to the person who has gained it and the difference is associated with many variables such as the identity of the person who has "committed" a difference. It allows the developers to come back and see previous versions of a file in case they have broken something. From a sociological point of view, this is a very interesting tool as it shows which person has contributed what and in which proportion.

7   Though the studied object was named free software upwards because only free software has the characteristics of a public good, I will use the term open source downwards. What follows focus on the way people coordinate in an opened structure and the chosen license has little impact on the subject. I do not deny that free software and open source software are different but the processes that will be described here are in my opinion not specific to free software but shared by open source software. The only important point is that the license should not impose any pre-existent hierarchical structure and that the repartition of power, whether formal or informal, should be the result of a perpetual bargaining between the members of the project. As stated by Friedberg (1993), power is always bargained but in open source software, the license rarely prevents fork, which allows some developers to chose exit rather than voice (Hirschman, 1970) without much cost and to start a new project where a new structure of power will be elaborated. Because they rely on the fact that power is informally distributed, conclusions drawn downwards would not apply to software released under a license that prevents forks and imposes its own organization.

8   A part of an open source project often neglected although it plays a major role in the success of the project as it is one of the main media to make the work public, to gain interest from the users and other developers.

9   http://www.videolan.org

10  The French educational system is quite different from Anglo-Saxon educational system; universities are wide open to anyone, whereas one can study in some schools called 'Grandes Ecoles' only if one passes selective tests. Because the access to these Grandes Ecoles is much harder than to university, they are more prestigious and often considered as being at the top of the system. The Ecole Centrale Paris (http://www.ecp.fr) is one of the best of these Grandes Ecoles and is as such one of the top scientific school in France.

11  However, these advantages have to be paid by the fact that it will be difficult to generalize the results of this case study. This will be discussed in conclusion.

12  http://statcvs.sourceforge.net

13   The sources for such a result were the THANKS and AUTHORS files of the VLC and the VLS, the persons that were quoted on the team page of the project (http://www.videolan.org/team) and those who give contributions in the mailing-lists.

14   The website was added in the CVS on October 27, 2002 in order to allow anyone who has a CVS to have access to it and modify it.

15   http://vlado.fmf.uni-lj.si/pub/networks/pajek

16   Such a number is calculated *after* having ignored a CVS commit of 107,521 modified lines. This corresponds to the move of some part of the code from one directory to a new one. According to the author of the commit, it should not be taken into account as no code has been added. It has therefore been ignored. Some other commits from the same author have also been ignored as they were devoted to remove double spacing. The score of the top developer calculated on the whole CVS was 52% of modified lines. I was not able to find other commits to ignore. The addition of license and documentation were taken into account as they constitute a real addition to the software and are part of it.

17   The more people have sought advice directly or indirectly from a person, the greater authority score he or she has.

18   Which is still the fifth score.

19   The friendship network was elaborated by asking people to design among the 53 developers who they consider as friends. The definition was voluntarily very strong as people were asked to check the box related to the person only if they consider that the person was one of the closest persons from them. The instructions seem to have been well understood as the answers seem to correspond to the ethnographic observation and there were never more than a few friends for each answer.

20   The betweenness centrality score is only as high that the person is indispesable to keep two persons in touch. It is calculated as:

$$C_{B_i} = \frac{\sum_{j<k, i \neq k, j \neq k} g_{jk}(i)}{g_{jk}}$$

i.e., the proportion of shortest paths (geodesics) between $j$ and $k$ that go through $i$. $g_{jk}$ is the ensemble of geodesics between $j$ and $k$ ; $g_{jk}(i)$ is a geodesic between $j$ and $k$ that goes through $i$. See Wasserman & Faust (1995) for more details. However, this score does not reflect the fact that the graph representing bonds of friendship is oriented. Friendship relationships were considered as being symmetric, the fact that one person declares the other his friend as being sufficient to consider that a friendship link between them existed. What is at

[21] stake is that one of the two people considers the relationships that unite them as friendship, not the fact that it is reciprocal.

[21] The nature of the conflicts and their solutions are also influenced by this kind of regulation but this goes too far to be in this article.

[22] This dichotomical distribution of tasks may be moderated, as the most technically skilled persons from the ECP are both among the main contributors to the project and endorse coordination of tasks.

[23] I share the intuition of Healy and Schussman (2003) that the closer to the core functions developers are the more important hierarchy is. But this is not only due to the fact that these functions are technically primordial: because these developers are in close contact, the network is much more structured than one which could group occasional contributors. The dynamic that ends in and perpetuate a particular hierarchical state is therefore a phenomenon which should be examined more closely, as it may reveal the nature and the variety of resources that are exchanged in order to reach that state.

[24] The Debian developers are aware of this and some of them show interest in elaborating the network that links the personal keys of developers (i.e., who offer who to join Debian). See http://people.debian.org/~edward/globe/earthkeyring/ which could provide a good start to study the phenomenon.

[25] In graph theory, a vertex is a good hub, if it points to many good authorities, and it is a good authority, if it is pointed to by many good hubs.

# APPENDIX

*Synthetic Results of the Network Analysis*

| Number | Student of the ECP | Technical expertise | Hub score in the advice network | Quoted as sought for advice | Authority score in the advice network | Quoted as friend | Betweenness centrality in the friendship network |
|---|---|---|---|---|---|---|---|
| 1 | Yes | 5 | 0.3407527 | 8 | 0.201308 | 12 | 0.0236097 |
| 2 | Yes | 6 | 0.1794858 | 0 | 0 | 3 | 0.0947544 |
| 3 | No | 2 | 0.1395097 | 4 | 0.1029956 | 1 | 0.0008752 |
| 4 | No | 6 | 0 | 0 | 0 | 0 | 0 |
| 5 | No | 6 | 0.0906997 | 0 | 0 | 0 | 0 |
| 6 | Past | 5 | 0.0943715 | 3 | 0.1005174 | 11 | 0.0035425 |
| 7 | No | 6 | 0.0490672 | 4 | 0.1108783 | 1 | 0 |
| 8 | Past | 5 | 0.0829895 | 3 | 0.0885726 | 10 | 0.0017557 |
| 9 | No | 3 | 0.0943715 | 4 | 0.0804054 | 1 | 0 |
| 10 | Yes | 3 | 0.2408508 | 9 | 0.2256629 | 9 | 0.0137979 |
| 11 | Yes | 4 | 0.1582746 | 0 | 0 | 5 | 0.0454545 |
| 12 | No | 3 | 0.1901184 | 2 | 0.0314699 | 1 | 0 |
| 13 | No | 6 | 0.0490672 | 0 | 0 | 0 | 0 |
| 14 | Yes | 2 | 0.3150331 | 7 | 0.1522243 | 18 | 0.0336772 |
| 15 | Yes | 3 | 0.1553381 | 7 | 0.1666198 | 6 | 0.0029551 |
| 16 | Yes | 4 | 0.0661207 | 0 | 0 | 4 | 0.0003157 |
| 17 | Past | 1 | 0.1483358 | 23 | 0.4356981 | 13 | 0.1767123 |
| 18 | Yes | 5 | 0.2443668 | 5 | 0.147656 | 5 | 0.0016204 |
| 19 | Yes | 2 | 0.1788027 | 11 | 0.2363794 | 15 | 0.0430132 |

# APPENDIX

*Synthetic Results of the Network Analysis (continued)*

| | | | | | | | |
|---|---|---|---|---|---|---|---|
| 20 | Yes | 6 | 0.1395594 | 1 | 0.0164587 | 4 | 0.0430132 |
| 21 | No | 6 | 0.1454884 | 0 | 0 | 0 | 0 |
| 22 | No | 4 | 0.0882767 | 0 | 0 | 0 | 0 |
| 23 | Yes | 1 | 0.1189504 | 21 | 0.4341018 | 7 | 0.0212058 |
| 24 | Yes | 4 | 0.3109421 | 8 | 0.1640062 | 13 | 0.0133441 |
| 25 | Yes | 4 | 0.0952353 | 0 | 0 | 2 | 0 |
| 26 | Past | 6 | 0.0490672 | 4 | 0.0898567 | 10 | 0.0123051 |
| 27 | Yes | 5 | 0.0943715 | 6 | 0.1572758 | 8 | 0.0105216 |
| 28 | Yes | 6 | 0.305013 | 1 | 0.0186641 | 9 | 0.1294141 |
| 29 | Yes | 6 | 0.1960486 | 1 | 0.0317148 | 8 | 0.0104299 |
| 30 | No | 3 | 0.0536649 | 0 | 0 | 1 | 0 |
| 31 | No | 1 | 0.1543112 | 10 | 0.2610486 | 1 | 0 |
| 32 | Yes | 5 | 0.1115307 | 2 | 0.0331766 | 4 | 0.0123106 |
| 33 | Past | 1 | 0.2856623 | 32 | 0.4718874 | 18 | 0.0649254 |
| 34 | Yes | 6 | 0.0443968 | 1 | 0.0099036 | 4 | 0 |

## Chapter VII

# Free Software Development:
## Cooperation and Conflict in a Virtual Organizational Culture

Margaret S. Elliott, University of California, Irvine, USA

Walt Scacchi, University of California, Irvine, USA

## ABSTRACT

*This chapter presents an empirical study of a free software development community and how its virtual organizational culture influences its work practices. Results show that beliefs in free software and freedom of choice, and values in cooperative work and community influence work practices and norms. The authors wish to convey the importance of understanding the deeply held beliefs and values of the free software movement by showing how a free software development community uses computer-mediated communication in the form of IRC (instant messaging), mailing lists, and summary digests to mitigate and resolve conflicts, build a community, reinforce beliefs, and facilitate teamwork. Results are intended to assist future contributors and managers of free/open source software development projects in understanding the social world surrounding free/open source software development.*

# INTRODUCTION

Free/open source software (F/OSS) development projects are growing at a rapid rate. The SourceForge website estimates 750,000+ users with 700 new ones joining every day and a total of 75,000+ projects with 60 new ones added each day. Thousands of F/OSS development projects have emerged within the past few years (DiBona, Ockman, & Stone, 1999; Pavlicek, 2000) leading to the formation of globally dispersed virtual communities (Kollock & Smith, 1999). Examples of F/OSS development projects are found in the social worlds that surround computer game development; X-ray astronomy and deep space imaging; academic software design research; business software development; and Internet/Web infrastructure development (Elliott, 2003; Elliott & Scacchi, 2003a, 2003b; Scacchi, 2002a, 2002b). Working together in globally distributed virtual communities, F/OSS developers communicate and collaborate using a wide range of web-based tools including Internet Relay Chat (IRC) for instant messaging, CVS for concurrent version control (Fogel, 1999), electronic mailing lists, and more (Scacchi, 2002b).

Proponents of F/OSS claim advantages such as improved software validity, simplification of collaboration, and reduced software acquisition costs. While some researchers have examined F/OSS development using quantitative studies exploring issues like developer defect density, core team size, motivation for joining free/open source projects, and others (Koch & Schneider, 2000; Mockus, Fielding, & Herbsleb, 2000, 2002), few researchers have explored the social phenomena surrounding F/OSS development (Berquist & Ljungberg, 2001; Mackenzie, Rouchy, & Rouncefield, 2002). Popular literature has described F/OSS developers as members of a "geek" culture (Pavlicek, 2000) notorious for nerdy, technically savvy, yet socially inept people, and as participants in a "gift" culture (Berquist & Ljungberg, 2001; Raymond, 2001) where social status is measured by what you give away. However, no empirical research has been conducted to study F/OSS developers as virtual organizational cultures (Martin, 2002; Schein, 1992) with beliefs and values that influence decisions and technical tool choices. In this chapter, we present the results of a virtual ethnography to study the work culture and F/OSS development work processes of a free software project, GNUenterprise (GNUe) (http://www.gnuenterprise.org). We identify beliefs and values of the free software movement (Stallman, 1999) associated with the work culture of the GNUe community, and we show the importance of computer-mediated communication (CMC) such as chat/instant messaging and summary digests in building community, resolving conflicts, and facilitating teamwork.

The free software movement promotes the production of free software that is open to anyone to copy, study, modify, and redistribute (Stallman, 1999). The Free Software Foundation (FSF) was founded by Richard M. Stallman (known as RMS in the F/OSS community) in the 1980s to promote the ideal of freedom and the production of free software based on philosophical convictions and on the concept that free source code is necessary for innovation to flourish in computer science (DiBona, Ockman & Stone, 1999). It is important to distinguish between the terms free software (Stallman, 1999) and open source (DiBona et al., 1999). Free software differs from open source in its philosophical orientation. Details regarding the Open Source Initiative (OSI) may be found at http://www.opensource.org. RMS believes that the difference between the free and open source software movements is in their values and ways of looking at the world. He explains it below:

*For the Open Source movement, the issue of whether software should be open source is a practical question, not an ethical one. As one person put it, "Open source is a development methodology; free software is a social movement." For the Open Source movement, non-free software is a suboptimal solution. For the Free Software movement, non-free software is a social problem and free software is the solution (RMS) (http://www.fsf.org/philosophy/free-software-for-freedom.html)*

A popular expression in the free software culture is "Think free speech, not free beer." The FSF promotes the use of the General Public License (GPL) for free software development as well as other similar licenses (http://www.gnu.org/licenses/license-list.html). While the majority of open source projects use the GPL, alternative licenses are suggested by the OSI (see http://www.opensource.org).

The free software movement has spawned a number of free software projects all adhering to the belief in free software and belief in freedom of choice (see http://www.gnu.org and http://www.fsf.org) as part of their virtual organizational culture. As with typical organizations (Martin, 1992; Schein, 1992), virtual organizations develop work cultures, which have an impact on how the work is completed. Subsequently, there is a need for better articulation of how cultural beliefs and values of the free software movement influence work practices of F/OSS development. Such an understanding would benefit both managers and developers of F/OSS development. In this chapter, we present empirical evidence from the study of the work culture of one free software community — GNUe. As with all qualitative research (Strauss & Corbin, 1990; Yin, 1994), we do not intend to portray a generalized view of all free software development projects. However, research has shown that many F/OSS development projects follow similar procedures (Feller & Fitzgerald, 2002; Scacchi, 2002b). Future research will show how closely the GNUe work culture resembles that of other free software projects.

In the next section, we present the GNUe project, followed by research methods. In subsequent sections we present the background, discuss the GNUe virtual organizational culture, and present the three case studies followed by a discussion. Finally, we end with practical implications, future research, and conclusions.

# GNUE PROJECT

GNUe is a meta-project of the GNU (http://www.gnu.org) Project. GNUe is organized to collect and develop free electronic business software in one location on the Web. The plans are for GNUe to consist of:

1) a set of tools that provide a development framework for enterprise information technology (IT) professionals to create or customize applications and share them across organizations;
2) a set of packages written using the set of tools to implement a full Enterprise Resource Planning (ERP) system; and
3) a general community of support and resources for developers writing applications using GNUe tools.

GNUe is an international virtual organization for software development (Crowston & Scozzi, 2002; Noll & Scacchi, 1999) based in the U.S. and Europe. This organization is centered around the GNUe Web portal and global Internet infrastructure that enable remote access and collaboration. As of the writing of this paper, GNUe contributors consist of six core maintainers (co-maintainers who head the project), 18 active contributors, and 18 inactive contributors. The six core maintainers share various tasks, including the monitoring of the daily Internet Relay Chat (IRC), accepting bug fixes to go into a release, testing software, documentation of software, and other tasks. Companies from Austria, Argentina, Lithuania, and New Zealand support paid contributors, but most of the contributors are working as non-paid participants.

# RESEARCH METHODS

This ongoing ethnography of a virtual organization (Hine, 2000; Olsson, 2000) is being conducted using the grounded theory approach (Strauss & Corbin, 1990) with participant-observer techniques. The sources of data include books and articles on F/OSS development, instant messaging (Herbsleb & Grinter, 1999; Nardi, Whittaker, & Bradner, 2000), transcripts captured through IRC logs, threaded electronic mail (email) discussion messages, and Kernel Cousins (summary digests of the IRC and mailing lists — see http://kt.zork.net). This research also includes data from email and face-to-face interviews with GNUe contributors and observations at Open Source conferences. The first author spent over 100 hours studying and perusing IRC archives and mailing list samples during open and axial coding phases of the grounded theory. During open coding, the first case study was selected as representative of the strong influence of cultural beliefs on GNUe software development practices. The selection of cases was aided by the indexing of each Kernel Cousin into sections labeled with a topic. For example, we read through all Kernel Cousins looking mainly at the indices only and found the following title "Using Non-Free Tools for Documentation" in http://kt.zork.net/GNUe/gnue20011124_4.html. Hyperlinks from this cousin pointed us to a similar case where non-free tools were being used for documentation of code. The third case was discovered during axial coding for Case Two. In the third case, a newcomer asks for help regarding the use of GNUe, and we show how cooperation and community building are facilitated by the use of IRC. The initial research questions that formed the core of the grounded theory are:

1) How do people working in virtual organizations organize themselves such that work is completed?
2) What social processes facilitate open source software development?
3) What techniques are used in F/OSS development that differ from typical software development?

We began this research with the characterization of open source software communities as communities of practice. A community of practice (COP) is a group of people who share similar goals, interests, beliefs, and value systems in a common domain of recurring activity or work (Wenger, 1998). An alternative way of viewing groups with shared goals in organizations is to characterize them as organizational subcultures (Martin, 2002; Schein, 1992; Trice & Beyer, 1993). As the grounded theory evolved, we discovered rich cultural beliefs and norms influencing "geek" behavior (Pavlicek, 2000). This led to us to the characterization of the COPs as virtual organizations having organizational cultures.

We view culture as both objectively and subjectively constrained (Martin, 2002). In a typical organization, this means studying physical manifestations of the culture such as dress norms, reported salaries, annual reports, and workplace furnishings and atmosphere. In addition, subjective meanings associated with these physical symbols are interpreted. In a virtual organization, these physical cultural symbols are missing, so we focus on unique types of accessible manifestations of the GNUe culture, such as website documentation and downloadable source code. We selected GNUe as a research site because it exemplified the essence of free software development, providing a rich picture of a virtual work community with downloadable free software as well as lengthy documentation—all facilitating a virtual ethnography (Hine, 2000). We took each IRC and Kernel Cousin related to the three cases and applied codes derived from the data (Strauss & Corbin, 1990). We used a text editor to add the codes to the IRC text logs using [Begin and End] blocks around concepts we identified, such as "belief in free software." In this way, we discovered the relationships shown in Figure 1. During the axial coding phase of several IRC chat logs, mailing lists, and other documentation, we discovered relationships between beliefs and values of the work culture and manifestations of the culture.

# BACKGROUND

In this section, we first discuss the organizational culture perspective and its application to virtual organizations. Next, we present research on conflict resolution in virtual communities.

## Organizational Culture Perspective

Much like societal cultures have beliefs and values manifested in norms that form behavioral expectations, organizations have cultures that form and give members guidelines for "the way to do things around here." An organizational culture perspective (Martin, 2002; Schein, 1992; Trice & Beyer, 1993) provides a method of studying an organization's social processes often missed in a quantitative study of organizational variables. Organizational culture is a set of socially established structures of meaning that are accepted by its members (Ott, 1989).

The substances of such cultures are formed from ideologies, the implicit sets of taken-for-granted beliefs, values, and norms. Members express the substance of their cultures through the use of cultural forms in organizations—acceptable ways of expressing and affirming their beliefs, values, and norms. When beliefs, values, and norms coalesce over time into stable forms that comprise an ideology, they provide causal models for explaining and justifying existing social systems. Researchers have theorized the application of a cultural perspective to understand IT implementation and use (Avison & Myers, 1995; Robey & Azevedo, 1994), but few have applied this to the workplace itself (Dubé & Robey, 1999; Elliott, 2000). More research is needed to understand the relationship between organizational culture and the development and use of IT in typical and virtual organizations.

In a virtual organization, cultural beliefs and values are manifested in norms regarding communication and work issues (if work-related communities like F/OSS development) and in the form of electronic artifacts—IRC and mailing list archives, and summary digests (Kernel Cousins). Most organizational culture researchers view work culture as a consensus-making system (Ott, 1989; Schein, 1992; Trice & Beyer, 1993). However, some researchers view organizational culture from two other perspectives (Martin, 2002): 1) differentiation where culture is viewed as resulting in inconsistencies, lack of consensus, and non-leader centered sources of culture content often manifested in differing subcultures, and 2) fragmentation where culture is viewed as having no shared values except that of the awareness of ambiguity. Martin (2002) proposes the use of all three perspectives on the same organizational culture, where appropriate, for a more in-depth look at work cultures. In this study, we focus on the integration perspective to characterize the unifying effects of the GNUe beliefs and values.

## Conflict Resolution in Virtual Communities

Researchers have attempted to understand conflict resolution in virtual communities (Kollock & Smith, 1996; Smith, 1999) in the areas of online communities and in the game world. Many others have studied conflict resolution in common work situations such as computer-supported cooperative work (CSCW) (Easterbrook, 1993). For our purposes, we are interested in virtual communities and how they resolve conflicts, so this discussion does not include studies on conflict management tools.

Smith (1999) studied conflict management in MicroMUSE, a game world dedicated to the simulation and learning about a space station orbiting the earth. There were two basic classes of participants: users and administrators. Disputes arose in each group and between the two groups regarding issues like harassment, sexual harassment, assault, spying, theft, and spamming. These problems can be attributed to differing views of MicroMUSE by participants and to diverse values, goals, interests, and norms of the group. Smith concluded that virtual organizations have the same kinds of problems and opportunities brought by diversity as real organizations do, and that conflict is more likely and more difficult to manage than in real communities. Factors contributing to this difficulty are: wide cultural diversity; disparate interests, needs and expectations; nature of electronic participation (anonymity, multiple avenues of entry, poor reliability of connections, and so forth); text-based communications; and power asymmetry among users.

Kollock and Smith (1996) explored the implications of cooperation and conflict in Usenet groups, emphasizing the importance of recognizing the free-rider problem. In a group situation where one person can benefit from the product or resource offered by others, each person is motivated not to contribute to the joint effort, instead free-riding on others' work. Success in a newsgroup is predicated by the active and ongoing contributions of its membership. If many members free-ride by lurking or by infrequently contributing to the newsgroup, a critical mass of participants may never occur. In addition, the authors suggest that bandwidth be used judiciously by taking care not to post extremely long articles or by not posting the same message to many newsgroups. Otherwise, participants might free-ride on the efforts of other members by using the available bandwidth without restraint while others post carefully. However, success of a Usenet group also depends on its members following the groups' cultural rules of decorum. We explore the topic of following cultural rules in the next section.

# GNUe VIRTUAL ORGANIZATIONAL CULTURE

The substance of a culture is its ideology—shared, interrelated sets of emotionally charged beliefs, values, and norms that bind people together and help them to make sense of their worlds (Trice & Beyer, 1993). While closely related to behavior, beliefs, values, and norms are unique concepts as defined below (Trice & Beyer, 1993):

- **Beliefs** - Express cause and effect relations (i.e., behaviors lead to outcomes).
- **Values** - Express preferences for certain behaviors or for certain outcomes.
- **Norms** - Express which behaviors are expected by others and are culturally acceptable.

In the GNUe study, we apply an integration perspective (Martin, 2002) to the GNUe community to show how beliefs and values of the free software movement tie the virtual organization together in the interests of completing the GNUe free software project (see Elliott & Scacchi, 2003a, for a detailed report of the GNUe study). We present the GNUe virtual organization as a subculture of the FSF inculcating the beliefs and values of the free software movement into its everyday work. As members of the FSF, free software developers share an ideology based on the belief in free software and the belief in freedom of choice. These beliefs are espoused in the literature on free software (Williams, 2002). The values of cooperative work and community are inferred from this research.

Figure 1 shows a conceptual diagram of the GNUe case study. The columns are labeled with terms derived from a grounded theory approach (Strauss & Corbin, 1990). The causal conditions consist of the beliefs (free software and freedom of

*Figure 1: Conceptual diagram of variables.*

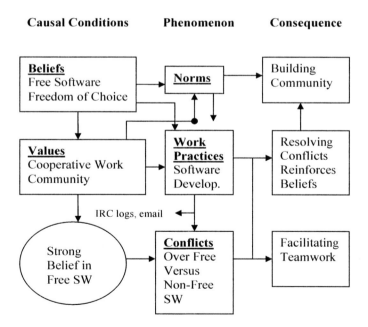

choice) and the values (cooperative work and community). The phenomenon is the free software development process (i.e., code reviews, CVS releases, etc.) influenced by the norms (open disclosure, informal management, and immediate acceptance of outsider critiques), which are a manifestation of beliefs and values. The interaction/action occurs on the IRC and mailing lists. It consists of: 1) the conflict over the use of a non-free tool to create a graphic diagram of the emerging GNUe system design; 2) the conflict over the use of a non-free tool to create GNUe documentation; and 3) acceptance of a newcomer. The consequences are: 1) building community; 2) resolution of conflicts with a reinforcement of the beliefs; and 3) strengthened teamwork. The beliefs, values, and norms are described below; the consequences are presented in the discussion section.

## Belief in Free Software

The belief in free software appears to be a core motivator of free software developers. GNUe developers extol the virtues of free software on its website and in daily activity on the IRC logs. The FSF website has many references to the ideological importance of developing and maintaining free software (see http://www.fsf.org). This belief is manifested in electronic artifacts such as the Web pages, source code, GPL license, software design diagrams, and accompanying articles on its website and elsewhere. The data analysis of the GNUe cases showed that this belief varies from moderate to strong in strength. For example, those who have a strong belief in free software may refuse to use any form of non-free software (such as a commercial text editor) for development purposes.

## Belief in Freedom of Choice

F/OSS developers are attracted to the occupation of F/OSS development for its freedom of choice in work assignments. Both paid and unpaid GNUe participants to some degree can select the work they prefer. This belief is manifested in the informal methods used to assign or select work in an F/OSS project. During a face-to-face interview with one of GNUe's core contributors, we asked about job assignments, and he responded with: "The number one rule in free software is 'never do timelines or roadmaps' (Derek, LinuxWorld conference, August 2002)." The belief in freedom of choice also refers to the ability to select the tool of choice to develop free software. Some F/OSS developers believe that a mix of free and non-free software tools is acceptable when developing free software, while others adhere to the belief that free software should be used exclusively.

## Value in Community

The beliefs in free software and freedom of choice foster a value in community. This value is evident in the IRC archives when newcomers join GNUe offering suggestions or pointing out bugs, and GNUe contributors quickly accept them as part of the community. In addition, one of the GNUe case studies shows that when a frequent contributor refuses to use a computer program that requires non-free software, fellow contributors join in an attempt to persuade him that a temporary use of non-free software should be acceptable.

## Value in Cooperative Work

The GNUe community's beliefs in free software and freedom of choice combined with the value in community foster a value in cooperative work. As with previous researchers (Easterbrook, 1993; Kollock & Smith, 1996; Smith, 1999), our results indicate that conflict arises during the course of cooperative work. GNUe contributors work cooperatively to resolve conflicts through the use of IRC and mailing lists.

## Open Disclosure Norm

Open disclosure refers to the open content of the GNUe website including the software source code, documentation, and archived records of IRC, Kernel Cousins, and mailing list interchanges. The GNUe contributors join others online via IRC on a daily basis and record the conversations for future reference.

## Informal Management Norm

The entire GNUe virtual organization is informal. There is no lead organization or prime contractor that has brought together the alliance of individuals and sponsoring firms as a networked virtual organization. It is more of an emergent organizational form where participants have in a sense discovered each other and have brought together their individual competencies and contributions in a way whereby they can be integrated or made to interoperate (Crowston & Scozzi, 2002). The participants come from different small companies or act as individuals who collectively move the GNUe software and the GNUe community forward. Thus, the participants self-organize in a manner more like a meritocracy (Fielding, 1999).

## Immediate Acceptance of Outsider Critiques Norm

In the GNUe organization, outsiders who have not visited the GNUe IRC before can easily join the discussion and give criticisms of the code or procedures. Sometimes this criticism revolves around the use of free versus non-free tools and other times it is related to attempts to fix bugs in the code. In either case, the GNUe maintainers who discuss these critiques respect and respond to outsiders' reviews with serious consideration without knowledge of reviewers' background.

# GNUe CASE STUDY

The GNUe case study consists of data analysis of three cases of GNUe software development—two involving conflict and one conflict-free case. Each case will be described briefly in this section. For a more detailed description of the first two cases, see Elliott and Scacchi (2003a).

## Case One - Use of Non-Free Graphic Tool for Documentation

In this section we present the first case study that reveals a trajectory of a conflict and debate over the use of a non-free tool to create a graphic on the GNUe website (http://www.gnuenterprise.org/irc-logs/gnue-public.log.25Nov2001). This exchange takes place on the IRC channel on November 25, 2001, and ends the next morning. This example illustrates the ease with which a newcomer comes onboard and criticizes the methods used to produce a graphical representation of a screenshot on the GNUe website. CyrilB, an outsider to GNUe, finds a graphic that was created using ADOBE Photoshop, a non-free graphical tool. He begins the interchange with a challenge to anyone onboard stating that "it is quite shocking" to see the use of non-free software on a free software project. He exhibits a **strong belief in free software**, which causes a debate lasting a couple of days. Table 1 displays the total number of contributors and the number of days of the conflict. Eight of the nine regular GNUe contributors were software developers, and one was working on documentation. The infrequent contributors drifted on and off throughout the day — sometimes lurking and other times involved in the discussion.

CyrilB's **strong belief in free software** leads to conflict among regular contributors: "I hope I'm wrong: it is quite shocking...We should avoid using non-free software at all cost, am I wrong?" **(Strong BIFS-1)**

Reinhard responds with a moderate view of belief in free software: "Our main goal is to produce good free software. We accept contributions without regarding what tools were used to do the work especially we accept documentation in nearly any form we can get because we are desperate for documentation."

Later, Neilt, who originally created the GNUe diagram using ADOBE Photoshop, joins the IRC, reviews the previous discussion on the archived IRC, and returns to discuss the issue with Reinhard and CyrilB. A lively argument ensues between Neilt and others with onlookers testing free software graphics tools in real-time. CyrilB becomes adamant that the choice should be free software as shown below:

<CyrilB>*We need people do be able to use free softwares...* **neilt:** *You are compromising our freedom by using non-free software: we can't modify and/or redistribute the source vector file.*

*Table 1: Contributors and duration of conflict in Case One.*

| Total Contributors | Regular Contributors | Infrequent Contributors | Number of Days |
|---|---|---|---|
| 17 | 9 | 8 | 2 |

Neilt responds by arguing with CyrilB over the utility in changing the graphic:

<Neilt> *otoh I see no reason to avoid non-free software either if this is really a freedom thing then we should be free to use whatever we want.*

Neilt feels that his **freedom of choice** is hampered by the strict adherence to using free software to develop an F/OSS system.

These exchanges illustrate how participants use the IRC medium to support and enable the cooperative work needed to resolve this issue. It also conveys the community spirit and cooperative work ethic that is a value in the GNUe work culture. They both agree to wait until CyrilB comes back to give more suggestions for an alternative. One of the norms of the work culture is **immediate acceptance of outsider contributions**. CyrilB's critique is considered very important to GNUe contributors and causes many to test free graphics packages in real-time for Neilt as the discussion unfolds. Eventually, Neilt agrees with CyrilB and others to change the graphic to one created with a free graphics tool. Consequences of the debate are a clarification and reinforcement of the beliefs and values of the GNUe community, and a recreation of a website graphic with free software to replace the original created with a non-free software tool.

## Case Two - Use of Non-Free Software for GNUe Documentation

The second case study explores project insider review of the procedures and practices for developing GNUe documentation (see http://www.gnuenterprise.org/irc-logs/gnue-public.log.15Nov2001 for the full three-day logs). Once again the debate revolves around polarized views of the use of non-free tools to develop GNUe documentation. In this case, chillywilly, a frequent contributor, balks at the need to implement a non-free tool on his computer in order to edit the documentation associated with a current release. Even though his colleagues attempt to dissuade him from his concerns by suggesting that he can use any editor—free or non-free—to read the documentation in HTML or other formats, Chillywilly refuses to back down from his stance based on a **strong belief in free software**. This debate lasts three days. Table 2 displays the number of contributors and their classification for participation in case two. This case exemplifies the fierce adherence to the **belief in free software** held by some purists in the free software movement and how it directs the work of the day. While the three- day debate reinforces cultural beliefs and values and builds community, at the same time, it ties up valuable time that could have been spent writing code or documentation.

In order to understand this example, some background information is needed. The GNUe core maintainers selected a free tool to use for all documentation called *docbook* (http://www.docbook.org). *Docbook* is based on an SGML document-type definition which provides a system for writing structured documents using SGML or

*Table 2: Contributors and duration of conflict in Case Two.*

| Total Contributors | Regular Contributors | Infrequent Contributors | Number of Days |
|---|---|---|---|
| 24 | 9 | 15 | 3 |

XML. However, as of November 15, 2001, several GNUe developers were having trouble with its installation. Consequently, they resorted to using *lyx* tool to create documentation (http://www.lyx.org) even though it required the installation of a non-free graphics packages (called *libxforms*). The following example shows the initiation by chillywilly of a debate that lasts several days:

> Action: chillywilly trout whips jamest for making lyx docs
> Action: jcater troutslaps chillywilly for troutslapping jamest for making easy to do docs
> **&lt;chillywilly&gt;** lyx requires non-free software
> **&lt;Maniac&gt;** lyx rules
> **&lt;chillywilly&gt;** should that be acceptable for a GNU project?
> **&lt;jcater&gt;** chillywilly: basically, given the time frame we are in, it's either LyX documentation with this release, or no documentation for a while (until we can get some other stinking system in place) pick one :)
> **&lt;chillywilly&gt;** use docbook then

As the day wears on, chillywilly continues his debate, continuously harassing fellow contributors. Reinhard agrees with chillywilly as do others, but in order to complete the documentation, they agree to use an interim solution. Chillywilly is so adamantly opposed to the use of non-free software that he references Richard Stallman as part of his reasoning—"**I will NOT install lyx and make vrms unhappy.**" This passage shows how RMS is considered by some to be the "guru" of the free software movement. In the GNUe culture, the mailing list is considered to be a more "formal" form of CMC, and it is a norm to save its use for more serious, technical problems, not for joking as seen on the IRC logs. However, chillywilly sends the following email to the mailing list causing a stir among fellow contributors:

*"OK, I saw on the commit list that you guys made some LyX documents. I think it is extremely \*\*\*that a GNU project would require me to install non-free software in order to read and modify the documentation. I mean if I cannot make vrms happy on my debian system them what good am I as a Free Software developer? Is docbook really this much of a pain? I can build html versions of stuff on my box if this is what we have to do. This just irks me beyond anything. I really shouldn't have to be harping on this issue for a GNU project, but some ppl like to take convenience*

*over freedom and this should not be tolerated... Is it really that unreasonable to request that we not use something that requires ppl to install non-free software? Please let me know." (chillywilly, mailing list)*

Later that day, Jcater sends a reply to chillywilly's message to the mailing list explaining the importance of his belief in freedom:

*"I would like to personally apologize to the discussion list for the childish email you recently received. It stemmed from a conversation in IRC that quickly got out of hand. It was never our intention to alienate users by using a non-standard documentation format such as LyX. ... LyX was chosen because it is usable and, more importantly, installable. After many failed attempts at installing the requirements for docbook, James and I made the decision that LyX-based documentation with the upcoming 0.1.0 releases was better than no documentation at all...*
*PPS, By the way, Daniel, using/writing Free software is NOT about making RMS happy or unhappy. He's a great guy and all, but not the center of the free universe, nor the motivating factor in many (most?) of our lives. For me, my motivation to be here is a free future for my son." (Jcater, mailing list)*

After posting this message, Jcater discusses this issue further with chillywilly and others over the IRC. Finally, fellow contributors persuade chillywilly that he can continue creating documentation in HTML or text until a free GNUe is available. Although he does not like this choice, he is reminded by mdean that he has a **freedom of choice** in which tool to choose:

<mdean> chillywilly: you have a choice – which is what is "really" important
<chillywilly> mdean: I choose GNU whenever I can, that is my choice.
<Mr_You> Sorry your choice is a frustrating one...
<chillywilly> whatever man I am burnt to a crisp.

Finally, chillywilly accepts the fact that fellow contributors are not willing to change the documentation tool to one based exclusively on free software.

## Case Three - Newcomer Asking for Help with GNUe Installation

In this example, mcb30 joins the IRC as a newcomer who wants to install and use GNUe business applications for his small business in England (http://www.gnuenterprise.org/irc-logs/gnue-public.log.16Nov2001). In addition, he offers his services as a contributor and immediately starts fixing bugs in real-time. This case is a good example of the community-building spirit of GNUe, since mcb30 is immediately accepted by frequent contributors especially because he posts significant bug fixes very rapidly.

> \<**mcb30**\> Is anyone here awake and listening?
> \<**reinhard**\> yes
> \<**mcb30**\> Excellent. I'm trying to get a CVS copy of GNUe up and running for the first(ish) time - do you mind if I ask for a few hints?
> \<**reinhard**\> shoot away :)
> \<**reinhard**\> btw what exactly are you trying to run?
> \<**reinhard**\> as "GNUe" as a whole doesen't exist (yet)
> \<**reinhard**\> GNUe is a meta-project (a group of related projects)
> \<**mcb30**\> OK - what I want to do is get **something** running so I can get a feel for what there is, what state of development it's in etc. - I'd like to con tribute but I need to know what already exists first!
> \<**reinhard**\> ok cool
> \<**reinhard**\> let me give you a quick overview
> \<**mcb30**\> I have finally (about 5 minutes ago) managed to get "setup.py de vel" to work properly - there are 2 bugs in it
> Mcb30 goes offline and continues to fix bugs. He then comes back and sug gests that he has a patch file to help:
> \<**mcb30**\> I've got a patch file - who should I send it to? jcater?
> \<**reinhard**\> jcater or jamest
> \<**mcb30**\> ok, will do, thanks
> \<**reinhard**\> mcb30: btw sorry if i tell you things you already know :)
> \<**mcb30**\> don't worry - I'd rather be told twice than not at all! :-)
> \<**reinhard**\> people appearing here in IRC sometimes have *very* different levels of information :) …
> \<**reinhard**\> mcb30: i will have to thank you
> \<**reinhard**\> mcb30: we are happy if you are going to help us

Later, mcb30 comes back to the IRC and posts code that he wrote to fix a problem, and several frequent contributors thank him and say that they wish they could hire him for pay. As with the first case, contributors immediately accept him into the "club."

# DISCUSSION

The three GNUe examples will be discussed in this section in relation to the three main themes found in the data: building community, conflict resolution, and facilitating teamwork.

## Building Community

Kollock (1996) suggests that there are design principles for building a successful online community, such as identity persistence. He draws upon the work of Godwin (1994) showing that allowing users to resolve their own disputes without outside interference and providing institutional memory are two principles for making virtual communities work. Applying these principles to the GNUe project shows that disputes are resolved simultaneously via IRC without management interference and recorded in IRC archives as a form of institutional memory. In the GNUe virtual community, the community is continuously changing with newcomers and infrequent contributors sporadically participating, yet the core maintainers are dedicated for long periods of time and help build a community by continuous participation (everyday, for some) in IRC discussions. Derek, a core maintainer, explains below how the IRC helps sustain a community:

*"Many free software folks think IRC is a waste of time as there is 'goofing off,' but honestly I can say it's what builds a community. I think a community is necessary to survive. For example, GNUe has been around for more than three years. I can not tell you how many projects have come and gone that were supposed to be competition or such. I put our longevity solely to the fact that we have a community."* (Derek, email interview, 2002)

## Conflict Resolution

In the two GNUe cases involving conflict, the issues were resolved via debate over the IRC and mailing lists. In the first case, the contributor who created the graphic with ADOBE Photoshop agreed to change it in the future using a free tool. In the second case, chillywilly stopped badgering his co-workers about the use of a non-free graphics package to complete documentation. An interesting finding from the research is that, for some F/OSS developers, there is tension between the **belief in free software** and the **belief in freedom of choice**. In Case One, CyrilB's strict adherence to the free software principle prompted Neilt, the creator of the graphic, to feel hampered in his freedom of choice (to select a graphics tool). In Case Two, chillywilly's strong beliefs in free software restricted his freedom of choice for a documentation tool. In resolving conflicts over the use of free versus non-free software tools, it appears that having a **strong belief in free software** can also inhibit an F/OSS developer's **freedom of choice**. This reveals how important the philosophical convictions of the **belief in free software** are to some F/OSS developers. In both cases, the conflicts were resolved without formal management techniques via the IRC exchanges. At the same time, the beliefs in free software are reinforced by people defending their positions and this, in turn, helps to perpetuate the community.

## Facilitating Teamwork

In the two cases involving conflict, as the day proceeded on the IRC, people were going offline to experiment with free software that would help to resolve the conflict (i.e., a free graphics package and a free text editor). Many infrequent contributors or newcomers, who were lurking and watching the problem unfold on the IRC, also offered suggestions for workable free tools. The real-time aspect of the work clearly facilitates the teamwork with people simultaneously working together to solve a technical problem. For example, in the conflict-free case, a newcomer, who was having trouble with the GNUe installation, was directed by a maintainer to the original author of the code who shortly thereafter joined the IRC and worked with mcb30 to fix the bugs. The real-time connection enabled a quick resolution to the newcomer's problems.

# PRACTICAL IMPLICATIONS

We have shown that the persistent recording of daily work using instant messaging (IRC) and Kernel Cousins can serve as an aid to community building. Managers of open source might benefit from incorporating these CMC media into their computing infrastructures. It assists employees in conflict resolution and also binds the groups together by reinforcing the organizational culture. As illustrated in the non-conflict GNUe example, the IRC serves as an expertise Q&A repository. The author of the software quickly emerged, and mcb30 was able to gain detailed knowledge of how the system works. In addition, the IRC enables real-time software design and debugging. As F/OSS development projects proliferate, managers should consider the benefits of using an IRC to facilitate software development and to help build a community.

# FUTURE WORK

We plan to continue with the analysis of GNUe data and to compare the results with other free software communities. Likewise, we expect to find similar beliefs and values in open source projects (e.g., Apache), and we plan to explore this phenomenon. In this way, we can assertain whether GNUe is in fact a unique culture (Martin, 2002), or whether other F/OSS software projects have similar software development processes. In a cursive look at other GNU projects' websites, we have found evidence of proselytizing of strong beliefs in free software (http://www.gnu.org/projects/projects.html. In addition, possible future research could be an analysis of the ongoing dispute in the Linux Kernel community about using Bitkeeper (non-free) versus CVS (free) as a configuration management system. Another potential topic of interest is to explore the hypothesis that strong beliefs in free software lead to more successful, productive F/OSS projects.

## CONCLUSIONS

Previous CSCW research has not addressed how the collection of IRC messaging, IRC transcript logs, email lists, and periodic digests (Kernel Cousins) can be collectively mobilized and routinely used to create a virtual organization that embodies, transmits, and reaffirms the cultural beliefs, values, and norms such as those found in free software projects like GNUe. Strong cultural beliefs among contributors in an F/OSS development community combined with persistent recording of chat logs help to build the GNUe community and to perpetuate the project. The beliefs in free software and freedom of choice create a special bond for the people working on free software projects. These beliefs foster the values of cooperative work and community building. Schein's (1990) theory of organizational culture includes revelation of underlying assumptions of cultural members that are on a mostly unconscious level. In the GNUe world, the underlying assumptions of the importance of cooperative work and community become an integral part of everyday work practices in the pursuit of a free ERP system. These beliefs and values enhance and motivate acceptance of outsiders' criticisms and resolution of conflict despite the distance separation and amorphous state of the contributor population.

## ACKNOWLEDGMENTS

The research described in this chapter is supported by grants from the National Science Foundation #ITR-0083075, #ITR-0205679, #ITR-0205724 and #ITR-0350754. No endorsement implied. Mark Ackerman at the University of Michigan Ann Arbor; Les Gasser at the University of Illinois, Urbana-Champaign; John Noll at the Santa Clara University; Chris Jensen, Mark Bergman, and Xiaobin Li at the UCI Institute for Software Research; and also Julia Watson at the Ohio State University are collaborators on the research project that produced this chapter.

## REFERENCES

Avison, D. E. & Myers, M. D. (1995). Information systems and anthropology: An anthropological perspective on IT and organizational culture. *Information Technology and People, 8*, (43-56).

Berquist, M. & Ljungberg, J. (2001). The power of gifts: Organizing social relationships in open source communities. *Information Systems Journal*, 11(4), 305-320.

Crowston, K. & Scozzi, B. (2002). Exploring strengths and limits on open source software engineering processes: A research agenda. Paper presented at the *2nd Workshop on Open Source Software Engineering*, Orlando, Florida, May 25, 2002.

DiBona, C., Ockman, S., & Stone, M. (Eds.) (1999). *Open sources: Voices from the open source revolution*. Sebastol, CA: O'Reilly & Associates Inc.

Dubé, L., & Robey, D. (1999). Software stories: Three cultural perspectives on the organizational practices of software development. *Accounting, Management and Information Technologies, 9*(4), 223-259.

Easterbrook, S. M. (Ed.) (1993). *CSCW: Cooperation or conflict?* London: Springer-Verlag.

Elliott, M. (2000). Organizational culture and computer-supported cooperative work in a common information space: Case processing in the criminal courts. (Unpublished Dissertation). Irvine, CA: University of California, Irvine.

Elliott, M. (2003). The virtual organizational culture of a free software development community. Paper presented at the *3rd Workshop on Open Source Software*, Portland, Oregon, May 3, 2003.

Elliott, M. & Scacchi, W. (2003a). *Free software: A case study of software development in a virtual organizational culture*, Technical Report # UCI-ISR-03-6. Irvine, CA: Institute for Software Research, University of California, Irvine. Available online at: http://www.ics.uci.edu/~melliot/reports/UCI-ISR-03-6.pdf.

Elliott, M. & Scacchi, W. (2003b). Free software developers as an occupational community: Resolving conflicts and fostering collaboration. *Proceedings 2003 International ACM SIGGROUP Conference on Supporting Group Work*, 21-30, Sanibel Island, FL.

Feller, J. & Fitzgerald, B. (2002). *Understanding open source software development*. New York: Addison-Wesley.

Fielding, R. T. (1999). Shared leadership in the Apache project. *Communications of the ACM, 42*(4), 42-43.

Fogel, K. (1999). *Supporting open source development with CVS*. Scottsdale, AZ: Coriolis Press.

Godwin, M. (1984). Nine principles for making virtual communities work. *Wired*, 2.06, 72-73.

Herbsleb, J. D. & Grinter, R. E. (1999). Splitting the organization and integrating the code: Conway's law revisited. *Proceedings of the 1999 International Conference on Software Engineering*, Los Angeles, CA, May 16-22, 85-95.

Hine, C. (2000). *Virtual ethnography*. London: Sage.

Koch, S. & Schneider, G. (2000). Results from software engineering research into open source development projects using public data. Wirtschaftuniversitat Wien, Austria: Working Paper #22. Available online at: http://citeseer.nk.nec.com/koch00result.html.

Kollock, P. (1996). Design principles for online communities. In the *Proceedings of the Harvard Conference on the Internet and Society*, May 20-22, 1996, Cambridge, MA. Also published in *PC Update 15*(5): 58-60 (June 1998). Available online at: http://sscnet.ucla.edu/soc/faculty/kollock/papers/design.html.

Kollock, P. & Smith, M. (1996). Managing the virtual commons: Cooperation and conflict in computer communities. In S. Herring (Ed.), *Computer-mediated communication: Linguistic, social, and cross-cultural perspectives*, pp. 109-128, Amsterdam: John Benjamins.

Kollock, P. & Smith, M. A. (1999). Communities in cyberspace. In M. A. Smith & P. Kollock (Eds.), *Communities in Cyberspace,* pp. 3-25. New York: Routledge.

Mackenzie, A., Rouchy, P., & Rouncefield, M. (2002). Rebel code? The open source 'code' of work. Paper presented at the *Open Source Software Development Workshop*, February 25-26, 2002, Newcastle-upon-Tyne, UK.

Martin, J. (2002). *Organizational culture: Mapping the terrain*. Thousand Oaks, CA: Sage Publications.

Mockus, A., Fielding, R., & Herbsleb, J. (2002). Two case studies on open source software development: Apache and Mozilla. *ACM Transactions on Software Engineering and Methodology,* 11(3), 309-346.

Mockus, A., Fielding, R. T., & Herbsleb, J. (2000). A case study of open source software development: The Apache server. In the *Proceedings of the 22nd International Conference on Software Engineering*, 263-272, Limerick, IR.

Nardi, B., Whittaker, S., & Bradner, E. (2000). Interaction and outeraction: Instant messaging in action. In the *Proceedings of the CSCW, 2000*, Philadelphia, PA, 79-88.

Noll, J. & Scacchi, W. (1999). Supporting software development in virtual enterprises. *Journal of Digital Information,* 1(4), Available online at: http://jodi.ecs.soton.ac.uk.

Olsson, S. (2000). Ethnography and Internet: Differences in doing ethnography in real and virtual environments. Paper presented at the *IRIS 23: 23rd Information Systems Research Seminar in Scandinavia Doing It Together,* August 12-15, 2001, Laboratorium for Interaction Technology, University of Trollhattan Uddevalla, Sweden.

Ott, J. (1989). *The organizational culture perspective*. Pacific Grove, CA: Brooks/Cole.

Pavlicek, R. G. (2000). Embracing insanity: Open source software development. Indianapolis, IN: Sams Publishing.

Raymond, E. S. (2001). The cathedral and the bazaar: Musings on Linux and open source by an accidental revolutionary. Sebastopol, CA: O'Reilly & Associates.

Robey, D. & Azevedo, A. (1994). Cultural analysis of the organizational consequences of information Technology. *Accounting, Management, and Information Technology,* 4(1), 23-37.

Scacchi, W. (2002a). *Open EC/B: A case study in electronic commerce and open source software development*, Technical Report. Irvine, CA: University of California, Irvine.

Scacchi, W. (2002b). Understanding requirements for developing open source software systems. *IEE Proceedings - Software, 149*(2), 24-39.

Schein, E. H. (1990). Organizational culture. *American Psychologist, 45,* 109-119.

Schein, E. H. (1992). *Organizational culture and leadership.* San Francisco, CA: Jossey-Bass.

Smith, A. D. (1999). Problems of conflict management in virtual communities. In M. A. Smith & P. Kollock (Eds.), *Communities in Cyberspace,* pp. 134-163. New York: Routledge.

Stallman, R. (1999). The GNU operating system and the free software movement. In C. DiBona, S. Ockman & M. Stone (Eds.), *Open Sources: Voices from the Open Source Revolution.* Sebastopol, CA: O'Reilly Press, 53-70.

Strauss, A. L. & Corbin, J. (1990). *Basics of qualitative research: Grounded theory procedures and techniques.* Newbury Park, CA: Sage Publications.

Trice, H. M. & Beyer, J. M. (1993). *The cultures of work organizations.* Englewood Cliffs, NJ: Prentice Hall.

Wenger, E. (1998). *Communities of practice: Learning, meaning, and identity.* Cambridge, UK: Cambridge University Press.

Williams, S. (2002). *Free as in freedom: Richard Stallman's crusade for free software.* Sebastopol, CA: O'Reilly & Associates.

Yin, R. K. (1994). *Case study research, design and methods,* 2nd ed. Newbury Park, CA: Sage Publications.

# SECTION IV:

# Simulating F/OSS Development – "Dynamic Swarms"

## Chapter VIII

# Dynamical Simulation Models of the Open Source Development Process

I.P. Antoniades, Aristotle University of Thessaloniki, Greece

I. Samoladas, Aristotle University of Thessaloniki, Greece

I. Stamelos, Aristotle University of Thessaloniki, Greece

L. Angelis, Aristotle University of Thessaloniki, Greece

G.L. Bleris, Aristotle University of Thessaloniki, Greece

## ABSTRACT

*This chapter will discuss attempts to produce formal mathematical models for dynamical simulation of the development process of Free/Open Source Software (F/OSS) projects. First, a brief overview for simulation methods of closed source software development is given. Then, based on empirical facts reported in F/OSS case studies, we describe a general framework for F/OSS dynamical simulation models and discuss its similarities and differences to closed source software simulation. A specific F/OSS simulation model is introduced. The model is applied to the Apache project and to the gtk+ module of the GNOME project, and simulation outputs are compared to real data. The potential of formal F/OSS simulation models to turn into practical tools used by F/OSS coordinators to predict key project factors is demonstrated. Finally, issues for further research and efforts for improvement of this first-attempt model are discussed.*

Copyright © 2005, Idea Group Inc. Copying or distributing in print or electronic forms without written permission of Idea Group Inc. is prohibited.

# INTRODUCTION

There have been a few studies attempting to define the Open Source Software (OSS) development process in general terms (Bollinger, Nelson, Self & Turnbull, 1999; Feller & Fitzgerald, 2000; McConnell, 1999; O'Reilly, 1999; Raymond, 1998; Wilson, 1999), and there have also been a few case studies of OSS projects: Linux (Godfrey & Tu, 2000), Apache WWW-server (Mockus, Fielding, & Herbsleb, 2000), FreeBSD (Jorgensen, 2001), GNOME (Capiluppi, Lago, & Morisio, 2003; Koch & Schneider, 2000). The latter studies presented some interesting qualitative data for the F/OSS development process, managerial issues, and programmer attitudes, as well as quantitative data regarding the total Lines of Code (LOC) added as a function of time, the defect density of the code produced, number of programmers and contributions per project module/task, average work-effort/time to submit a contribution (code change, defect correction, code testing), and other statistical measures. Despite the fact that these studies have produced interesting results validating or disproving certain hypotheses regarding F/OSS development on a per case basis, there is not sufficient *global* understanding nor a precise definition of the open source development process—the results show both similarities and clear differences in processes and outputs among different projects, but there is no adequate explanation of presented facts based on more general principles. In many cases, the authors offer descriptive explanations based on plausible assumptions but, as there is no general model to quantify their claims together with their possibly complicated interactions, the validity of such explanations cannot be directly demonstrated.

Therefore, there is a need to move from descriptive models based on special cases to a more general *quantitative* mathematical model that would hopefully be used as a demonstrating tool of real case results. Most importantly, this model could serve as a *predicting* tool of key F/OSS project factors, such as project failure/success, dynamical evolution of source code, defect density/architectural quality, expected number of programmers involved, and distribution of work effort to distinct project modules and tasks.

Previous studies have shown that the dynamical evolution of the above key factors is quite sensitive to a) the type of software developed, and b) the specific technical management framework of an F/OSS project. Therefore, the model should be general enough so that, by a straightforward adjustment of model parameters, it is able to simulate various types of F/OSS projects under alternative managerial scenarios.

The "predictive power" of such a model could be viewed as follows: by first calibrating the model parameters against available historical data from a certain time period within the development phase of an F/OSS project, the model should be able to approximately reproduce the *future evolution* of the *same* F/OSS project.

This chapter will discuss attempts to produce formal mathematical models for dynamical simulation of the development process of F/OSS projects. Whereas several such models and corresponding computer simulation studies exist for the traditional (closed source) software development process, it is only very recently that such attempts have started to appear in international literature regarding the F/OSS development process. We will briefly describe closed source simulation models and their practical applications in software engineering. We will then introduce a general framework for generic F/OSS dynamic simulation models, compare and contrast it to traditional closed source simulation models, and discuss its possible use in describing the F/OSS development process and its possible practical applications for the F/OSS community. Then, we will introduce a specific simulation model and demonstrate it against the Apache case study (Mockus et al., 2000) and evolution of the gtk+ module of the GNOME project (Koch & Schneider, 2000; http://bulunga.dat.escet.urjc.es/gnome-cvs/index.php). As F/OSS dynamical simulation opens an entirely new field of research, we will discuss future possibilities and suggestions for improvements and further work.

# BACKGROUND

## Towards the Creation of a Simulation Model of F/OSS Development Dynamics

In this section, we first review briefly simulation modeling for traditional closed source software projects. We discuss mainly a generic structure for such models and the ways these models may be used for project management. We then proceed by discussing the requirements of a simulation model for the F/OSS case by describing a general framework (Antoniades, Stamelos, Angelis, & Bleris, 2003) that may be used as a basis for the design of such models. It is evident that, as there exist major differences between the closed and F/OSS approach, respective simulation models are also quite different in nature, structure, and accuracy. Then, we discuss plausible ways of exploiting F/OSS simulation models by project coordinators. Finally, we present briefly a very recent approach to OSS simulation modeling (Dalle & David, 2003) that follows a different point of view from the general framework we present here.

## Closed Source Project Simulation Modeling

*Closed Processes and Models*

As it is known, closed source software projects are developed according to more or less well-defined software processes. Examples of such development cycle models are the traditional *waterfall* model, the *prototyping* model, the *spiral* model and, more recently, Rational Unified Process (RUP) (Pfleeger, 2001). Initially, development models were quite rigid, but gradually more flexible and adaptable models were proposed and applied.

In general, three modeling methods may be distinguished, namely, the analytical, the continuous, and the discrete-event methods. Various analytical models have been proposed in software engineering for predicting effort, time (Boehm, 1981; Center for Software Engineering, 1997), reliability (Fenton, 1997), etc., when producing software. They are capable of estimating specific aspects of software, but lack the necessary flexibility to model interactions and variations within the project.

In parallel, various simulation approaches have been also proposed to model software development practices. Continuous simulation is the most common choice of the researchers, although discrete-event simulation is also proposed. The approach of Abdel-Hamid and Madnick (1991) based on System Dynamics is probably the most known approach. This kind of approach uses dynamic processes to simulate the software process. A descriptive model, a graph for example, is built of the various activities that are involved in software production, with each node representing a factor that affects software production. For example, one node can represent the fraction of the experienced staff, while another, the actual fraction of a person's days on the project. These nodes are connected with directed arrows that represent how changes in one factor affect changes in another. Nodes can be grouped together to represent a bigger part of the software development process, such as human resource management. The next step after the identification of the nodes and the relations between them is to quantify those relations and assign a probability distribution to those nodes that change in time. After the model is built, its dynamic view enables us to simulate the process.

Perhaps the most straightforward way to understand the application of simulation on closed source software modeling is to look at the approach presented in Doncelli (2001). In that paper, a two-level, hybrid approach is proposed. At a high level of abstraction, discrete-event simulation seems to be more appropriate. At that level, requirement definition, component interaction, project artifact exchanges, deliverable appearance and consumption, and macro-management activities, based on discrete decisions, take place. Consequently, a discrete-event queuing network may adequately model high-level process dynamics.

At a lower level, i.e., at the level of artifact production (for example, code generation), a combination of analytical and continuous approach looks quite suitable. At this level, the production activities to be modeled are of a more continuous nature, presenting dynamics that resemble those of a production line, and therefore, analytical, equation-based models or continuous simulation models are more suitable. A discrete-event approach at this level of detail might produce overly cumbersome, difficult to manage models.

Certain process characteristics are typical inputs to closed source software simulation. For example, a group of persons possessing a specific role within the project may be defined. For instance, the number of programmers is often used. Such a figure may be known and fed to the model by its user as a single number or as a short range of numbers, e.g., the user may determine that a group of 4-6 persons will code one system function. Other simulation inputs or parameters may be schedules (e.g., delivery date), requirements rate of arrival, personnel experience and turnover rates, complexity of the system to be produced, etc. For a review of recent developments in software simulation refer to the Special Issue on Software Simulation, *Journal of Systems and Software* (Vol.59, 2001).

## A Generic Structure for F/OSS Simulation Modeling

We now describe a general framework that should be followed (perhaps not exclusively) by generic F/OSS simulation models and point out the extra difficulties that have to be confronted relative to analogous models of the "traditional" (closed source) process.

1. Unlike closed source projects, in F/OSS projects the number of contributors (programmers) greatly varies in time, cannot be directly controlled, and cannot be predetermined by project coordinators. As any qualified programmer can freely contribute to the project at any time, on any task and as often as he or she desires, the number of distinct individual contributors varies based on the interest that the specific F/OSS project attracts. *Therefore, an F/OSS model should a) contain an explicit mechanism for determining the flow of new contributors as a function of time, and b) relate this mechanism to specific project-dependent factors that affect the overall "interest" in the project. These project-dependent factors should be identified and parameterized (quantified).*

2. In any F/OSS project, any particular task at any particular instant can be performed either by a *new* contributor or an *old* one who decides to contribute again. In addition, it has been shown that almost all F/OSS projects have a dedicated team of programmers (core programmers) that perform most of the contributions, especially in specific tasks (e.g., code writing), while their interest in the project (judged by how often they contribute in the course of time) stays approximately the same (e.g., Mockus et al., 2000). *Therefore, the F/OSS simulation model must contain a mechanism that determines the number of contributions that will be undertaken per category of contributors (e.g., new, old, or core) at each time interval.*

3. In F/OSS projects, there is also no direct central control over the number of contributions per task type or per project module. Anyone may choose any task (e.g., code writing, defect correction, code testing/defect reporting, functional improving, etc.) and any project module to work on. The allocation of contributions per *task type* and per project *module* depends on the following sets of factors:

    a. *Factors pertaining to programmer profile* (e.g., some programmers may prefer code testing to defect correcting). These factors can be further categorized as follows:

i. factors that remain constant in time (e.g., the aptitude or preference of a programmer in code writing), and
ii. factors that vary with time (e.g., the overall interest of a programmer to contribute to any task or module may vary based on frequency of past contributions).
b. *Project-specific factors* (e.g., a contributor may wish to write code for a specific module, but there may be nothing interesting left to write for that module).

Therefore, *the F/OSS model should (a) identify and parameterize the dependence of programmer interest to contribute to a specific task/module on (i) programmer profile, (ii) project evolution, and (b) contain a quantitative mechanism to allocate contributions per task type and per project module.*

4. In F/OSS projects, because there is no strict plan or task assignment mechanism, the total number of Lines of Code (LOC) written by each contributor varies significantly per contributor and per time period, again in an uncontrolled manner. Therefore, project outputs such as LOC added, number of defects, or number of reported defects are expected to have a *much larger statistical variance* than in closed source projects (e.g., Koch & Schneider, 2000). This fact is not only due to the lack of strict planning, but also to the much larger numbers and diverse profiles of contributors that participate in an F/OSS project. *Therefore, the F/OSS simulation model should determine delivered results of particular contributions in a stochastic manner, i.e., drawing from probability distributions. This is a similar practice to what is used in closed source simulation models, with the difference being that probability distributions here are expected to have a much larger variance.*

5. In F/OSS projects there is no specific *time* plan or deadlines for project deliverables. Therefore, the number of calendar days for the completion of a task varies greatly and must be drawn from probability distributions with relatively large variances. Also, delivery times should depend on project-specific factors such as the amount of work needed to complete the task. For example, writing 1,000 LOC should, on average, take more time than writing 200 LOC, while discovering all defects in a source file containing 10,000 LOC should take, on average, more time than the same task for a file containing only 100 LOC. Therefore, *task delivery times should be determined in a stochastic manner on the one hand, while average delivery times should follow certain deterministic rules, on the other.*

A fact that becomes immediately apparent from the discussion in Points 1-3 above is that the core of any F/OSS simulation model should be based upon a specific *behavioural model* that must be properly quantified (in more than one possible way) in order to model the behaviour of the "crowd" of project contributors in deciding: a) whether to contribute to the project or not, b) which task to perform, c) to which module to contribute, and d) how often to contribute. *The behavioural model should then define the way that the above four aspects of programmer behaviour depend on a) programmer profile (both static and dynamic), and b) project-specific factors (both static and dynamic).*

The formulation of a behavioural model must be based on a set of *qualitative rules*. Fortunately, previous case studies have already pinpointed such rules either by questioning a large sample of F/OSS contributors or by analyzing publicly available data in F/OSS project repositories. There is probably no single behavioural model that can fit contributor behaviour in all types of F/OSS projects. However, as previous case studies identified many common features across several F/OSS project types, one certainly can devise a behavioural model general enough to describe at least a large class of F/OSS projects.

An extra degree of freedom that comes in when designing a behavioural model is the variety of ways that a set of qualitative rules may be quantified; there can be an infinite number of specific equations that describe a specific qualitative rule. Selecting a suitable equation is largely an arbitrary task in the beginning; however, a particular choice may be subsequently justified by the model's demonstrated ability to fit actual results. Once the behavioural model equations and intrinsic parameters are validated, then the model may be applied to other F/OSS projects.

## Application of an F/OSS Simulation Model
*General Procedure*

In Figure 1 we show the structure of a generic F/OSS dynamic simulation model. Just as in any simulation model of a dynamical system, the user must specify on input a) values to *project-specific* time-constant *parameters*, and b) *initial conditions* for the project *dynamic variables*. These values are not precisely known at project start. One may attempt to provide rough estimates for these values based on results of other (similar) real-world F/OSS projects. *However, these values may be readjusted in the course of evolution of the simulated project as real data becomes available.* The way to readjust these values would be to try to fit actual data from an initial historical period of the project to the results of the simulation for the same period. By applying this continuous re-adjustment of parameters (*backward propagation*), the simulation should become more accurate in predicting the future evolution of the project. If this is not the case, it means that a) either some of the behavioural model qualitative rules are based on wrong assumptions for the specific type of project studied, or b) the values of behavioural model *intrinsic* parameters (*project-independent*) must be re-adjusted.

Initial values of dynamic variables are known at project start. Dynamic variables can be the number of contributors, the number of LOC written for each module, the number of source files for a module, the number of modules itself, the defect density, the number of reported defects, activity in each task type, etc.

*Figure 1: General structure of an OSS dynamic simulation model.*

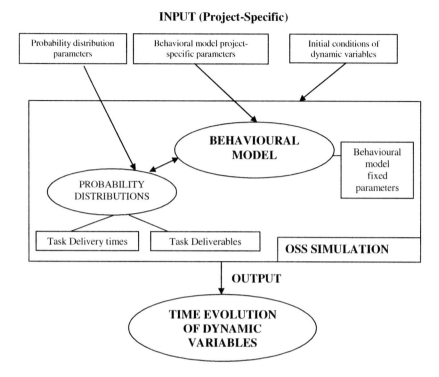

Due to the stochastic features of the simulation model, several computer runs must be performed in order to obtain average values and variances of the dynamic variables.

*Calibration of the Model*

The adjustment of behavioural model *intrinsic* (project-*in*dependent) parameters is the *calibration procedure* of the model. According to this procedure, one may introduce arbitrary values to these parameters as reasonable "initial guesses." Then one would run the simulation model, re-adjusting parameter values until simulation results satisfactorily fit the results of a real-world F/OSS project (*calibration project*) *in each time-window of project evolution*. More than one similar type of F/OSS projects may be used in the calibration process.

*Validation of the Model*

Once the project-*in*dependent parameters of the behavioural model are properly calibrated, the model may be used to simulate other F/OSS projects according to the procedure outlined in the following chapters.

## Practical Use of F/OSS Simulation Models

Before proceeding with the whereabouts of an F/OSS simulation model, let us consider its plausible practical uses, in the same way we have done with closed source models. We provide such a tentative list, although the reader might suggest further uses.

- *Prediction of F/OSS project evolution.* Project coordinators may obtain a picture of plausible evolution scenarios of the project they are about to initiate. Moreover, they may calibrate a generic model to their own project, using data from its initial stages (e.g., first 1-2 years) and anticipate future development with increased accuracy. In such a way, an F/OSS project profile (size, quality, population of programmers) may be maintained and updated as frequently as desired. Software *users* may also be interested in such prediction, as it would indicate when the software will most likely be available for use. This also applies to organizations, especially if they are interested in pursuing a specific business model that is based on this software.
- *F/OSS project risk management.* Much as a closed source project, F/OSS projects are risky, in the sense that many, not easily anticipated factors may negatively affect their evolution. Simulation models may help in quantifying the impact of such factors, taking into account their probability of occurrence and the effect they may have, in case they occur.
- *What-if analysis* (a typical usage mode for simulation models). F/OSS coordinators may try different development processes, coordination schemes (e.g., core programming team), tool usage, etc., to identify the best possible approach for initiating and managing their project.
- *F/OSS process evaluation.* F/OSS dynamics are not yet fully understood. Moreover, the nature of F/OSS guarantees that, in the future, we will observe new types of project organization and evolution patterns. Researchers may be particularly interested in understanding the dynamics of F/OSS development and simulation models may provide a suitable tool for that purpose.
- *Education of students and professionals in F/OSS* (e.g., in the case of hybrid projects). F/OSS is being introduced in the curricula of various academic institutions as a novel way of developing and managing software. The contrast with the traditional approach of doing software has a tremendous educational value for students of informatics and guarantees that such courses, along with extreme programming, agile methods, etc., will flourish in the near future. F/OSS project simulators appear as natural learning tools, providing to novice students/professionals a readily available picture of how an F/OSS project evolves and which factors steer such evolution.

An interesting approach in F/OSS simulation was presented recently by Dalle and David (2003). In this paper, the authors aim to develop a stochastic simulation model in order to describe how people involved in the F/OSS development community decide how to allocate programming resources both within and among various F/OSS projects. The core of the model relies on the reward that is gained when someone (a F/OSS developer) contributes to a project. Based on behavioural models, the authors try to determine a) the way F/OSS developers decide to which module to contribute and b) the processes that drive the evolution of certain modules to a mature stage of development. Making assumptions — among others — like: "Launching a new project is more rewarding than contributing to an existing one," "Contributing to early releases is more rewarding than contributing to late ones," and "The creation of new projects/modules is related to existing ones, to which they add functionality," Dalle and David developed a dynamic stochastic (random graph) model. In this model, at any moment in time, a specific "agent" (i.e., an F/OSS developer) must choose how to allocate a fixed level of development effort (i.e., adding new functionality and/or bug fixing) to a particular module of a project. Without quoting here any particular mathematical formula, the authors result in a function called "social utility function," that depicts the assumptions mentioned above. The model also seems to confirm the well-known "release early" rule that Eric Raymond first formulated (Raymond, 2002).

The Dalle & David approach is similar to ours in that it also utilizes a 'behavioural model' for prescribing how programmers decide on which task and which module to work. Our approach, however, moves a step further, attempting to also reproduce specific outputs of a F/OSS project (such as LOC, defect density, and number of committers) in real-time.

# AN EXAMPLE SIMULATION MODEL OF THE F/OSS DEVELOPMENT PROCESS

## Definitions

*Project modules:* A project module consists of a collection of code files that have the same general specifications.

*Project tasks:* Project tasks are the different type of actions that can be performed by individual contributors. The tasks considered are:

1. Design and submit the first release version of a new module. We call this task *new module submission* and index all quantities related to it by index **S**. This task may refer to a submission of an alternative or more advanced version of an existing (old) module. This occurs frequently during the evolution of the most popular F/OSS projects.
2. Correct defects that were previously reported for a specific file. We will call this task *debugging* and index all quantities related to it by index **B**.

3. Test a specific source file and report defects. We will call this task *testing* and index all quantities related to it by index **T**.
4. Add functionality and/or improve an existing source file. We will call this task *functional improving* and index all quantities related to it by index ***F***.

## Model Equations

In order to quantify the above dynamical procedure, three major sets of model equations are employed:

1. *Equation Set 1*: It yields the number of tasks initiated at project day *t* per task type, per project module.
2. *Equation Set 2*: It yields the specific output of each task (number of LOC, number of defects, defect corrections, etc.)
3. *Equation Set 3*: It yields the time period needed for each task to be completed.

The above sets of equations have been described in detail in Antoniades, Stamelos, Angelis and Bleris (2003), thus we limit the discussion here to the most significant points.

*Equation Set 1*

Equation 1.1: Determination of initial "interest" in the project.

We assume that the number of individuals $E(t)$ that tentatively decide to contribute to the project starting at day *t* depends on (i) the "overall quality" of the project, and (ii) the profile of the programmers that either have worked on the same project before day *t* or are new to the project. The "overall quality" of the project is determined by all those project-specific factors that stimulate the interest of prospective programmers, leading them to decide to devote personal effort and time on any one contribution. Previous studies have pinpointed such factors as are summarized in Antoniades et al. (2003).

Taking into account these factors, a time-dependent *overall quality factor* $Q(t)$ defined as:

$$Q(t) = \begin{cases} 1 + \log_{10}(Q_1(t)) + Q_o, & Q_1 > 1 \\ Q_o(1 - 10^{-Q_o Q_1(t)}), & Q_1 \leq 1 \end{cases} \quad (1)$$

$$Q_1(t) = w_R R(t) + w_L \langle \Delta L(t) \rangle + w_A \langle A(t) \rangle + J(t)$$

where $R(t)$ the *percentage* increase *per unit time* in the LOC, from the *last production release* to the one before the last, $<\Delta L(t)>$ the *cumulative time average rate of change in total* LOC in the current development release *from one day to the next*, $<A(t)>$ the *cumulative time average of the Activity* in the project, which is defined as the *total* number of tasks that were terminated on day $t$ and $J(t)$ the *"interest boost factor,"* which is zero for all $t$, except for the day(s) when there is an extra boost in interest (e.g., due to the public announcement that a renowned programmer participates in the project). $Q_o$, $w_R$, $w_L$ and $w_A$ are time-constant, project-specific parameters that determine the scale (*calibration*) of $Q(t)$ for a specific F/OSS project. We will discuss the way to determine their values later.

Except for the overall project "quality" $Q(t)$, the number of individuals that show an interest to initiate a task depends also on the programmer's profile. Project contributors are divided into two categories: "*core*" contributors and "*normal*" contributors. It is assumed that up to a certain number of man-days $s_o$ spent on the project by any contributor, interest to contribute again *increases*, whereas beyond this number ($s > s_o$) *interest decreases*.

Based on the above discussion, $E(t)$ is given by the following equation:

$$E(t) = Q(t)\Lambda(t) + N'_{core}(t) \qquad (2)$$

where $N'_{core}(t)$ is the number of *available core programmers* at day $t$ and $\Lambda(t)$ is a factor determining the variation in programmer interest according to the time they spent on past contributions, as mentioned above.

The core programmers are assumed always to show the same interest in the project independent of its quality and their past contributions (last term in equation 2).

In order to determine $E$ at $t=t_o$, where $t_o$ is the first day of the project after which simulation will apply, we impose the condition $Q \equiv 1$ at $t=t_o$, assuming that we use the first $t_o$ days of the real-world project to calibrate project-specific, time independent parameters mentioned above. In order to do this, we let $E(t_o)=N_o$, where we define $N_o$ as the average number of tasks actually performed per day as measured in the first $t_o$ days of the project. Then, we run the simulation from $t=0$, and fix $Q_o$, $w_R$, $w_L$ and $w_A$ so that simulation results *fit actual results* for the first $t_o$ days as closely as possible.

*Equation 1.2: Determination of number of tasks initiated for each task type and each project module.*

The $E(t)$ individuals who have decided to contribute to the project starting at day $t$ will then have a closer look at what in particular they can do. It is not necessary for all of them to actually initiate a task, as they may lose interest after they browse the project site in search for an interesting assignment.

Denoting by upper index $j=S,B,T,F$, the task type, first lower index $i=1,2,...,M$, the particular project module and second lower index $k=1,2,...,S_i$, possible *alternatives* of module $i$ that an individual may contribute to, we define (t) as the number of tasks actually initiated for project module $i$, alternative $k$ (of module $i$)[1] and task type $j$ at day $t$.

In Antoniades et al. (2003), we proposed the following master equation to determine the $P_{ik}^j$'s *for every project day t*:

$$\frac{P_{ik}^j(t)}{E(t)} \quad I_{ik}^j(t) \quad I_{ik}^j(t) \quad \underset{(p,q,r)\;(j,i,k)}{{}^p_{qr}(t)} \tag{3}$$

where $I_{ik}^j$ is the *total interest factor* for task type *j*, module *i* and alternative *k*, defined as:

$$I_{ik}{}^j \equiv \Gamma_{ik}^j - \gamma_{ik}^j \tag{4}$$

where $\Gamma_{ik}^j(t)$ the *interest increase factors* and the $\gamma_{ik}^j(t)$ *interest reduction factors* for task type *j*, module *i* and *alternative k* at time *t*. The particular way $\Gamma_{ik}^j$ and $\gamma_{ik}^j$ depend on project-specific time variables was also defined in Antoniades et al. The precise mathematical formulae are reported in http://sweng.csd.auth.gr/TR/report#001.pdf. For example, for task type **T** (testing), $\Gamma_{ik}^T$ (interest increase factor) is *inversely* proportional to the number of times alternative *k* of module *i* has been tested since its last update. On the other hand, $\gamma_{ik}^T$ (interest decrease factor) is *larger* the more times a given alternative *k* has been tested since its last update, and *smaller* the larger the total number of tests performed in the past for the *entire* module. The behavioural rule assumed here is that the more times an alternative *k* of module *i* has been tested, the less interested a potential contributor would be to test it again, and also, if potential contributors see that many tests are being performed for a given module *i* (including all its alternatives), the more interested they would be to test it themselves.

The basic assumption behind (3) is that a proportion of the users who turn down tasks other than *j*, *i*, *k* will finally turn to task *j*, *i*, *k* with a probability proportional to the total interest in *j*, *i*, *k*. The values of the $\Gamma$'s and $\gamma$'s are properly normalized so that the sum of all tasks that will actually be initiated at day *t* for all task types, modules and alternative modules in the project is *at most* equal to $E(t)$.

*Equation Set 2*

Equation Set 2 determines the deliverable quantities of each of the initiated tasks by drawing from probability distributions. We use *lognormal distributions*, since relevant quantities are all positive. The following *time-constant project-specific* quantities must be provided as initial input to the simulation model:

$\bar{L}_i^S, \sigma_{L_i^S}$ : The mean and standard deviation for the number of LOC added as *initial* contribution for a *new* module, *alternative* to module *i*, which appears in the development release.

$\bar{L}_i^F, \sigma_{L_i^F}$ : The mean and standard deviation for the number of LOC added per contribution for a functional improvement (update) of an existing source file in module *i*.

$\bar{d}_i, \sigma_{d_i}$ : The mean and standard deviation of the initial defect density (number of defects per LOC) of code written and submitted by a contributor for module *i* in a single check-in.

$\bar{T}_i, \sigma_{T_i}$ : The mean and standard deviation of the estimated *fraction* of actual defects in a source file of module *i* that are reported by a contributor undertaking a testing task.

One can obtain reasonable *initial* estimates for the above parameters by using real results from a known F/OSS and periodically re-adjusting their values at later project times by the *backward-propagation* procedure described in section, *Application of an F/OSS Simulation Model*.

## Equation Set 3

Finally, the delivery time of each initiated task is also drawn from lognormal probability distributions. The following time-constant *project-specific* quantities must be provided as initial input to the simulation model:

$\bar{t}_i^S, \sigma_{t_i^S}$ : The mean and standard deviation for the time (in days) to write one LOC for a file in module *i*.

$\bar{t}_i^B, \sigma_{t_i^B}$ : The mean and standard deviation for the time (in days) to correct a single defect in 1,000 LOC.

$\bar{t}_i^T, \sigma_{t_i^T}$ : The mean and standard deviation of the time (in days) to test the most recent version of the project's development release and report detected defects for a file in module *i*, when the file contains 1,000 LOC and the development release $M \times 1,000$ LOC ($M$ is the number of modules of the project).

Equation Sets (1)-(3) fully determine the dynamics of the system. Due to the stochastic character of Equation Sets 2 and 3, each run has to be repeated an adequate number of times with different random number generator seeds. Averages and standard deviations of output variables must be calculated at each time step.

# SIMULATION STUDIES AND RESULTS

In order to adequately validate the above model one has to a) calibrate the behavioural model intrinsic parameters (project-*in*dependent) using results from one or more case studies, and b) simulate other F/OSS projects comparing simulated data with actual data. Although, there is relative data in all case studies that could be used for these purposes, *no single case study contained all the information needed in order to adequately calibrate or validate the entire model.* This is quite expectable, as none of the existing studies was conducted having any particular simulation model in mind. For example, the GNOME case study contained a lot of data for key project factors' time averages, but lacked information on the averages or dynamic evolution of defect density, activity per *task type*. The Apache case study, on the other hand, contained much more information for cumulated activities per task type but a) reported no results regarding the evolution of relevant quantities as a function of time, b) lacked other necessary data such as statistical information about deliverables per single check-in (LOC, defects, reports, etc.), and information about the evolution of separate project modules.

*Therefore, it was not possible to attempt a full-scale calibration and validation of the proposed model in the present work.* However, in Antoniades et al. (2003), solely for the purpose of producing at least an initial demonstration of the model's ability to fit actual data and reproduce some of the reported *qualitative* features of F/OSS development process, we managed to side-track some of the difficulties mentioned above by using data mainly from the Apache case study combined with data from the GNOME case study that were reasonably assumed to be similar for both projects. For certain project parameters, for which we had no data from either project, we had to make our own plausible assumptions. Finally, we produced simulation results that compare very well for the respective results given in the Apache case study.

## *Results after 1,094 Project Days and Comparison to Apache Case Study*

Details and justification of the values given to project-specific parameters for application to the Apache case study were presented analytically in Antoniades et al. (2003). A summary of these inputs is presented in Table 1, along with respective inputs for the gtk+ case study that we will present in the next section. Simulations were realized by a custom C application. One hundred runs with different random number generator seeds were performed, each one for 2,000 time steps and averages taken for each dynamical variable. All 100 runs finished in about seven minutes on a regular PC with Pentium IV 1.8GHz. The interactive calibration procedure showed that the Apache case study data were best fitted by assuming that 6.5% of normal contributors by priority are interested in creating a new source file, 3.89% of them in debugging, 53.7% in testing, and 35.9% in functional improving.

*Table 1: Input parameters for Apache and gtk+ simulations.*

| Parameters | Apache | Source/Justification | Gtk++ | Source/Justification |
|---|---|---|---|---|
| $\bar{L}_i^S, \sigma_{L_i^S}$ | (4000,2000) LOC | Assumption. Data not available in Apache case study. | - | Deduced value from http://bulunga.dat.escet.urjc.es/gnome-cvs/index.php for the first 1,000 days. |
| $\bar{L}_i^F, \sigma_{L_i^F}$ | (36, 40) LOC/commit | Average LOC difference per commit deduced from Apache case study (equals total LOC written divided by reported number of type F tasks). | (15, 97) LOC/commit | Reasonable assumption. No data available for GNOME or gtk+. |
| $\bar{d}_i, \sigma_{d_i}$ | (2, 3) defects/1KLOC | Reasonable assumption. Data not available in Apache case study. | (2, 3) defects/1KLOC | Reasonable assumption. Data not available in GNOME or gtk+ public data. |
| $\bar{T}_i, \sigma_{T_i}$ | (50%, 25%) | Reasonable assumption. Data not available in Apache case study. | (50%, 25%) | Exact deduced value from GNOME case study (average time a single contributor takes to submit one LOC). |
| $\bar{t}_i^S, \sigma_{t_i^S}$ | (0.11, 0.02) days/LOC | Deduced value from Apache case study (average time a single contributor takes to submit 1 LOC). | (0.0117, 0.03872) | No data available. Use same values as in Apache, reduced roughly in proportion to the ratios between $\bar{t}_i^S, \sigma_{t_i^S}$ in the two projects. |
| $\bar{t}_i^B, \sigma_{t_i^B}$ | (35, 47) days | Deduced value from Apache case study (average time a single contributor takes to correct one defect). | (3, 4) days | No data available. Use same values as in Apache reduced roughly in proportion to the ratios between $\bar{t}_i^S, \sigma_{t_i^S}$ in the two projects. |
| $\bar{t}_i^T, \sigma_{t_i^T}$ | (1, 2) days | Deduced value from Apache case study (average time a single contributor takes to test a portion of the Apache code). | (0.1, 0.2) | Actual average number of commits per day for gtk+ for the first 1,000 days. |
| $N_o$ | 23.3 | Adjusted value so that Equation 1 is properly calibrated. | 25 | Adjusted value so that Equation 1 is properly calibrated. In the present version of the model $N_o$ obtains its actual value and calibration is achieved through $Q_o$. |
| $Q_o$ | - | In the version of the model presented in Antoniades et al. (2003) $Q_o$ was not present in Equation 1. Calibration was achieved solely by adjusting $N_o$. The actual $N_o$ for Apache was ~9.8. | 0.089 | |
| $(w_R, w_L, w_A, J)$ | (0.00227, 0.00227, 0.0, 0.0) | Arbitrary weights given to respective terms in Equation 1. | (0.00227, 0.00227, 0.0, 0.0) | Arbitrary weights given to respective terms in Equation 1. |
| $N_{core}$ | 15 | Reported value from Apache case study (contributors that wrote 88% of source code). | 7 | Reduced value from gtk+ data as obtained from http://bulunga.dat.escet.urjc.es/gnome-cvs/index.php. It equals the number of committers that clearly produced much more code than all others. |
| $s_o$ | 120 days | Reasonable assumption. No data available in Apache case study. | 246 days | Exact reported average value in Koch and Schneider (2000) for the entire GNOME project for the mean number of days a contributor spent on the project in the first three years. It is assumed that the same value holds for gtk+. |
| Functional completion threshold | 400 KLOC | Reasonable assumption based on the fact that Apache had nearly reached completion in the first three years. | 2,000 KLOC | Reasonable assumption for gtk+. |

*Table 2: Apache simulation results compared to real results from Mockus et al. (2000).*

| Variable | Simulation (average ± st. dev.) | Real Data |
|---|---|---|
| Total number of LOC after 1094 days (3 years) | 220,907.0 ± 31.6 LOC | 220 KLOC |
| Average Defect density in the first 1094 days | 2.71 ± 0.42 defects/KLOC | 2.64 defects per KLOC |
| Residual defect density (i.e., actual reported defects that were not corrected after 1094 days) | 0.03 ± 0.09 defects/KLOC | Not available |
| Average Number of reported defects per day | 33.9 ± 11.05 | Not available |
| Total activity in task type B (in the first 1094 days) | 715 ± 26.7 tasks | 695 tasks |
| Total Activity in task type T | 4,040 ± 60.9 tasks | 3,975 tasks |
| Total Activity in task type F + type S | 5,991 ± 83.2 | 6,092 |
| Total Activity in all task types | 10,747 ± 106.5 | 10,762 |
| Number of individual contributors | 489 ± 21.8 | 388 |

Previous case studies reported programmer interest per task type in the same order as what the present simulation yielded, namely, testing, functional improving/new source file creation, and debugging.

Table 2 compares average values of certain key project variables between simulation and reported results in the Apache project. Standard deviations of the reported quantities as calculated after 100 simulation runs are also reported.

*Temporal Evolution of Project Variables of the Apache Case Study*

In Figures 2 and 3, the dynamical evolution of project variables is shown for 2,000 days. The Apache case study reported results for the first three years (1,094 days) of the project, but we continued the runs in order to look at the simulated evolution at later times. For all figures except 2b, average data for the 100 runs is further "smoothed" by taking *running* (window) time averages within a running window of 30 days.

Figure 2a shows the evolution of the total number of LOC difference (for all project modules). By LOC "difference" we mean actual (uncommented) LOC that is purely added to the development release from one day to the next. The bold line shows the average of the 100 runs tried, and the two dashed lines show the bounds of one standard deviation above and below average. We see that in the first 310 days the project already reaches half the size (110 KLOC) of the number of LOC after 1,094 days. Only about 35 KLOC are added from Day 1095 until Day 2,000. *This means that project size growth rate rises at the first stages and slows down towards later stages of the F/OSS project.* This fact is more clearly demonstrated by Figure 2b, which shows the average rate of adding LOC each project day. The rate reaches a maximum around Day 100 and subsequently drops. *This behaviour has been observed with certain strands (core) of the GNOME project and certain modules of the LINUX project.* In fact, the slowing down of software projects is a well-known fact for both closed source projects (e.g., Boehm, 1981) and F/OSS projects (e.g., Godfrey & Tu, 2000; Koch & Schneider, 2002).

*Figure 2: Simulation results for the Apache project: (a) Cumulative LOC difference vs. time for the Apache project. The bold line is the average of the 100 runs. The gray lines are one standard deviation above and below the average. The dashed vertical line shows the day of the project until data was reported in the Apache case study (Mockus et al., 2000). (b) Total number of LOC added per day. (c) Density of (unfixed) defects as a function of time.*

(a)

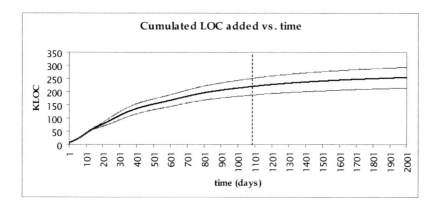

(b)

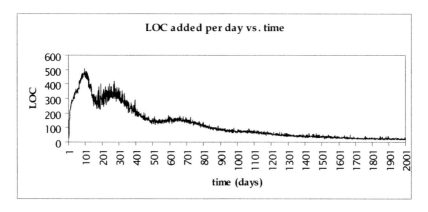

*Figure 2: Simulation results for the Apache project (continued).*

(c)

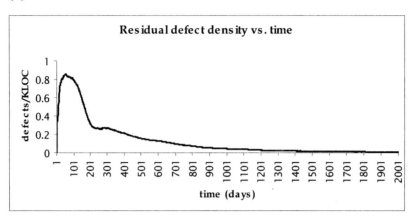

*Figure 3: Number of committers vs. time. (a) Simulation results for the Apache project. (b) Respective plot with actual results for the Mutt project from Capiluppi et al. (2003). The relevant curve for comparison with our simulation results is the curve for "contributors."*

(a)

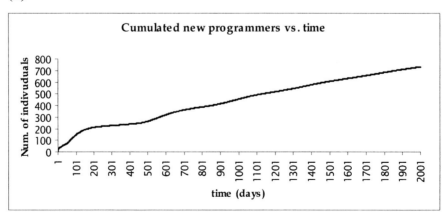

*Figure 3: Number of committers vs. time. (continued)*

(b)

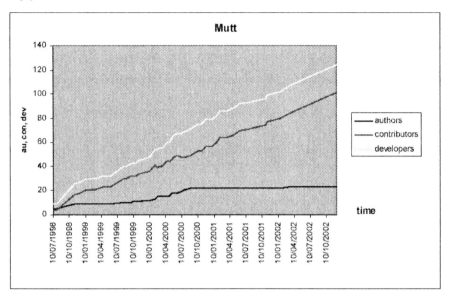

Finally, Figure 2c shows the evolution of residual defect density, i.e., defects per KLOC that are left unfixed. We see that the density rises rapidly in the beginning, when a lot of code is added, and defect correction activity cannot keep up. Fortunately, the model predicts that defect density will drop less than 0.1 defects per KLOC that are left after Day 1,000. This agrees with the *Apache*, *FreeBSD*, and *Linux* case studies that state that defect correction is quite effective in F/OSS projects.

Finally, Figure 3a shows the simulation results for cumulative number of committers in time for the Apache project. At Day 1094, there is a total of $489 \pm 21.8$ individual programmers that have performed at least one task for the project. Compared to the actual number for Apache, which is 388, this is indeed larger, but the disagreement is surprisingly satisfactory if one considers that $Q$ was calibrated using only the reported *three-year period* time average values for the evolution of project variables in the Apache case study and that many other data used for adjustment of model parameters was either assumed or picked from other cases studies. The step-like increase in number of committers has been noticed in several studies of other F/OSS projects, e.g., in Capiluppi et al. (2003). Figure 3b shows an example plot of actual data on the "Mutt" project taken from Capiluppi et al. The definitions of "Authors, contributors and developers" shown in the figure can be found in Capiluppi et al.; the definition consistent with what we have called "committers" in this chapter is the one the above authors used for the term "contributors."

## Demonstration of Model against Evolution of gtk+ Module in GNOME Project

In this chapter, we also present simulation results for the gtk+ module of the GNOME project. For gtk+, we gathered data from http://bulunga.dat.escet.urjc.es/gnome-cvs/index.php, a website that graphically reports dynamic evolution of committers, commits, and LOC along with other interesting statistics based on CVS analysis of several F/OSS projects. For data not present in the aforementioned site, we used data from the GNOME case study of Koch and Schneider (2000), from which we managed to deduce estimates for averages and standard deviations of delivery times necessary for Equation Set 3. Both sources lacked data regarding activity for individual task types (code writing, defect correction defect reporting), thus we used the same values that we had available from the Apache case study.

In the present simulation, we are examining the evolution of a single module (gtk+). Therefore, we disabled assignments of task type S in the simulation so that there is no change in the number of modules during the simulation. Further, we used the exact same equations (with the same time-constant parameter values) for the behavioural model in Equations 1 and 3 as in the Apache case study simulations. Table 1 shows the values of project-specific parameters used for Equation Sets 2 and 3. Notice that gtk+ programmers were more productive in code writing than in the Apache project (more LOC were written on the average for shorter time periods). Also, the estimated threshold for "functional completion" of gtk+ was set to 2,000 KLOC, much higher than in the Apache (see discussion in the next section). $N_o$ was set equal to 25, the actual average value of commits per day in the first 1000 days of gtk+ development. Subsequently, $Q_o$ (the project-specific "quality factor" calibration parameter) was interactively adjusted to a final value of 0.089 so that cumulated LOC difference and number of commits fit actual data for the first 1,000 days as closely as possible. Finally, the runs were continued until Day 2,000. One hundred repetitions were performed and averages taken. Table 3 shows simulation

*Table 3: gtk+ simulation results compared to real data from http://bulunga.dat.escet.urjc.es/gnome-cvs/index.php (data taken on April 29, 2003).*

| Variable | Simulation (average ± st. dev.) | Real Data |
|---|---|---|
| Total number of LOC after 1,000 days | 390±93 KLOC | 350 KLOC |
| Total number of LOC after 2,000 days | 951±185 KLOC | 950 KLOC |
| Total number of commits by Day 1,000 | 24,600 ± 311 | 24,500 |
| Total number of commits by Day 2,000 | 59,880 ± 344 | 60,000 |
| Number of individual committers by Day 1,000 | 459 ± 27 | 125 |
| Number of individual committers by Day 1,000 | 975 ± 40 | 250 |

results compared to actual data from http://bulunga.dat.escet.urjc.es/gnome-cvs/index.php.

Figure 4a shows the cumulative LOC difference (i.e., LOC added minus LOC deleted) for 2,000 days of gtk+ evolution. At Day 1,000, about 390±93 KLOC were written according to simulation results, compared to about 350 KLOC in reality. At Day 2,000, 951±185 KLOC were written according to simulation results, the value being 950 KLOC in reality. The rise in LOC difference is almost linear, showing that gtk+ has not reached a plateau in development yet, contrary to the Apache case study. This is basically due to the fact that we set the "functional completion" cut-

*Figure 4: LOC evolution in gtk+ module of GNOME project. (a) (Simulation results) Cumulative LOC difference vs. time. The bold line is the average of the 100 runs. The gray lines are one standard deviation above and below the average. The dashed vertical line shows approximately the day of the project until data was reported in the GNOME case study (Koch & Schneider, 2000). (b) (Simulation results) Cumulative LOC difference vs. time for three individual runs. Bold curve: least $\chi^2$ curve from actual results. Dashed curve: Maximum LOC run. Gray curve: Minimum LOC run. (c) (Real results from http://bulunga.datescet.urjc.es/gnome-cvs/index.php, downloaded on April 29, 2003) Cumulative LOC difference vs. time © Grupo de Sistemas y Comunicaciones (Universidad Rey Juan Carlos), Spain. In order to compare with simulation results in (a), time axis of (a) must be shifted to the left by about 350 days (see discussion in text).*

(a)

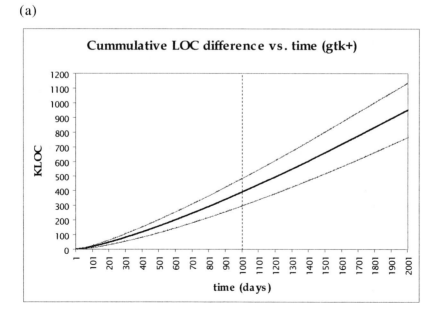

*Figure 4: LOC evolution in gtk+ module of GNOME project (continued).*

(b)

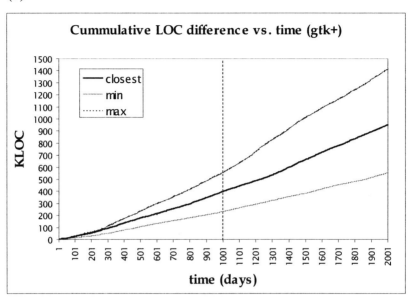

(c)

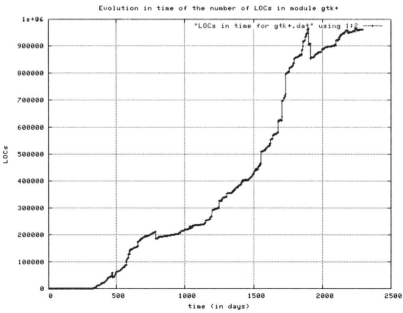

*Figure 5: Number of committers vs. time: (a) Simulation results for the gtk+. Bold line is average after 100 runs. Thin line is result of a single (least $\chi^2$) Run. (b) Equivalent plot for actual results from the GnuParted project taken from Capiluppi et al. (2003). The relevant curve for comparison with our simulation results is the curve for "contributors."*

(a)

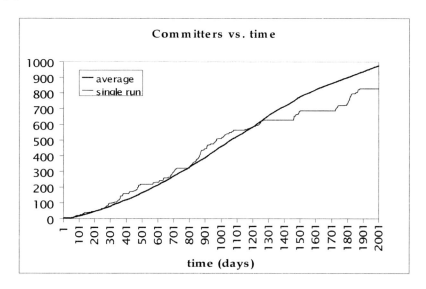

(b)

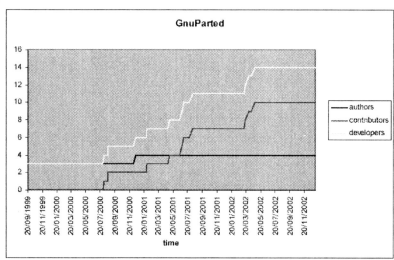

off at a high value. The dashed lines show one standard deviation above and below the average curve.

Figure 4b shows the cumulative LOC difference for three individual simulation runs: a) the run that had the least $\chi^2$ deviation from the actual LOC difference curve reported in http://bulunga.dat.escet.urjc.es/gnome-cvs/index.php ("closest"), b) the run that resulted in the fewest LOC after 2,000 days ("min"), and c) the run that resulted in the most LOC after 2,000 days ("max"). The large standard deviations in LOC difference are expected due to the large standard deviations of the LOC submitted in a single commit by a single programmer. Finally, Figure 4c shows the actual LOC difference curve for the gtk+ module as reported in http://bulunga.dat.escet.urjc.es/gnome-cvs/index.php. (The actual LOC difference curve, as reported on the mentioned website, is shifted 350 days to the right; Day zero, for simulation, corresponds to Day 350 for the actual plot.)

Figure 5a shows simulation results for cumulative number of committers vs. time. The bold curve is the average of all runs and the plain curve the result of a single run ("closest" run). The almost linear increase compares well with respective real results in http://bulunga.dat.escet.urjc.es/gnome-cvs/index.php. As a comparison to the step-like shape of the second curve, we show the respective plot for the GnuParted project taken from Capiluppi et al. (2003).

Figure 6 shows the cumulative time variation in number of commits (for task type F only). It rises linearly to about 24,600 commits (actual value for gtk+ is ~25,000) at day 1,000 and up to 60,000 (actual value for gtk+ is) commits at day 2,000.

*Figure 6: Commits for task type F (functional improving) vs. time. Simulation results for gtk+ project. Average of 100 runs.*

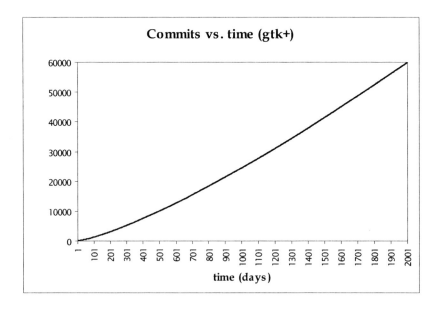

The almost linear rise in number of commits is also seen in the real data plotted in http://bulunga.dat.escet.urjc.es/gnome-cvs/index.php. However, the actual number of *committers* is almost four times higher in simulation results than in real data (see Table 3). Compared to the almost exact agreement in the number of commits, this fact suggests that, in simulation, individual committers appear to be about four times less productive in code writing than actual gtk+ committers. This discrepancy is partly due to a "time resolution" problem and partly due to the fact that the gtk+ real data on committers refer only to individuals that applied *changes* to the source code. In the present model, testing tasks (task type T) produces no change in source code, but still occupies a great number of individuals that must be drawn from the general "pool" of programmers. The "time resolution" problem refers to the fact that the simulation model must have a discrete time scale. In the Apache simulation, our basic unit of time was one calendar day, whereas in the gtk+ simulation it was 1/10 of a day (due to the shorter delivery times). When a delivery time shorter than the basic unit is selected from the time probability distribution, it must inevitably be rounded off to a higher time scale resulting in "loss" of effort and thus leading to a reduced average productivity of individual programmers.

## Sensitivity of Results to Model Parameters

Simulation results are quite sensitive to the values of parameters $N_o$ and $Q_o$. The reason is that they both determine, by large, the flow of new contributors into the project (see Equation Set 1). Since the rate of growth is exponential in the initial project phases—as correctly predicted by the present model—a slight change in $N_o$, *especially in the early stages of project evolution*, will greatly affect the future evolution of the total number of LOC. However, by letting $N_o$ be equal to the average number of commits per day in an initial period of the project, one can use $Q_o$ to fit the actual cumulated LOC difference and the actual number of commits as closely as possible in the same time period. The accuracy of the model may be concluded if, by keeping these values fixed, it succeeds in reproducing the future evolution of the project.

The most critical parameter that determines the time scale after which the development in a particular module will reach a plateau (i.e., the module is nearing functional completion) is the value given to the "LOC threshold" parameter for each module $i$ in the project. The value of this parameter must be estimated according to the known specifications of the module and represents a rough estimate of the *maximum* size (in LOC) of the module so that it may be considered "functionally complete." For example, for a module like gtk+ it is reasonable to assume that this threshold must be set high, since possible functionality is limited (theoretically) only by programmer imagination. For projects like Apache, on the other hand, where maximum expected functionality is naturally much more limited, this threshold may be set much lower. Of course, it is not at all the case that the evolution of a project with a high functional completion threshold will necessarily evolve up to this threshold without the risk of reaching a plateau much earlier. According to the present model, evolution depends on other dynamically changing factors, such as residual defect density of the code written, frequency of production releases, the initial flow of contributors, and how long the interest of each contributor is maintained during the lifetime of the project.

# CONCLUSIONS AND POSSIBILITIES FOR FUTURE DEVELOPMENT

We introduced a general framework for the production of OSS dynamic simulation models. We then presented a first effort to produce a specific simulation model for the OSS development process and attempted to produce some indicative simulation results, applying the model to the Apache case study and to the gtk+ module of the GNOME study. The present model contains enough generality to allow its application to multi-module projects.

Unfortunately, existing case studies do not contain the complete set of data necessary for a full-scale calibration and validation of the present model. Despite this fact, qualitatively, the simulation results demonstrated the super-linear project growth at the initial stages, the saturation of project growth at later stages where a project reached a level of functional completion (Apache), the "step-like" increase in number of contributors, and the effective defect correction, facts that agree with known studies.

Due to the lack of adequate literature data, simulation results presented here cannot be considered at all as full-scale validation of the model, a task that would require future F/OSS case studies to be conducted in parallel with the application of the simulation model for all the necessary data to be available. For example, future case studies should report commits per each task type and per each project module as a function of time. This way, the full-scale dynamics of the model will be exploited revealing, for example, possible complicated interactions among different modules and tasks.

One of the most evident intrinsic limitations of the present F/OSS simulation models comes from the very large variances of probability distributions. On output, this leads to large variances in the evolution of key project variables, a fact that naturally limits the predictive power of the model. A remedy for this is to attempt a less coarse description of these probability distributions, for example, by treating different categories of contributors separately or applying correlations between time distributions in Equation Set 3 and project output distributions in Equation Set 2. Another solution would be to apply a continuous re-adjustment of model parameters at shorter past time windows, trying to predict the evolution in the next (short) future time window. Of course, the above solutions would require a more detailed recording of real-world data, which we hope to stimulate by our present work.

Despite the aforementioned intrinsic and extrinsic limitations, these first attempt simulation runs fairly demonstrated the model's ability to capture reported qualitative and quantitative features of F/OSS evolution, indicating that the present work is a first promising effort that could stimulate further research in designing alternative OSS simulation models within the general framework described in the following chapters. Future case studies on F/OSS real-world projects conducted with the purpose of collecting all the necessary data needed for accurately calibrating and validating F/OSS simulation models is first priority research that will reveal the full potential of such models in practical applications.

## ACKNOWLEDGMENTS

Figure 4c is copyrighted by Grupo de Sistemas y Comunicaciones (Universidad Rey Juan Carlos), Spain.

The authors wish to thank A. Capiluppi, P. Lago and M. Morisio for permission to reproduce Figures 3b and 5b in this chapter

## REFERENCES

Abdel-Hamid, T. & Madnick, A. (1991). *Software Project Dynamics: an Integrated Approach*. Upper Saddle River, NJ: Prentice Hall.

Antoniades, I. P., Stamelos, I., Angelis, L., & Bleris, G.L. (2003). A novel simulation model for the development process of open source software projects. *Software Process: Improvement and Practise, 7*(3-4), 173-188.

Boehm, B. (1981). *Software Engineering Economics*. Englewood Cliffs, NJ: Prentice Hall.

Bollinger T., Nelson, R.O., Self, K.M. & Turnbull, S. J. (1999). Open-source methods: Peering through the clutter. *IEEE Software 16*(4), 8-11.

Capiluppi, A., Lago, P., & Morisio, M. (2003). Models for the evolution of OS projects. In Proceedings of the 9th International Software Metrics Symposium, Sydney, Australia, September 3-5, p.65.

Center for Software Engineering (1997). *COCOMO II Model Definition Manual*, Computer Science Dept., USC Center for Software Eng., Los Angeles, CA.

Crowston, K. & Scozzi, B. (2002). Open source software projects as virtual organizations: Competency rallying for software development. *IEE Proceedings – Software Engineering*, 149(1), 3.

Dalle, J. M. & David, P. M. (2003). The allocation of software development resources in "open source" production mode, SIEPR Discussion Paper No. 02-27, Stanford Institute For Economic Policy Research, Stanford, CA.

Doncelli, I. (2001). Hybrid simulation modeling of the software process. *Journal of Systems and Software*, 59(3), 227-235.

Feller, J. & Fitzgerald, B. (1999). A framework analysis of the open source software development paradigm. *ICIS*, p.58, Retrieved November 25, 2000 from: http://citeseer.nj.nec.com/feller00framework.html.

Fenton, N.E. (1997). *Software metrics: A rigorous and practical approach.* London: International Thomson Computer Press.

Godfrey, M. W. & Tu, Q. (2000). Evolution in open source software: A case study. In *Proceedings of the International Conference on Software Maintenance (ICSM '00)*, San Jose, CA, October 11-14, p. 131.

Jorgensen, N. (2001). Incremental software development in the FreeBSD open source project. *Information Systems Journal*, 11, 321.

Koch, S. & Schneider, G. (2000). Results from software engineering research into open source Development projects using public data. Diskussionspapiere zum Taetigkeitsfeld Informationsverarbeitung und Informationswirtschaft Nr. 22, Wirtschaftsuniversitaet, Wien.

Koch, S. & Schneider, G. (2002). Effort, cooperation and coordination in an open source software project: GNOME. Information Systems Journal, 12(1), 27-42.

McConnell, S. (1999). Open source methodology: Ready for prime time? *IEEE Software*, 16(4), 6-8.

Mockus, A., Fielding, R., & Herbsleb, J. (2000). A case study of open source software development: The Apache server. In *Proceedings of the International Conference on Software Engineering*, 263-272, Limerick, IR.

O'Reilly, T. (1999). Lessons from open source software development. *Communications of the ACM*, 42(4); 33-37.

Pfleeger, S. L. (2001). *Software Engineering, Theory and Practice.* Upper Saddle River, NJ: Prentice Hall.

Raymond, E.S. (1998). The cathedral and the bazaar. Retrieved November 25, 2000 from: http://www.tuxedo.org/esr/writings/cathedral-bazaar.

Rus, I., Collofello, J.,& Lakey, P. (1998). Software process simulation for reliability strategy assessment. In *International Workshop on Software Process Simulation Modeling (ProSim'98)*, Silver Falls, OR.

Special Issue on Software Simulation (2001). *Journal of Systems and Software* 59(3), December 2001.

Wilson, G. (1999). Is the open source community setting a bad example? *IEEE Software*, 16(1), 23-25.

# ENDNOTE

[1] By "alternative $k$" to a project module $i$, we mean a separate project module with the same specifications as the original, which is developed as a separate strand of the original within the project. For example, the gtk+ module in GNOME project is an alternative to gtk; the "nedit" text editor in UNIX (X-windows) is an alternative to the "edit" text editor.

## Chapter IX

# Modeling the Free/Open Source Software Community:
## A Quantitative Investigation

Gregory Madey, University of Notre Dame, USA

Vincent Freeh, North Carolina University, USA

Renee Tynan, University of Notre Dame, USA

## ABSTRACT

*In this chapter we summarize the latest results from an ongoing study examining Free/Open Source Software (F/OSS) Development communities as self-organizing systems. Using publicly available data about projects, developers, and their relationships at F/OSS hosting sites such as SourceForge, we have found the existence of several power-law relationships, which is consistent with the contention that F/OSS communities are self-organizing systems. The F/OSS community is modeled as a collection of ad hoc, social networks consisting of heterogeneous agents, self-organizing into projects and clusters of projects. A computer simulation of the F/OSS community model is developed using SWARM, an agent-based simulation toolkit. Empirical data is used to parameterize the simulation, which in turn is used to investigate a social psychological model of communication and team effectiveness in F/OSS projects.*

Copyright © 2005, Idea Group Inc. Copying or distributing in print or electronic forms without written permission of Idea Group Inc. is prohibited.

# INTRODUCTION

Our investigation aims to increase the understanding of the Free/Open Source Software (F/OSS) movement by providing a quantitative investigation of the network properties of the community. In some ways, the F/OSS movement is a prototypical example of a self-organizing system (Axelrod & Cohen, 1999; Barabasi, 2002; Barabasi, Albert, & Jeong, 2000; Faloutsos, Faloutsos, & Faloutsos, 1999; Holland, 1998; Huberman & Adamic, 1999; Kuwabara, 2000; Madey, Freeh, & Tynan, 2002a, 2002b, 2002c), but it also possesses some unique properties that may affect the development of the network.

The lack of central planning or control in F/OSS projects challenges conventional economic assumptions, turns conventional software engineering and project management principles inside out, threatens traditional proprietary software business strategies, and presents new legal and governmental policy questions regarding software licensing and intellectual property. Understanding F/OSS is far from an academic enterprise—F/OSS is a major component of the IT infrastructure enabling global e-commerce. Free/open source software includes BIND, sendmail, Apache, Linux, INN, GNU utilities, MySQL, PostgreSQL, and Perl, all critical elements of the Internet.

In this chapter we describe a social network investigation of almost 60,000 F/OSS projects at SourceForge (2003), a web-based project support site sponsored by VA Software. With permission, we collected data on developers and projects over time at SourceForge. We analyzed the data using cluster analysis to learn more about the structure of the developer-project network, and then used the data to create a model of the network for agent-based simulations. We ran simulations of the network using the model to validate the model and to discover emergent properties of the network that can only be observed by studying the network growth over time.

We find that both project size and the number of projects on which developers are working can be modeled with the power-law relationship, providing empirical evidence for the claim that the F/OSS community is a self-organizing system. We also find that the cluster size of connected developers fits the power law, if the largest and most connected cluster, comprising almost 35% of the developers, is removed, and we discuss the possible causes behind this dual structural nature of the network. Finally, extending Barabasi's construct of a network fitness component (Barabasi, 2002), we find that a dynamic lifecycle fitness parameter for projects is necessary to best model the project data at SourceForge.

We begin with a discussion of social network theory and the utility of using simulation modeling to understand self-organizing systems. We then describe our data collection, cluster analysis results, model development and simulation results, followed by a discussion of the theoretical and practical ramifications of our results and directions for future research.

## MODELING SOCIAL NETWORKS

Why should we invest the effort to do a quantitative simulation of the F/OSS network? The rationale behind such an investigation is that the F/OSS community is a social network that possesses several prototypical features of complex systems, systems that prior investigations have shown possess temporal and emergent properties that can be discovered only through modeling the system as a whole over time. For example, Axelrod (1984) found that certain types of "guarded cooperation" emerged as the most effective strategies for maximizing long-term joint outcome of dyads in a community, a result that could not have been obtained without simulation modeling.

Social network theory seeks to understand the network properties of people in relation to one another. Social network theory models persons as vertices or nodes of a graph and their relationships as edges or links of the graph (Barabasi, 2002; Jin, Girvan, & Newman, 2001; Wasserman & Faust, 1994; Watts, 1999; Watts & Strogatz, 1998). Thus two persons are directly connected if they have a relationship (e.g., friendship) with each other; they then are one edge away from one another, a distance of one. More distant relationships are modeled as paths through the graph; a "friend of a friend" is two edges away. For example, displayed in Figure 1 is a diagram of a social network composed of four individuals, labeled A through D and represented as vertices of the graph. In this social network, individuals A through C are fully connected by edges, representing a "circle of friends" or a clique. Vertices C and D are connected by one edge, and since this is the only path between C and D, that edge is called a weak tie. The number of edges attached to a vertex is called its index value. Vertex C, called a hub (or later in this chapter, a linchpin), is critical in social networks because its removal would break the network into two disconnected components. Hubs are also important because they often share the edges that weakly tie cliques together, and those weak ties have been shown to be most important in the spread of information through a network (Barabasi, 2002; Granovetter, 1973; Watts, 1999). Several studies reveal an interesting phenomenon present in many of these social networks; most persons are very few links from any other person—the Small World Phenomenon (Watts, 1999; Watts & Strogatz, 1998). This idea was popularized in the play (and movie) *Six Degrees of Separation* (Guare, 1990) that claims that all persons in the world are at most six friendship links away.

Collaboration networks are variations of social networks where the relationships are collaborations, e.g., actors in movies (Tjaden, 1996; Watts, 1999) or co-authors on research papers (Barabasi et al., 2001; Newman, 2001). Often entire populations are connected into one large cluster with high characteristic clustering coefficients (Watts, 1999). Highly prolific actors or authors are linchpins in collaborative networks. Linchpin actors or researchers play key roles in bridging disparate groups into one large cluster.

Social networks and collaboration networks have another interesting property in common; they are often self-organizing systems, forming patterns of connectiv-

*Figure 1: Components of a social network.*

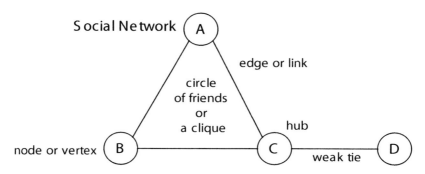

ity that emerge in a bottom-up process based on local interactions. Such systems displaying self-organizing behaviors and emergence based primarily on local interactions are the subject of study called complex systems (Cowan, Pines, & Melzer, 1999; Johnson, 2001; Kelly, 1994).

Social networks, collaborative networks, and other self-organizing systems (e.g., the Internet, corporation sizes, cities, economic systems, word usage in languages, ecosystems, lines-of-code in programs) often have another interesting property; they have highly skewed distributions, which under a log-log transformation results in a linear relationship. This is called a power-law relationship. Power-law relationships have been reported for the Internet (Albert, Jeong, & Barabasi, 1999; Barabasi & Albert, 1999; Barabasi et al., 2000; Faloutsos et al., 1999; Huberman & Adamic, 1999), sizes of U.S. firms (Axtell, 2001), city size distributions (Pumain & Moriconi-Ebrard, 1997), ecosystems (Jorgensen, Mejer, & Nielsen, 1998), word rank in languages and writing (Schroeder, 1991) and other systems for which emergent properties are of interest. One major hypothesis about the source of power-law relationships is nonrandom attachment of nodes (called preferential attachment by Barabasi, 2002; Barabasi et al., 2001; Newman, 2001). Preferential attachment refers to the fact that some vertices tend to attract new edges over time in an evolving network with greater probability than other vertices. Unlike random networks, where all vertices have equal probability of attracting new edges as the network grows, in collaboration networks the probabilities tend to be in part proportional to the index of the vertex, resulting in a "rich-get-rich" phenomenon. They also may have different intrinsic "fitness" that can outweigh the attractiveness for new edges caused by index value of the graphs' vertices (Barabasi, 2002). This property helps model the fact that "young upstarts" can attract new edges with greater probability that older vertices that typically have greater index values because they have been part of the network for a longer time.

We analyze the free/open source software phenomenon by modeling it as a collaborative social network. The developers are vertices of a network, and joint

*Figure 2: The Free/Open Source Software (F/OSS) community modeled as a collaborative social network.*

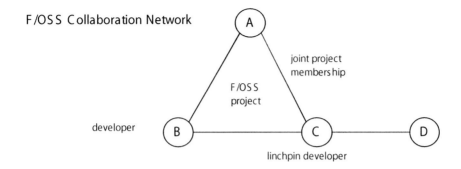

membership on an open source project is a collaborative link between the developers, as shown in Figure 2.

The F/OSS development community is highly decentralized and is a volunteer effort where developers freely join projects that they find appealing—all attributes of typical self-organizing systems. We hypothesize that the open source community displays power-law relationships in its structure and grows with a preferential attachment process modified by a fitness property. Our empirical analysis of structural data collected from SourceForge suggests that this is the case. These results, along with additional unique properties of the F/OSS community discovered using agent-based simulation and data-mining techniques are described in the next section.

## DATA COLLECTION

Because free/open source software is typically developed by global virtual teams, most projects use the Web and the Internet to facilitate their work. This provides research opportunities for the acquisition of data directly from online data sources, as has been used by several prior studies, including Mockus, Fielding, and Herbsleb (2002), Ghosh and Prakask (2000), and Krishnamurthy (2002). We collected data on F/OSS developers and projects from January 2001 to March 2003 at SourceForge.net, a large project management website supporting F/OSS development with project management tools, bug tracking software, mail list services, discussion forums, and version control software (SourceForge, 2003). We believe that SourceForge is representative of a wide cross-section of the F/OSS community. While SourceForge does not support many large high-profile projects that maintain their own developer sites (e.g., Apache, Perl, sendmail, and Linux), some large projects have moved to SourceForge, including Samba, and there are many smaller projects that have joined SourceForge.

Each project in SourceForge has a unique project number, and each developer is assigned a unique ID when he or she registers with SourceForge. We collected data

*Table 1: Typical data (anonymized) retrieved from SourceForge associating projects and developers.*

```
9001 | dev378
9001 | dev8975
9001 | dev9972
9002 | dev27650
9005 | dev31351
9006 | dev12509
9007 | dev19395
9007 | dev4622
9007 | dev35611
9008 | dev7698
```

linking developers to the projects on which they are working, over time. The data thus consisted of a table of records with two fields, project number and developer ID. Because projects can have many developers and developers can be on many projects, neither field is a unique primary key, and thus, the composite key composed of both attributes serves as the primary key. After the data collection, the data was completely anonymized by an algorithm that shuffles the ID values.

To collect the data, we created a web crawler to traverse the SourceForge website monthly, with the permission of SourceForge. All project home pages in SourceForge have a similar top-level design, with most of them dynamically generated from a database. A simple shell script fetches each project's developer page. It parses the HTML and extracts the names of the developers. A python program parses the HTML source, giving as output one line for each developer, containing the project number and the developer's ID. For March 2003, the table generated with this method exceeded 110,000 records in the form shown in Table 1. Thus, in the example data displayed in Table 1, there are six projects ranging in size from one developer to three developers. All the data was stored in a relational database by month. The combined data for all months of collected data exceeds two million records.

## RESULTS AND ANALYSIS

We submitted the data to three types of analyses. First, we examined the cluster properties of the network using developers as nodes, with an edge existing between nodes if both developers are on the same project. This representation is analogous to using research paper authors as nodes and joint authorship as a link in a collaboration network (Barabasi et al., 2001; Newman, 2001). As can be seen in Figure 3, a randomly selected small cluster from one monthly data set from

*Figure 3: Social network consisting of one connected cluster, 24 developers, five projects, and two Linchpin developers (dev[58] and dev[46]).*

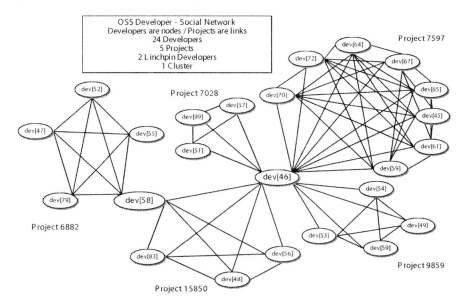

*Figure 4: Distribution of developer index values on linear axes (left) and log axes (right).*

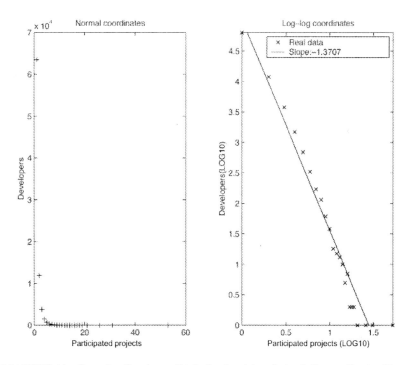

SourceForge, linchpin developers (Developer 58 and Developer 46), or developers who link two projects together, play an important role in linking clusters together (Granovetter, 1973). Our analysis showed that the developer collaboration network at SourceForge fits the power-law model, as determined by ordinary least squares (OLS) regression in log-log coordinates. As shown in Figure 4, the distribution of the number of projects per developer has a power-law distribution, with the solid line the OLS regression line through the data (adjusted R-squared greater than .95). The data displayed in Figure 4 was collected from SourceForge in March 2003. Equivalent results are seen in all monthly data for distributions of the number of projects developers join.

The second analysis used projects as nodes and developers as edges, with two projects linked together if they shared a common developer. Project size frequency showed equivalent results to the number of projects developers join, fitting the power-law and with R-squared greater than .95 displayed in Figure 5.

We also conducted an analysis on the size of clusters, with a cluster defined as a connected component with a path between all developers. Cluster analysis of

*Figure 5: Power-law distribution on size of projects determined by number of participating developers.*

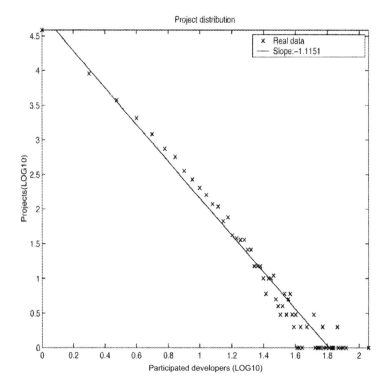

the SourceForge data from March 2003 identified the presence of one large cluster consisting of 28,394 developers, about 34.6% of the developers at SourceForge. The next largest cluster was of size 75, with sizes ranging down to one (i.e., those developers on single-member projects). Projects with only one developer occurred most frequently (in addition, developers on only one project was the most frequent value for the number of projects joined by a developer, with a large intersection between the two groups). Identification of clusters, or connected groups of developers, was implemented using a version of Prim's spanning tree algorithm (Corfmen, Leiserson, Rivest, & Stein, 2001).

*Figure 6: Cluster analysis on SourceForge developer data. One large cluster (far right) and remaining clusters with a power distribution of sizes.*

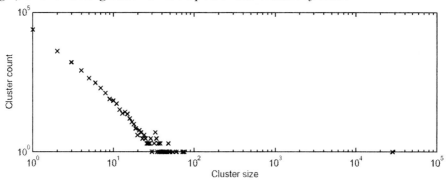

*Figure 7: Relative cluster sizes. Major cluster comprises about 34.6% of all developers.*

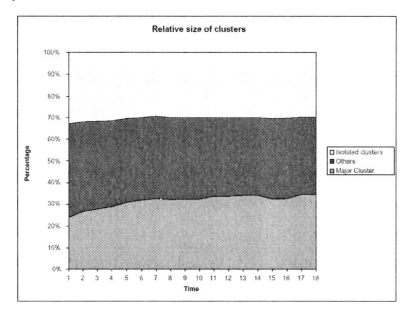

The large super-cluster was an outlier relative to the rest of the data points. If this cluster is removed, the cluster components fit the power-law, with a log-log transformation giving a good fit using regression (adjusted R-squared = .93) shown in Figure 6.

In Figure 7, we display the relative sizes over the months of gathered data, measured in numbers of developers, of the major cluster (the largest cluster), the isolated clusters (consisting of a single project with a single developer), and all other clusters. Note that the major cluster has grown over time but is slowing relative to the rest of the social network.

## DISCUSSION OF THE ANALYSIS

The analysis provides support for the contention that the F/OSS community is a self-organizing system, and it also yielded an unexpected finding regarding the structure of the community. Several different types of analyses on the F/OSS data obey the power-law, which gives support to the hypothesis advanced by many qualitative researchers that the F/OSS community operates as a self-organizing system. When one examines the size of connected projects and developers, however, a second phenomenon emerges. It appears that there may be a dual nature to the structure of the F/OSS community, at least at this point in time. While the less well-connected clusters fit the power-law, suggesting that part of the network is operating as a self-organizing system, there is a substantial percentage of the network (34.6% in March 2003) that is behaving differently, and that cluster does not fit the power-law pattern of the rest of the network data.

To our knowledge, this type of phenomenon has not been reported for other self-organizing systems, which may be due to the fact that our data set is large enough for us to be able to do a cluster analysis on cluster size. There are many possible explanations for this dual nature to the network structure that we are observing in the network. The F/OSS community does not operate in a vacuum, and in fact it may operate in some respects as a "shadow" network, with its structure influenced by the structure of the outside networks to which some of the developers and projects belong. For example, the commercial software industry can be expected to exert an influence on the structure of the F/OSS network by influencing the training and background of the developers, as well as influencing the reward and incentive structure for working on a particular type of project. Many F/OSS projects have developers that are full-time employees of commercial software firms assigned to the projects; e.g., IBM is reported to have as many as 350 employees on various F/OSS projects including Apache, Linux, Jikes, and Samba, and has invested over $1 billion USD in Linux-related development (Hochmuth, 2002).

Several follow-on research questions suggest themselves to improve understanding of the meaning and significance of this dual structure. Do projects and developers in the large cluster differ in observable ways from projects and developers that fit the power-law? At what point in the development of the network does the super-cluster emerge, and what is the percentage of the whole will to which it will converge? What relationship does this network structure have to threshold effects that have been observed in real-world network phenomenon such as the spread of computer viruses or vulnerability to denial-of-service network attacks? If there really is a dual nature to the F/OSS network, we need to recognize and understand it as such, as the network and behavioral dynamics observed in the super-cluster may differ significantly from that part of the network that is organizing itself as a power-law distribution. While we think this phenomenon is likely to be due to idiosyncratic properties of the F/OSS community and the fact that it is a shadow network and subject to strong external pressures, another possibility that must be examined is whether data in other self-organizing systems shows this dual nature when it becomes large enough or at a certain point in its development.

## AGENT-BASED MODELING OF F/OSS

Agent-based modeling is a technique for understanding the temporal dynamics and emergent properties of self-organizing system by simulating the behavior of individual components of the network over time (Axelrod, 1984; Epstein & Axtell, 1996; Resnick, 1994; Schelling, 1978). We use empirically collected data to generate models of the F/OSS using social network theory (see Figure 8). We then use that model to provide specifications for an agent-based simulation in which we grow an artificial SourceForge over time (Epstein & Axtell, 1996; Goldspink, 2000, 2002). We can then compare the outcome of the simulation to the data to assess and improve the model. In addition, using multiple iterations of the simulation given different starting conditions, we can discover invariant properties of the network/simulation that would not otherwise be observable.

Although the rules describing the local interactions of the components or agents of a self-organizing system (in our case, F/OSS projects and developers) may be few and simple, often unexpected and difficult to predict global properties emerge. Many investigators of such systems have found that they can only be understood through modeling and, specifically, through agent-based simulation (also called iconological, individual-based, and structural modeling) (Eve, Horsfall, & Lee, 1997; Harvey & Reed, 1997; Kiel & Elliot, 1997; Smith, 1997).

An understanding of a complex system can be obtained by discovering the rules and mechanisms that control agent interactions, by discovering invariant global properties, and conditions that lead to stability, periodic behavior, or chaotic behavior. Prediction may not be possible because of sensitive dependence to initial conditions, dynamic coupling, and feedback. Rather, the goal of these simulations is to develop an understanding of how and why the elements of the system are able to produce emergent behavior (Axelrod, 1997a, 1997b; Harvey & Reed, 1997; Holland, 1998). Possible emergent behaviors in these complex systems might include

*Figure 8: Coupled agent-based modeling and simulation help gain a better understanding of the F/OSS phenomenon.*

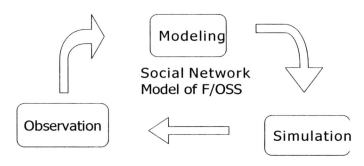

adaptation, learning, memory, cooperation, and the persistence of self-sustaining temporal patterns.

To conduct an agent-based simulation, researchers specify types of agents. In our simulations, we used only one type of agent for each simulation, F/OSS developers. Future research could, however, specify different types of developers and different types of projects (e.g., game projects, programming language projects, Internet projects, etc.). Next, the researchers specify behavior rules for the agents. These rules can be either deterministic (e.g., "If X happens, agents of type Y will do Z") or probabilistic (e.g., "If X happens, agents of type Y will have a 20% chance of doing Z"). The model can also contain information about the environment in which the agents are interacting.

In our simulation, we model at every time period (1 day) whether the developers in our artificial SourceForge community will 1) start a new project, 2) join another existing project, or 3) quit a project. In addition, at every time period we model whether a new developer will join the community, either by starting a new project or joining an existing project. We use the data we collected at SourceForge to provide parameters for our simulation (e.g., the growth of the number of developers and projects at SourceForge over time, the ratio of new project creation between current developers and new developers, the evolution of projects as measured by number of developers, downloads, lines of code, bug reports, and posting on discussion lists, rates at which developers leave projects, project termination rates, etc.).

The simulations are built using the Java programming language and the agent-based modeling library Swarm (Minar, Burkhart, Langton, & Askenzi, 1996; Swarm Development Group, 2000; Terna, 1998). The Swarm library enables discrete event, multi-agent simulations. In these simulations, agents are organized into "swarms." Several swarms can be nested into another swarm in a hierarchical manner. Thus,

OSS developers can be grouped into a "project swarm," which in turn can be grouped into a "cluster swarm," which would be part of the entire "OSS development swarm." Both the Swarm library and the Java language are object-oriented, providing the object-oriented programming benefits of attribute and behavior encapsulation, information hiding, and inheritance (Epstein & Axtell, 1996; Minar et al. 1996; Swarm Development Group, 2000; Terna, 1998). We are using modeling and simulation in concert with empirical data collection and analysis, as shown in Figure 8.

We use the empirically collected data to generate models of F/OSS using social network theory. Next, we use that model as a specification for a Java/Swarm agent-based simulation in which we grow an artificial SourceForge. Using multiple iterations of the simulation given different starting conditions (e.g., different random seeds), we discover invariant properties of the simulation.

These properties, along with other simulation output, are compared against the data we have on the "real-world' SourceForge. Agreement between the real and artificial suggests that the model may be correct, although it does not prove it. Disagreement between the real and artificial suggests that the model (or simulation) may be flawed or incomplete and provides clues on where to look in the empirical data for hints on how to correct or refine the model. This process of coupling model and simulation to real data has provided us with additional insights into the F/OSS community. We conducted our agent-based simulations of SourceForge in an iterative manner, starting with the simplest model, checking its fit against the data, and improving the model. Using this method we obtained the following results:

1) *Preferential attachment improves the model.* Our first simulation used random attachment of developers to projects, which fails to replicate the power-law observed in the real data. Adding preferential attachment does improve the fit of the model. By experimenting with different simulation implementations, we discovered that preferential attachment can be implemented independently for a) developers choosing a project yielding a power distribution on project sizes, and b) developers preferentially choosing to join an additional project, yielding a power distribution of developer index size. Figures 9 displays a comparison between 1) the actual values of SourceForge's average degree and diameter over time, and 2) the average degree and diameter of the network created in our simulation under the assumption of random joining (attachment) of developers to projects. (In our analysis, we identify networks that grow with random attachment as Erdos-Renyi graphs or ER for short.) Likewise, Figure 10 displays a comparison between 1) the actual values of SourceForge's average degree and diameter over time, and 2) the average degree and diameter of the network created in our simulation under the assumption of preferential joining (attachment) of developers to projects based on project size (or degree in our network representation). (In our analysis, we identify networks that grow with preferential attachment as Barabasi-Albert graphs or BA for short.)

By comparing Figures 9 and 10, we see a much better fit for these network parameters (average degree and average diameter) in the simulations that assume preferential attachment.

2) Simply adding preferential attachment did not provide a model that could properly model the "young-upstart" phenomenon, where a new project or new developer quickly passes older projects or developers. Barabasi (Barabasi, 2002) found that to model links to Internet sites a "fitness factor" was necessary. Fitness modifies the probability of a new edge, meaning that some websites are born more fit than others and are more likely to attract new links. Adding a fitness factor to our model did improve the fit of the model, although there were discrepancies.

3) In our search to understand these discrepancies, we made the discovery that the best fitness factor for the SourceForge data is not static but rather dynamic as a function of project age and life-cycle. In other words, these data are better modeled with a fitness factor that changes over time. For example, a new project may start with an attractive fitness factor, but its fitness levels over time and then decreases.

*Figure 9: Comparison between 1) the empirical values of SourceForge's average degree and diameter over time, and 2) the average degree and diameter of the network created in our simulation under the assumption of random attachment (ER) of developers to projects.*

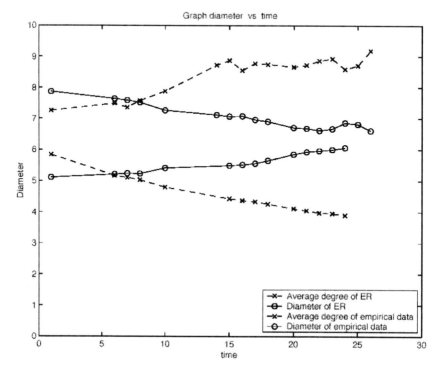

*Figure 10: Comparison between 1) the empirical values of SourceForge's average degree and diameter over time, and 2) the average degree and diameter of the network created in our simulation under the assumption of preferential attachment (BA) of developers to projects.*

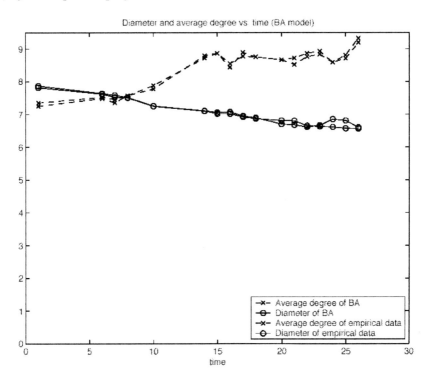

## CONCLUSION

We investigated the potential for learning more about the F/OSS community by 1) using SourceForge to learn more about developer and project evolution over time, 2) modeling the F/OSS community using social network theory, 3) discovering that the frequency of developer index, project sizes, and cluster sizes (excluding the one large component) all have a power-law distribution, 4) simulating the evolution of "artificial" F/OSS communities using agent-based modeling, and 5) iterating a coupled observation-modeling-simulation cycle to discover properties of the F/OSS community. Using the projects at SourceForge, the F/OSS community appears to be a highly fragmented social network, with the largest connected component comprising approximately 35% of all developers. Also, supporting this observation is the observed power-law distribution in the SourceForge community that has as the most frequent value, projects with only one developer and developers on only one project with a relatively large intersection of the two. Future work includes 1) studying Savannah and other F/OSS sites (under way), 2) computing and analyzing additional social network metrics such as clustering coefficients, network diameter, and average index values, and 3) continuing to refine the model and simulation using the process displayed in Figure 8.

## ACKNOWLEDGMENTS

This research was partially supported by the U.S. National Science Foundation, CISE/IIS-Digital Society & Technology, under Grant No. 0222829. We acknowledge the assistance of Patrick McGovern, Director of SourceForge.net and the contributions of University of Notre Dame students Yongqin Gao, Chris Hoffman, Carlos Siu, and Nadir Kiyanclar for their assistance with data collection, data analysis, and programming on this project. Finally, we wish to thank the two anonymous reviewers of this chapter for their most helpful suggestions.

## REFERENCES

Albert, R., Jeong, H., Barabasi, A. L. (1999). Diameter of the World Wide Web. *Nature, 401*, 130-131.

Axelrod, R. (1984). *The evolution of cooperation*. New York: Basic Books.

Axelrod, R. (1997a). Advancing the art of simulation in the social sciences. *Complexity, 3*(2), 16-22.

Axelrod, R. (1997b). *The complexity of cooperation: Agent-based models of competition and collaboration*. Princeton, NJ: Princeton University Press.

Axelrod, R. & Cohen, M. (1999). *Harnessing complexity: Organizational implications of a scientific frontier*. New York: The Free Press.

Axtell, R. L. (2001). Zipf Distribution of U.S. firm sizes. *Science, 293*(5536), 1818-1820.

Barabasi, A. L. (2002). *Linked: The new science of networks*. Boston, MA: Perseus Books.

Barabasi, A. L. & Albert, R. (1999). Emergence of scaling in random networks. *Science, 286*, 509-512.

Barabasi, A. L., Albert, R., & Jeong, H. (2000). Scale-free characteristics of random networks: The topology of the World Wide Web. *Physica A*, 69-77.

Barabasi, A. L., Jeong, H., Neda, Z., Ravasz, E., Schubert, A., & Viscek, T. (2001). *Evolution of the social network of scientific collaborations*. Retrieved April 10, 2001, from: xxx.lanl.gov/arXiv:cond-mat/0104162v1, from xxx.lanl.gov/arXiv:cond-mat/0104162v1.

Corfmen, T., Leiserson, C., Rivest, R., & Stein, C. (2001). *Introduction to algorithms* (2nd ed.). Cambridge, MA: MIT Press.

Cowan, G., Pines, D., & Melzer, D. (Eds.). (1999). *Complexity*. Boulder, CO: Westview Press.

Epstein, J. M., & Axtell, R. (1996). *Growing artificial societies: Social science from the bottom up*. Cambridge, MA: MIT Press.

Eve, R., Horsfall, S., & Lee, M. (Eds.). (1997). *Chaos, complexity and sociology.* Thousand Oaks, CA: Sage Publications.

Faloutsos, M., Faloutsos, P., & Faloutsos, C. (1999). *On power-law relationships of the Internet topology.* Paper presented at the *SIGCOMM'99*, August 31-September 3, Cambridge, MA.

Ghosh, R., & Prakask, V. V. (2000). Orbiten free software survey. *First Monday, 5*(7).

Goldspink, C. (2000). Modelling social systems as complex: Towards a social simulation meta-model. *Journal of Artificial Societies and Social Simulation, 3*(2).

Goldspink, C. (2002). Methodological implications of complex systems approaches to sociality: Simulation as a foundation for knowledge. *Journal of Artificial Societies and Social Simulation, 5*(1), 1-19.

Granovetter, M. (1973). The strength of weak ties. *American Journal of Sociology, 78,* 1360-1380.

Guare, J. (1990). *Six degrees of separation.* New York: Vintage Books.

Harvey, D., & Reed, M. (1997). Social science as the study of complex systems. In L. D. Kiel & E. Elliot (Eds.), *Chaos theory in the social sciences: Foundations and applications* (pp. 295-323). Ann Arbor, MI: University of Michigan Press.

Hochmuth, P. (2002). IBM's open source advocate. *Network World,* , December 12. Retrieved on January 27, 2004 from: http://www.nwfusion.com/power/2002/frye.html.

Holland, J. (1998). *Emergence: From chaos to order.* Reading, MA: Addison Wesley.

Huberman, B. A. & Adamic, L. A. (1999). Growth dynamics of the World Wide Web. *Nature, 401,* 131.

Jin, E. M., Girvan, M., & Newman, M. E. J. (2001). *The structure of growing social networks.* Unpublished manuscript, Santa Fe.

Johnson, S. (2001). *Emergence.* New York: Scribner.

Jorgensen, S. E., Mejer, H., & Nielsen, S. N. (1998). Ecosystem as self-organizing critical systems. *Ecological Modeling,* 261-268.

Kelly, K. (1994). *Out of control.* Reading, MA: Addison-Wesley.

Kiel, L. D. & Elliot, E. (1997). *Chaos theory in the social sciences: Foundations and applications.* Ann Arbor, MI: University of Michigan Press.

Krishnamurthy, S. (2002). An empirical examination of 100 mature open source projects. *First Monday, 7*(6).

Kuwabara, K. (2000). Linux: A bazaar at the edge of chaos. *First Monday, 5*(3), 1-68.

Madey, G., Freeh, V., & Tynan, R. (2002a). Agent-based modeling of open source using Swarm. Paper presented at the *Americas Conference on Information Systems (AMCIS2002),* August 8-11, Dallas, TX.

Madey, G., Freeh, V., & Tynan, R. (2002b). The open source software development phenomenon: An analysis based on social network theory. Paper presented at the *Americas Conference on Information Systems (AMCIS2002),* August 8-11, Dallas, TX.

Madey, G., Freeh, V., & Tynan, R. (2002c). Understanding OSS as a self-organizing process. Paper presented at *The 2nd Workshop on Open Source Software Engineering at the 24th International Conference on Software Engineering (ICSE2002),* May 19-25, Orlando, FL.

Minar, N., Burkhart, R., Langton, C.,Askenzi, M. (1996). The Swarm simulation system: A toolkit for building multi-agent simulations. Retrieved on January 27, 2004 from: http://citeseer.nj.nec.com/minar96swarm.html.

Mockus, A., Fielding, R.T., & Herbsleb, J. (2002). Two case studies of open source software development: Apache and Mozilla. *ACM Transactions on Software Engineering and Methodology, 11*(3), 309-346.

Newman, M. E. J. (2001). *Clustering and preferential attachment in growing networks.* Unpublished manuscript, Santa Fe.

Pumain, D. & Moriconi-Ebrard, F. (1997). City size distributions and metropolization. *GeoJournal, 43*(4), 307-314.

Resnick, M. (1994). *Turtles, termites, and traffic Jams.* Cambridge, MA: MIT Press.

Schelling, T. (1978). *Micromotives and macrobehavior.* New York: W. W. Norton.

Schroeder, M. R. (1991). *Fractals, chaos, power laws.* New York: W. H. Freeman and Company.

Smith, T. (1997). Nonlinear dynamics and the micro-macro bridge. In R. Eve, S. Horsfall, & M. Lee (Eds.), *Chaos, Complexity and Sociology,* pp. 52-78. Thousand Oaks, CA: Sage Publications.

SourceForge. (2003). http://sourceforge.net/. Retrieved March 2003.

Swarm Development Group. (2000). Brief overview of Swarm. Retrieved on January 27, 2004 from: http://www.swarm.org/swarmdocs/set/book149.html.

Terna, P. (1998). Simulation tools for social scientists: Building agent-based models with Swarm. *Journal of Artificial Societies and Social Simulation, 1*(2), 1-12.

Tjaden, B. (1996). The Kevin Bacon Game. Retrieved July, 2001, from: http://www.cs.virginia.edu/oracle/.

Wasserman, S.& Faust, K. (1994). *Social network analysis: Methods and applications.* Cambridge, UK: Cambridge University Press.

Watts, D. (1999). *Small worlds.* Princeton, NJ: Princeton University Press.

Watts, D. & Strogatz, S. H. (1998). Collective dynamics of small-world networks. *Nature, 393,* 440-442.

# SECTION V:

# F/OSS Development Interacting with Commercial and Public Organizations

# Chapter X

# Benefits and Pitfalls of Open Source in Commercial Contexts

Jiayin Hang, Siemens Business Services GmbH & Co. OHG, Germany

Heidi Hohensohn, Siemens Business Services GmbH & Co. OHG, Germany

Klaus Mayr, IFS IT GmbH, Germany

Thomas Wieland, University of Applied Sciences Coburg, Germany

## ABSTRACT

*This chapter intends to show how companies can benefit from open source software and its development culture and how the open source communities could, in turn, be stimulated and accelerated. One of the first major steps for businesses that plan to act in this context is to accept that open source projects have their own communication culture. After explaining this fact, we illustrate its relevance on the basis of a case study in which an open source framework was used to build a commercial product. The decision-making process and the lessons learned from it point out some guidelines, particularly for companies that offer projects rather than products. As there are, however, more parties involved than just the developers when OSS is discussed as a business opportunity, we also classify the different players in the software business such as distributors, system integrators, and software/hardware vendors. Findings on roles and their motivations and restraints, partially based on a survey carried out within our research project, point up this categorization. The authors hope that*

Copyright © 2005, Idea Group Inc. Copying or distributing in print or electronic forms without written permission of Idea Group Inc. is prohibited.

*this overview of the benefits and pitfalls will encourage more companies to make use of and invest in the open source way to develop and deploy software.*

## INTRODUCTION

Open source has become an established model for software development. It is no longer hidden among obscure Internet mailing lists populated by purely idealistic and mostly academic programmers. It is today well known in the IT world, albeit not always fully understood. Books like this one do not need to start explaining what the basic ideas of open source software (OSS) development are, but can focus on trying to grasp the appeal of this model and its main antipodes: open source vs. proprietary source, free-of-charge vs. commercial. However, OSS is not just a menace for the business world, as some commercial software companies sometimes propagate it. OSS also offers great opportunities for enterprises. In this chapter, we want to show how companies can benefit from OSS and its development culture and how the open source community could, in turn, be stimulated and accelerated by adopting some best practices from classical commercial development.

In this chapter, we start with a short discussion about the uncertainties of OSS releasing companies and emphasize the importance of understanding the open source culture for businesses that plan to act in this context. We then describe one case study performed by one of the authors in which an open source framework was used to build a commercial product. The decisions in this project and the lessons learned from it are explained in detail. The second major part is an overview of the different players, their motivations, and restraints in the software business such as distributors, system integrators, and software/hardware vendors. Some results are based on a survey carried out within our research project.

Throughout this chapter, we will call software "open source software" if it complies with the Open Source Definition published by the Open Source Initiative (Open Source Initiative, 2003). This definition comprises ten clear and strict rules that a piece of software and its distribution license have to fulfil in order to be called "open source" justifiably. For classical OSS projects like Linux, Apache, or GNU, these requirements are a matter of course. But for companies that have become interested by the cheering press reports and just want to "jump on the OSS train," they can represent considerable barriers.

### Uncertainties of OSS Releasing Companies

One example of the ten requirements of the OSI definition is that any discrimination against a specific field of endeavour is forbidden. The consequence is that a company that releases this software cannot prevent other companies from using it for business and profit. Since it is also required that an OSS license must not place restrictions on other software with which is shipped, the originator must even tolerate his or her open source software being bundled and sold with commercial software.

Licenses like the GPL do not oblige the creator to publish derived works, but if the enlarged or new version is published, the sources have to be published as well (Free Software Foundation, 1999). There are a number of OSS licenses, however, , that allow the incorporation of OSS codes in commercial software systems, e.g., the BSD license. The basic question of software producers is: Can I still make profit with my software when putting it out as open source? The answer is as simple as the question: You can make profit, but you will make in a different way, especially with value-added services and higher level products built on the open source software base. One example is Digital Creations Inc., who turned its proprietary content management system *Zope* into open source. It has gained much more popularity and market share in the meantime and succeeded in increasing the sales volume of its service offering related to Zope (Tippmann, 2001).

Other prevailing uncertainties are related to legal aspects, especially with respect to licenses. The free character of the code and the commercial (but vital) interests of the company in terms of licenses look as if they are hardly compatible. But at the moment no one seems to know if the problem is inherent in the license and business structure or in the accustomed way of considering the source itself as an object of merchandise instead of only its use. Anyway, if considerable support from open source programmers is envisaged for a particular software development project, only a free and open license should be chosen.

So before releasing some software into the open source space, one should acquire a profound understanding of what open source is and what the "do's and don'ts" of the community are. Some smaller companies who do not want to build their entire business on such a product but just realize that releasing the code of one particular piece of software may bring them some advantages, are usually very uncertain about the consequences of such a step. One main fear is that open source shows a "viral" character, infecting the company's other products, too. Some people are afraid that they might not be able to sell any software if they release one as OSS. This is completely wrong, as many examples from IBM to Sun Microsystems show. In any case, the original creator of the software definitely keeps the right to decide under which conditions his software may be distributed. This includes also the right to put one product under an open source license and another under a commercial license. The GNU General Public License (GPL), however, indeed has the intention to transfer its license to derived works. If software is directly linked (in the sense of object code linking) with GPL software, it is considered such a derived work, and if published, it has to be published under GPL and in source code (Free Software Foundation, 1999). But in practice, there is seldom the need for direct linking with GPL software. Most common OSS systems to which commercial software is linked are released under the Library GPL (LGPL). This license explicitly allows linking with closed source and requires open source disclosure only for direct modifications of the libraries themselves.

## Understanding Open Source Culture

A more practical conflict is that the attitude, motivations, and expectations of commercial developers and OSS developers are usually very different. OSS projects are built on contributions by random developers, frequent releases, feature selection according to willingness and qualifications of developers, decentralization, and peer review (Raymond, 1999). Typical business practices like schedules, time-line planning, task distribution by a project manager, determination of feature set, and date of a release by a product manager are widely unknown (and ignored or even seen with disdain) among OSS developers. Companies that want to avail themselves of the OSS community have to learn these rules of the game. So, apart from any licensing issues, releasing software as open source usually means first a fundamental change of the development culture of the company. Therefore, the first step for companies realizing software, using OSS, or establishing their business around OSS has to be a deeper understanding of this sometimes conflicting culture—and vice versa, if an OS company wants to conquer commercial areas on a wider scale. This culture is widely known as the "geek culture" (Pavlicek, 2000). Since OSS developers are working "just for fun," due to their technical interest or sometimes because they offer consultancy services around it, they do not feel any obligation to the originator of the software. They generally do not accept when someone is trying to impose his or her development and management processes on them. For the other side, this means that the originating company has more or less to acknowledge the OSS culture and to create appropriate interfaces to the community alike (cf. Seifert & Wieland, 2003).

But the step into the open source universe affects many more business practices than the development process alone. A good example is the handling of patents and standards. Open source should mean "open" in every sense, allowing only the use of open standards and formats (Holmes, 2000). Usually OSS needs to be free of any parts that are protected by a patent. It is, however, possible to file patents to any part of a piece of software or its algorithms, but they must be publicly and freely available so nobody outside the company would run the risk of patent infringement by just using and distributing the software. As we will see later, this restriction can also be an advantage in the sense that concepts that are widespread as OSS may soon become "de facto" standards.

OSS projects with a completely unstructured, chaotic organization are hardly likely to succeed—just the same way management shortcomings threaten commercial projects. OSS does not always mean success. In contrast, there are many more unsuccessful projects that never get above the alpha stage and never produce useful software than successful ones, as the statistics of the project mediator Source-Forge show (McGovern, 2003). Among the OSS projects there is a "survival of the fittest"—the best designed (and maybe organized) projects get the most input (contributors) and attention (users). Although a stringent project management alone is not sufficient for a successful end product, it seems to be necessary, as Mockus,

Fielding, and Herbsleb reported (2000, 2002) for Apache and Mozilla. But remember that this project management is much more reliant on people and personal communication than on processes and tools. Nevertheless, OSS projects can benefit from a couple of insights and best practices from commercial development. This mutual fertilization will be the red thread for the rest of this chapter.

## The NOW Project

Much of the work described in this chapter was done in the context of a publicly funded research project called "NOW." It explores how commercially oriented enterprises can benefit from the software engineering practices of the open source community. NOW is the German acronym for "Utilization of open source concept in business and industry." The project consortium consists of three industrial members (Siemens Corporate Technology, Siemens Business Services, and 4Soft) and one university partner (Technical University of Munich).

The primary goal of this project is to point out an effective model of how a joint endeavour between the open source community and a "traditional" software company works effectively. The project therefore investigates the problem area from various angles:

- *Business models*. One result of the project will be working business models for interested companies. Clearly, there are already working models, but they do not apply on a development project level. Aside from "common" open source models, for example apparent in the Linux industry, small- and medium-sized companies need new business models to utilize the potential power of open source software projects in their own business processes.
- *Software engineering methodologies*. Open source software development differs from prevalent in-house software engineering processes. In order to combine both NOW analyzes the differences and communities and develops a suitable model for this type of software design.
- *Community management*. An important question to be answered is how to find the right partners in the open source community for an open source project and how to manage the project. Therefore, a community platform based on portal technologies—of course, open source—is being created to serve as virtual meeting point between interested companies and open source developers. This portal will also contain material or links to material of interest for creating and managing open source projects. In addition, project results will be presented and discussed there continuously and tools supporting development and employment processes will be tested.

The project is funded by the German Federal Ministry of Education and Research (BMBF) and runs from 10/2002 to 03/2005.

# TOWARDS AN OPEN SOURCE PROJECT METHODOLOGY: A CASE STUDY

In this section we try to sketch how the contrary philosophies described previously can be reconciled. For that purpose, we briefly summarize the experiences that we made in a case study at 4Soft that was undertaken within the context of the research project NOW.

4Soft was founded in 1999, has since that time grown to fifteen employees, and specializes in developing trendsetting IT-strategies and architectures, iterative development processes, and model-based software engineering. The case study was meant to simulate a real-world scenario in which a software company evaluates ways to successfully use and enhance open source software. For this reason, three experienced Java developers were assigned to a development team (one of which was the project leader and is co-author of this chapter) and opposed to another 4Soft employee "playing" the role of some customer. Starting in mid October 2002 and lasting for about five man-months, this case study resulted in about 800 lines of code and a very good understanding of the "do's and don'ts" of communication with an open source community. Besides using open source software in the scope of the technical infrastructure of the company, this case study was the first contact of 4Soft to open source software in progress and the thriving community beyond. In the following, we present some more insights in the activities and results of the associated activities.

The case study began with a workshop for setting the goals of a successful open source strategy at 4Soft. After several discussions, it became clear that along with more or less general marketing and technically oriented goals, open source software development could play an essential role in supporting existing and acquiring new projects.

The challenging task here was twofold—we aimed at developing *something new*, and we wanted to show *how* such a development can be successfully organized within a commercial project environment. Despite the advantages that we associated with the usage of open source, the question of how to make profit with the software that we were going to develop constantly remained open. Furthermore, there was also a constant skepticism against the risks that came along with the complexity of the open source components that we used and that we were going to extend. And, last but not least, we recognized the dilemma of developing a *product* within a company that is focused on *projects*. On one hand, and guided by the expectations of the open source community, our software engineers planned to develop a very generic tool that could be used in lots of different contexts for a long period of time and that would finally be very attractive for the open source community. On the other hand, and much more dominating, the development was stimulated by the specific and temporary needs of the customer. He was faced with the problem of modelling and modifying highly complex processes that contained thousands of nested activities and transitions, but no commercial product matched his requirements for flexibly defining different views onto this process graph. Therefore, we planned to build our own solution on the basis of existing open source.

Just as expected, managing the project and reconciling the interests of software engineers, researchers, and management was very difficult. The technical challenges were very hard, too. No matter which open source component we chose, we had to cope with big pieces of unknown, complex, and partially undocumented software. Moreover, since our initial requirements were not precise enough, they needed to be refined and reworked according to the restrictions that the underlying software imposed (Scacchi, 2001). Last but not least, during our work, an increasingly active open source community proceeded with an ongoing development that our development team was uncertain to follow. In order to manage our project, we successfully established an iterative process as shown in Figure 1.

The overall process was separated into two phases, where in the first phase (analysis) we chose a mixture between a top-down approach (oriented on our customer's requirements) and a bottom-up approach (oriented on the underlying OSS packages). Within a very limited time, we had to find the best candidates that met our basic requirements and gave the best perspective for unknown future customer requirements.

In the second phase (implementation), we proceeded in an iterative manner. The outer loop here illustrates the release cycles, each of which started with two actions: selection of new feature requests and definition of a plan by which the tasks were assigned to the developers. In the inner loop, we were concerned about iteratively

*Figure 1: Iterative development process used in case study project.*

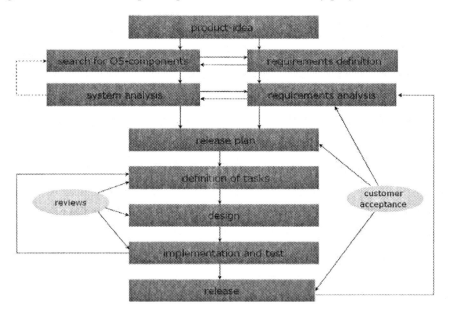

working on the tasks and subtasks that we created. After having described each task precisely, we concentrated on defining the appropriate test cases and detailed out the design for the code to be written, implemented, and tested. As in *Extreme Programming* (Beck, 1999), a "test first approach" was identified as useful for reworking and restructuring written code. However, instead of installing a pair programming procedure as Beck recommends for such refactoring tasks, we installed pairs for reviewing the corresponding design and implementation documents.

Before turning to the crucial point (evaluation), some of the stages of the above sketched process need to be explained in more detail.

## Elaboration of a Product Idea

After having settled on our general strategy that open source should help to support existing and acquire new projects, we came to the point to decide in which way this could be achieved. The product idea that we had (developing a process modelling tool) was already mentioned. The criteria that led to this decision should now be made clear.

Of course, the process modelling tool that we developed was *currently needed*. But we also expected that we could flexibly adjust its graph visualization features easily to a lot of other application areas. The tool that we planned to develop was required to be *useful also beyond our current needs*. Finally, we recognized that as long as we did *not* rely on a component that was published with the GPL license, we had a *free choice to preserve the secrets of the source code* that we linked to such an OSS component. Due to the LGPL license, no restrictions were imposed on a commercial usage.

It is also interesting to mention some of the other motivations to which lower priorities were assigned. The more technically oriented people in our company (also beyond the borders of our project team) came up with questions that targeted the *attractiveness* of the result. They encouraged us to develop something really innovative and were curious about the opportunities to find an OSS community that could help us with developing such a tool. They also took into account the status and features of underlying open source software that they already had in mind and that needed to be improved. Certainly, the quality and extensibility of such a piece of software become more and more important when implementation starts. However, in the early phase of elaborating a product idea such facts were of minor importance.

## Selecting an Open Source Component

After having settled on the product idea along with a set of initial requirements, we came to the point of finding the right component. Due to time restrictions, we were able to evaluate only a very limited number of systems. Furthermore, in each case the evaluation could not go very deep; so there remained a clear risk of having chosen the wrong software. In order to minimize these risks, we worked out a detailed checklist for evaluating the respective components. The essential criteria were:

- Functionality;
- Stability;
- Quality of architecture and design;
- License; and
- Community support.

The functionality of the underlying software was checked by comparing asserted features with required ones. In the next step, we evaluated the stability of the selected components. This was easily achieved by installing the latest version and giving it a try. Getting a good impression of the quality of design and architecture first seemed to be a very difficult and complex task. Surprisingly, however, by briefly reading the available documentation or performing more or less detailed code inspections, we were able to define a clear ordering of the candidates. In a final step, on the basis of the information that we had gained, we tried to estimate the amount of work that we expected for an implementation of the missing features. In order to achieve this and to avoid possibly redundant work, we did not hesitate to contact the respective open source communities. The help and competent advice that we got finally provided the key criteria that led us to select JGraph as our premier choice (JGraph, 2003).

JGraph is an open source component based on the Java Swing library and distributed under the LGPL. It visualizes an arbitrary business model, enables "drag & drop" on the view level and—like other Swing components—it provides access to the underlying objects.

## System Analysis and Requirements Analysis

The most difficult phase in our project began when we started in parallel to follow two different tasks: a more detailed list of requirements and system analysis. Clearly there were tight interdependencies between these tasks, and the essential problem that we had to solve was to find the right compromises. In order to verify that we were on the right path, we were forced to quickly build an early release. We presented a simple multiple-window application as our first prototype that was continuously developed further by iteratively refining the system according to upcoming requirements. It would have been useful to present the features of the JGraph component itself in advance (maybe in comparison with the demos of the other components). Alas, the impatience in awaiting the first release had increased in the meantime, while the developers could only tell that they were experimenting with the components that they were going to use.

## Release Planning and Development Issues

Unlike commercial projects that typically start with an analysis of the requirements and that proceed with a very detailed design phase, we were faced initially with a huge amount of code that with which we had to cope, In this situation, it would have been a great risk to independently build up our own detailed design while having hardly understood the underlying components. Hence, our first goal was to get a running prototype of the system that we were planning to implement. Instead of throwing away this prototype, however, we improved it step-by-step. Incrementally, we proceeded with an analysis of the old system and the development of the new one. For performing these tasks, we quickly recognized the advantages of feature request tracking systems that are often used by open source communities. Each release with a clearly and a priori defined set of features was considered a milestone, and the risk of exceeding our fixed time schedule was controlled continuously. Help from the outside experts was sometimes required, too. We were surprised by how quickly we got competent support. Typically, our support requests were answered within a few days, mostly within a few hours or even minutes. Of course, we could not rely on this, but we found it much more helpful to get an occasional help quickly than a guaranteed help within a much longer period of time

## Lessons Learned

Finally, we were successful both in developing a new tool on the basis of a previously unknown piece of OSS and in working together with an unfamiliar open source community. Nevertheless, in the end we realized that some of the virtues that we had regarded as best practices for commercial software development projects are not valuable in a mixed context, i.e., when OSS components and commercial components should be combined.

The experts of the open source community constantly offer substantial support. Without this support, it would have been impossible to fix a couple of intricate problems that we met during development. This support was offered to us for free and without any kind of contract. This means it is both commercially attractive and technically inspiring to stay in contact with such excellent software engineers, and our company will proceed on the successful path cooperating with people from appropriate open source communities. The unwritten laws for such cooperation, namely, giving back an elaborated piece of software for having used a basic one, are very well understood. However, the discussion about which pieces of software, licenses, and business model are best in this situation has not been finalized at the time of writing. But, as our case study clearly indicates, deriving and providing the right explanation models here is heavily needed. These guidelines should be helpful, especially for small companies like 4Soft or for companies that offer projects instead of products.

# OPEN SOURCE IN COMMERCIAL CONTEXT

Experiences as described above illustrate the difficulties in the adaptation process between OSS culture and commercial understanding and demands in software development. Nevertheless, a lot of resources have been invested. As is the case in most of the OSS-projects, this project was driven by technical aspects; however, to ensure real benefits, the business perspective has to be taken into account early enough—even beyond the scope of potential users and companies having OSS portfolios. All different involved parties have to be analyzed to develop sufficient models of OSS handling to support the OSS trend and its tendencies to gain a more and more important role in commercial areas.

Therefore, a central task of NOW is to investigate roles and their interaction to identify the potential benefits and ways to gain them. This explicitly includes the benefits of the community itself. As a first step, roles and motivations of all parties have been analyzed. As the demand of users—especially from industry and government—has caused the current OSS boom, their restraints and concern will be considered in particular. In support of this, an explorative survey has been carried out.[1] In the ongoing project, these first steps will be the basis for developing business models and suggestions for procedures.

A substantial discussion would fill a whole book. Accordingly only a brief summary of these aspects can be given.

## Definition of the Different Roles in the Open Source Process

To make use of the advantages that software development in an open source style provides and to make use of the results already available, business models for OSS have to be developed that constitute a "win-win" situation for all parties involved in the open source process. For example, one goal is to find out how companies can benefit from the open source approach. New business models have to be developed that allow companies to take OSS into their portfolio. Simultaneously, the reward of the developing community has to be secured.

By analyzing the open source process, five different roles can be identified that are involved in the open source development. For a better understanding of the open source phenomenon, these five roles should be differentiated based on their functions and interrelations. These five parties are: Developers, Distributors, System Integrators/Consulting Companies, Software and Hardware Companies, and Users.

In reality, some people and companies are covering several roles such as developing software as a part of the community, using open source products, and maybe at the same time offering services in the field of integration or consulting. But this does not affect the necessity of understanding each single role.[2]

- Developers: Private developer, developer in a company, academic developer.
- Distributors: Bundling and packaging of software, offering releases in a product style way.
- System Integrators: Providing services, carrying out advisory services and customization of software for customers, integration of software on behalf of the customers.

- Software and Hardware Companies: Selling software and hardware.
- Users: Private person, software developer, company or institution.

During the initial stage of OSS, mainly two roles existed—developers of OSS and users of OSS. Most of the time even these roles were played by the same people. With the growth and maturity of the segment, some intermediary roles arose: distributors, system integrators, and software and hardware companies.

Sometimes users can take OSS directly from the community—depending on their own abilities. But as the market grows, more and more customers are also non-IT specialists; therefore, in most of the cases, intermediaries are needed.

The motivation for each role has to be analyzed in order to identify intentions, demands, restrictions, and objections. By starting at this stage, win-win interactions, business models, etc., can be defined.

In the research project NOW, an online survey was carried out as an additional source of information to the already available studies and papers.[3]

## Motivations of Different Roles for Committing to the Open Source Phenomenon

What motives do the different parties have for their commitment to open source software? All five roles have different motivations.

### *Motivation of OSS Developers*

Since open source software is free of charge, the motives of OSS developers are not profit or substantial income; instead, there is a wider spectrum of human motivations (Hertel, Niedner, & Herrmann, 2003). The incentives often found in literature are fun, success, technical curiosity, intellectual challenge (Hars & Ou, 2000), prestige, acknowledgement, and appreciation (Hars & Ou, 2001), self-realization, learning, personal conviction, and idealism (Raymond, 1999). Some people also work on OSS projects to improve their personal qualification and offer consulting services later (Seifert & Wieland, 2003).

In the NOW survey, the two reasons "fun" and "personal conviction and idealism" were the most selected items by the respondents, followed by "technical curiosity," "success," and "intellectual challenge"; the motive "prestige and appreciation" scored the fewest number of points.

Nevertheless, one has to bear in mind that most of the developers from academia and industry are paid employees (Hars, 2001).

### *Motivation of Distributors*

The motivation of distributors is common commercial interest. Their interest lies in the establishment of open source software. For commercial success, they need input from the community and a growing market segment of users. They depend on this segment—more than software companies with a mixed portfolio of classical and open source-based software. Therefore, marketing activities and advertising are important to increase the publicity level of open source software. There are distributors for Linux operating systems, other OSS applications, and documentations/books about OSS. In addition, they offer different services around OSS that include full services along the software value chain (Wichmann, 2002a).

*Motivation of System Integrators*

The motivation of system integrators for engaging in open source software is their desire to differentiate themselves from competitors, who only offer proprietary software in their solution packages, and to meet the growing demand of customer requests for open source-based solutions. They want to fill a void in the market that has been neglected so far. As the market matures, the number of non-IT specialist customers who need such an intermediary grows. Another reason system integrators dealing with open source software is the assumption that with OSS a greater profit margin can be realized than with proprietary software. With open source software, system integrators can gain more independence from big commercial software companies regarding product and price policy. Another advantage of open source software is free availability of its source code, which means that system integrators can achieve a better customization of the software according to the individual customer needs.

*Motivation of Software and Hardware Vendors*

There are a lot of possible motives of software and hardware vendors for their commitment to open source software. In the following, some objectives are stated:

- Software and hardware vendors use the open source concept as a business strategy to compete with their business rivals and market leaders (Hecker, 1999).
- With the open source business model, more profit can be realized than with the commercial business model because of independency.
- An unprofitable proprietary software product could be turned profitable again by converting it into an open source software product (Behlendorf, 1999).
- To gain market share or to weaken competitors/market leaders (Behlendorf, 1999).
- Hope of enhancement of software with the help of OSS developers. To achieve this, they try to gain prestige/reputation in the OSS community.
- As a complementary product or incentive to promote sale of commercial software (Hecker, 1999).
- As a complementary product or incentive to promote the sale of commercial hardware (Raymond, 1999).
- To employ less own development resources.

- Own unprofitable software could be maintained free of charge from OSS community.
- Because open source software becomes a trend-appearance in the IT industry. Open source software gains importance in economy and industry.

As part of the NOW survey, the software and hardware vendors were asked for their motivations/objectives for engaging in open source software. The most frequently selected motives in the responses to the survey were the software and hardware companies' use of OSS as a business strategy to compete with business rivals and market leaders, and their desire to minimize their own development resources. The second most frequently selected objectives in the survey were the hope that their software product would be improved through the OSS developers, and the use of OSS as a complementary product or incentive to promote the sale of commercial hardware. Further motives stated by the respondents were that through converting a commercial software into open source software an unprofitable software could be turned profitable again, and to gain market shares or to weaken competitors or market leader. The objectives that were not taken into consideration at all by the respondents were the possibility to realize more profit with the open source business model than with the commercial business model, and the more indirect endeavour to increase their prestige/reputation in the OSS community.

## *Motivation for Users of OSS*

Open source software has a lot of advantages in comparison to proprietary software. The most important reasons stated by users for the application of OSS instead of commercial software are: higher cost savings potential; the Total Cost of Ownership (TCO)[4] of open source software is lower than that of proprietary software[5] (BMWi, 2001); no license fees; better price-to-performance ratio; higher stability (Wichmann, 2002b)'; higher performance[6] (University of Dortmund, 2002); better functionality; open and/or modifiable source code; better quality (University of Dortmund, 2003); better security (Raymond, 1999); higher maturity of software (Raymond, 1999); openness/flexibility of interfaces; more independence from the product and price policy of commercial software companies (Wichmann, 2002b); less dependence on a single provider; and better maintenance and support.[7]

In a survey carried out by META Group (2003) in 2002, 188 companies were asked for the most important reasons for their application of Linux. The most important reason stated by the respondents was the cost savings of open source software due to the loss of license fees and due to the fact that the calculated costs of ownership of open source software are much lower than those of proprietary software. This argument was stated by 81%. In second place, with 59% of the respondents was the argument that OSS has higher stability, reliability, and availability. Flexibility and ability of integration were stated by 54%. Other stated reasons were ability of administration (37%), security (33%), compatibility (22%), open/modifiable source code (15%), and performance (11%).

In the survey of the NOW Project, the result was similar. The main reason for their decision in favor of OSS was the cost savings that can be realized. Due to the loss of license fees, a better price-to-performance ratio can be achieved with OSS. Another advantage that was often stated by the respondents was open and/or modifiable source code of OSS. With an open source code, the software is modifiable and easier to assimilate to individual customer needs. Other very important aspects: openness and flexibility of the interfaces of OSS (thereby, OSS is better compatible with other software and hardware), and the desire to have more independence from the product and price policy of big commercial software companies. The above mentioned criteria nearly all received the same rating of 50%. The following criteria achieved the second highest rating, ranging from 20-30%: higher stability of OSS, higher performance, and better functionality. Additionally, OSS has better quality and it is more secure in contrast to proprietary software, e.g., because with OSS, bugs and errors in the source code can be discovered and removed more quickly by IT administrators and consequently the down time of the operating system can be reduced.

Those motivations are drawing an impressive picture of positive objectives. But OSS has encountered a point where this triumphal trend entering the industrial world can be stopped by "simple" aspects such as unsolved legal issues, liability, and warranty, which have a major impact on commercial interest—sometimes more than technical advantages.

*Prevailing Restraints and Concerns of OSS Users*

The driving force bringing open source to wider areas of industrial user scenarios is the end-user himself—companies considering this option as an alternative to traditional software. The different intermediaries and suppliers will follow such a request—if it is profitable for them as well. The developers in companies are steered by those decisions. But if those considerations end up in a more and more commercial deployment of open source, academia and private developers will ask for a participation—which does not need to be monetary.

But will there be a wide industrial demand of open source—more than Linux and some other well-known communication components?

Despite all the advantages of OSS, many users and companies still have a reserved and retentive attitude towards it. Management, in particular, still has a lot of restraints and objections towards OSS. In the following, the major problems and risks mentioned by companies concerning open source software are enumerated. They are divided into three groups. The first group of problems affects both end-users and intermediaries. The second group mainly refers to the end-users, and the third group applies primarily to the intermediaries (distributors, system integrators, and software provider companies).

Main restraints affecting both end-users and intermediaries:

- Legal aspects, warranty, and liability questions, patent law, and copyright (who takes the responsibility concerning warranty and liability for damages caused by open source software) (BMWi, 2001).
- License model (more developer- than user-oriented license models).

The additional disadvantages for end-users with OSS are: insufficient documentation (Raymond, 1999), no guarantee of ongoing support and regular update (Lieder, 2003), bad usability and user-friendliness (DiBona, Ockman & Stone, 1999) such as the lack of comfort for installation and configuration, insufficient compatibility with other software and hardware, absence of daily maintenance and support security (Lieder, 2003), high initial costs for migration from proprietary software to OSS, insufficient qualification of employees for OSS (Wichmann, 2002b), lack of availability of drivers, and problems with hardware support.

Additional problems and concerns of intermediaries are:

- Business models that are promising and profitable still have to be found.
- They are afraid to give away their new and innovative technologies to their competitors and thus to reduce their competitive advantage by publishing the source code of the software—or even giving back the advancements to the community.
- Bad marketing support.

One of the main questions today is the warranty—who can be prosecuted? This is of major interest for the using companies as well as the intermediaries. A lack of specific laws and regulations brings this case to common regulations of liability and warranty. But this means a wide range of interpretation and dependency on national laws.

As long as this uncertainty prevails, the following questions will often be dominated by legal aspects: Which alternative is cheaper, more efficient, more productive, and by which option can a lower Total Costs of Ownership be achieved? How can open source software promote and accelerate the software development? Shall the company distribute its software products as open source software or as commercial software?

In the NOW survey, the legal aspects also turned out to be the first priority problem for users. Warranty and liability still seem to be an unsolved problem, and nobody seems to be responsible for damages caused by OSS. The second biggest concern is the lack of OSS applications in the desktop area. Until now, OSS is strongly represented in the server-area (like Linux), but in the desktop area, it is still underrepresented, and there is a great deficiency in office applications, customer relationship software, personal information manager, and management information systems. Other disadvantages are the lack of compatibility with other software and deficits in the qualification of personnel, consequently training is necessary. The aspects insufficient documentation, bad usability, and missing user-friendliness were also criticized.

Recapitulating this means legal aspects have to be solved. Most other aspects are a question of investment—such as development of desktop applications, revision of code and compiling releases, documentation, etc. Intermediaries will be found—new companies offering this or "bricks-and-mortar" companies discovering a new business segment—when the business opportunities are clear. That means a clear demand with a relevant market volume, clear product and service elements to provide, and a sustainable market structure—including the developing community.

Elaborated, revised, and tested business models covering the different roles and especially their interactions are still missing. The legal basis and interdependencies have to be clarified, and acceptable concepts have to be designed taking into account the requirements of all roles. The system integrators are just defining their position in this market segment. They are needed to fill the gap between distributors and/or open communities to make open source practically usable in common business.

## CONCLUSIONS

OSS has a lot to offer to software portfolios in commercial projects and can contribute to the improvement of software development strategies. The culture of OSS projects is usually characterized by leadership through competence, but also by acceptance of random contributors. Strict planning of time and budget as is common and necessary in commercial projects is widely unknown. Case studies show that this is one of the main challenges for integration of OSS components into commercial development.

But the OSS hype ensnares commercial vendors and users to underestimate the necessary steps to capitalize from this. The current resentments mostly do not go back to real threats but, rather, to doubts and uncertainty. A lot of questions have to be answered and a lot of solutions—mainly un-technical—have to be derived. As our study shows, legal aspects are regarded as the biggest problem for users. Warranty and liability still seem to be an unsolved problem, and nobody seems to be responsible for damages caused by OSS. Surprisingly, the second biggest concern is the lack of enterprise OSS applications in the desktop area. With increasing adoption of OSS in business and industry, both problems will certainly be targeted and managed. Interesting business models, also for system integrators, software and hardware companies, and distributors, are at least available.

## ACKNOWLEDGMENTS

Parts of this work were supported by the German Federal Ministry of Education and Research (BMBF) in project "NOW – Nutzung von Open Source in Wirtschaft und Industrie." The authors would like to thank BMBF for the funding, as well as the other partners, namely, Siemens AG and Technical University of Munich, for the constructive and cooperative work and fruitful discussions.

## REFERENCES

Beck, K. (1999). *Extreme programming explained*. Reading, MA: Addison-Wesley.

Behlendorf, B. (1999). Open source as a business strategy. In C. DiBona, S. Ockman & M. Stone (Eds.), *Open sources: Voices from the open source revolution*, Sebastopol, CA: O'Reilly.

BMWi (2001). *Open-source software: A guideline for small and middle-sized enterprises*. 1st edition. German Federal Ministry for Economy and Technology.

DiBona, C., Ockman, S., & Stone, M. (1999). *Open sources: Voices from the open source revolution*, Sebastopol, CA: O'Reilly.

Feller, J. & Fitzgerald, B. (2002). *Understanding open source software development*, London: Addison-Wesley.

Free Software Foundation (1999). Licenses of free software. Retrieved March 31, 2003, from: http://www.gnu.org/licenses/.

Hars, A., & Ou, S. (2000). Why is open source viable? A study of intrinsic motivation, personal needs and future returns. In *Proceedings of the 2000 Americas Conference on Information Systems (AMCIS 2000)*, August 10-13, Long Beach, CA, pp. 486-490.

Hars, A. & Ou, S. (2001). Working for free? - Motivations for participating in open source projects. In *Proceedings of the 34th Hawaii International Conference on System Sciences (HICSS-34)*, January 3-6.

Hecker, F. (1999). Setting up shop: The business of open source software. Retrieved March 19, 2003, from: www.hecker.org/writings/setting-up-shop.html.

Hertel, G., Niedner, S., & Herrmann, S. (2003). Motivation of software developers in open source projects: An Internet-based survey of contributors to the Linux kernel. *Research Policy, 32*, 1159-1177, special issue on OSS.

Holmes, W. N. (2000). The evitability of software patents. *Computer*, 33(3), 30-34.

House, R. (1999). Uniting the open-source and commercial software worlds. Retrieved March 21, 2003, from: http://www.sci.usq.edu.au/staff/house/ipl/ppunite.htm.

JGraph (2003). The JGraph Software Project. Retrieved March 30, 2003, from: http://jgraph.sourceforge.net/

Lieder, H. (2003). Die Sicherheit von Software wird nicht allein durch deren Quellcode bestimmt. W*irtschaftsinformatik* 45(4), 478-479.

McGovern, P. (2003). SourceForge Sitewide Update, June 20. Email correspondence to all SourceForge users.

META Group (2003). Linux – operating system landscape in change, results of a META Group study. Germany [press conference, January 30, 2003].

Mockus, A., Fielding. R., & Herbsleb, J. (2000). A case study of open source software development: The Apache server. In *Proceedings of the 22nd International Conference on Software Engineering (ICSE 2000)*. 263-272. Limerick, Ireland: ACM Press.

Mockus, A., Fielding, R., & Herbsleb, J. (2002). Two case studies of open source software development: Apache and Mozilla. Technical report, Avaya Labs. Retrieved July 25, 2003, from: http://www.research.avayalabs.com/techreport/ALR-2002-003-paper.pdf.

Open Source Initiative (2003). The open source definition (Version 1.9). Retrieved March 14, 2003, from: http://www.opensource.org/docs/definition/php.

Pavlicek, R. (2000). *Embracing insanity: Open source software development*. Indianapolis, IN: Sams Publishing.

Raymond, E.S. (1999). *The cathedral and the bazaar: Musings on Linux and open source by an accidental revolutionary.* Sebastopol, CA: O'Reilly

Sandred, J. (2001). *Managing open source projects.* New York: Wiley Computer Publishing

Scacchi, W. (2001). Understanding the requirements for developing open source software systems. *IEE Proceedings Software*, Paper number 29840.

Seifert, T. & Wieland, T. (2003). Prerequisites for enterprises to get involved in open source software development. In *Proceedings of 1st Workshop on Open Source Software in an Industrial Environment at Net.ObjectDays 2003,* September 22-25, Erfurt, Germany. Retrieved August 20, 2003, from: http://www.netobjectdays.org/pdf/03/papers/ws-ossie/457.pdf.

Sharma, S., Sugumaran, V., & Rajagopalan, B. (2002). A framework for creating hybrid-open source software communities. *Information Systems Journal*, 12, 7-25.

Sieckmann, J. (2001). Bravehack: Technische, wirtschaftliche und gesellschaftliche Aspekte von freier Software und Open Source; ihr Wesen, ihre Geschichte, ihre Organisationen und Projekte. Retrieved March 19, 2003, from: http://www.bravehack.de/html/nodel.html.

Tippmann, D. (2001). Open source und zope: Eine Einführung in Freie Software. Retrieved July 31, 2003, from: http://userpage.fu-berlin.de/~danitipp/daniel/opensource.html.

University of Dortmund (2002). Entrepreneurial evaluation of open source software. An online-survey carried out by the University of Dortmund in cooperation with MATERNA GmbH. Retrieved April 30, 2003, from: http://www.it-surveys.de/itsurvey/pages/studie_oss_executive_summary.html.

Wichmann, T. (2002a). *Free/Libre Open Source Software: Survey and Study, Basics of Open Source Software Markets and Business Models*. FLOSS Final Report – Part 3. Berlin: Berlecon Research.

Wichmann, T. (2002b). *Free/libre open source software: Survey and study, Use of open source software in firms and public institutions, Evidence from Germany, Sweden and UK. FLOSS Final Report – Part 1*. Berlin: Berlecon Research.

Wieland, T. (2001). Open source im Unternehmen. In A.von Raison & R. Schönfeldt (Eds.): *Linux im Unternehmen*. Heidelberg: dpunkt.verlag.

Wieland, T. (2000). Linux als Geschäftsfaktor. Linux Enterprise, 2. Retrieved March 24, 2003, from: http://www.drwieland.de/articles/Linux_Geschaeftsfaktor.html.

# ENDNOTES

[1] The survey had an explorative approach. The findings have to be classified as hints but lack statistical validity because of the low rate of return. Thus statistics have been left out here.

[2] Additional roles like, e.g., platform or portal providers for open source communities, etc., can be considered. But to understand the interrelations and options of the open source market, the analysis has focused on the five roles as described.

[3] Most of the surveys and studies are focused on end-users, some on developers. The intermediary roles are usually not discussed. Some studies like the Meta Group Study mentioned are not focusing on open source in general but on Linux.

[4] The Total Costs of Ownership (TCO) contains the costs for implementation, operation, and support of the software.

[5] This aspect is controversial, because there are also studies stating that OSS is more expensive than proprietary software under special constraints.

[6] This aspect is controversial, because there are also studies stating the opposite.

[7] A remarkable aspect, because it is listed under pros as well as cons of open source.

# Chapter XI

# Experiences Enhancing Open Source Security in the POSSE Project

Jonathan M. Smith, University of Pennsylvania, USA

Michael B. Greenwald, University of Pennsylvania, USA

Sotiris Ioannidis, University of Pennsylvania, USA

Angelos D. Keromytis, Columbia University, USA

Ben Laurie, AL Digital, Ltd., USA

Douglas Maughan, Defense Advanced Research Projects Agency, USA

Dale Rahn, University of Pennsylvania, USA

Jason Wright, University of Pennsylvania, USA

## ABSTRACT

*This chapter reports on our experiences with POSSE, a project studying "Portable Open Source Security Elements" as part of the larger DARPA effort on Composable High Assurance Trusted Systems. We describe the organization created to manage POSSE and the significant acceleration in producing widely used secure software that has resulted. POSSE's two main goals were, first, to increase security in open source systems and, second, to more broadly disseminate security knowledge, "best practices," and working code that reflects these practices. POSSE achieved these goals through careful study of systems ("audit") and starting from a well-positioned technology base (OpenBSD). We hope to illustrate the advantages of applying OpenBSD-style methodology to secure, open-source projects, and the pitfalls of melding multiple open-source efforts in a single project.*

# INTRODUCTION

**Posse** - *a group of people summoned by a sheriff to aid in law enforcement.*

A variety of reasons, ranging from marketplace ignorance to a perceived trade-off between usability and security, have driven modern operating systems into the undesirable role of a potential lever with which system security can be breached. The use of *any* common operating system platform across an organization can make this lever effective, independent of the organization, its security policy, and security practices.

This problem has been exacerbated by the commercial success of the Internet over the last decade, as the Internet's "end-to-end" (Clark, 1988; Saltzer, Reed, & Clark, 1984) design implicitly relies on host security as the basis of security for the overall system. An example of this reliance and its consequence is the advent of Distributed Denial of Service (DDoS) attacks, effected by multiple computers bombarding one or more target hosts with traffic and disabling these targets.

As the commercial marketplace, and to a large degree the government marketplace, have converged towards a common platform (the dominant commercial operating system, Microsoft Windows), these organizations increasingly rely on the platform to be trustworthy, whether it is so or not. Further, the use of the Internet and computer systems in the functions of all of these organizations has made systems software, as a whole, "critical infrastructure." At the same time, a single point of vulnerability and failure has been created for systems dependent on this software.

## The Open Source Alternative

Concurrent with the growth of the Internet, an alternative software development paradigm began emerging. This paradigm had roots in the research UNIX community and its USENET, with some philosophical roots later added with the "Free Software" principles of Stallman. The mid-1960s MULTICS (Daley & Dennis, 1968; Organick, 1972) project, part of the U.S. Defense Advanced Research Projects Agency (DARPA)-supported Project MAC (Fano & David, 1965) at MIT, gave rise to the original UNIX system (Ritchie & Thompson, 1974, 1978; Thompson, 1978) (the name UNIX is in fact a pun on MULTICS) as a reaction to MULTICS system complexity. Unfortunately, in rejecting much of MULTICS, the UNIX system was not able to avail itself of the extensive effort devoted to developing protection models and security kernels (Schroeder, 1975; Schroder, Clark, & Saltzer, 1977) for

MULTICS. McKusick, Bostic, Karels, and Quarterman (1996) provide historical details on the emergence of UNIX.

UNIX, as an important consequence of its university base, boasted platform portability of much of the software and easy availability. These, in turn, meant that UNIX became the dominant platform for experimental operating systems research, and the availability of several good books explaining the system internals (Bach, 1986; Lions, 1977a, 1977b; McKusick et al., 1996) meant that the system could be taught. The result, entering the 1990s, was a substantial number of people who understood the ins and outs of most of the operating system. Thus, as the PC became the dominant platform in the mid-1990s, UNIX became the dominant model for "open source" operating systems projects, where system source was fully available for examination and modification. The dominant commercial platform, Microsoft's Windows, is not UNIX based; it has accreted (Cusumano & Selby, 1997) features and technologies starting with a simple microcomputer software platform.

UNIX-based platforms have presumed "shared use" since their inception, were early platforms for network software deployment and refinement, have sizeable and talented user communities, and are available to all for scrutiny. There is a belief in this community (Raymond, 1999) that "many eyes" lead to faster discovery and repair of flaws in software. While "open source" *enables* scrutiny (Raymond, 1999), it does not *cause* it.

The following (quoted with permission) was posted to the "Robust Open Source" mailing list by Peter Gutmann:

*I can provide a data point on this based on a disk encryption device driver I wrote about 8-9 years ago. For various reasons too boring to go into here, I never released the source code (AFAIK it's the only thing I've ever written where I haven't published the source). At various times I'd get people sending me mail asking me why I hadn't released the code so it could be reviewed. When I offered to send it to them, they replied that they didn't want to review it themselves, they expected someone else to review it for them. That is, even the people who went so far as to express an interest in the source code admitted they'd never look at it (and furthermore that they'd be quite happy to have some complete stranger tell them it was OK based on the claim that they'd reviewed it)... As an experiment I also planted a comment which should raise eyebrows in some code I released years ago and which is fairly widely used just to see if I'd get any reaction from anyone... No one has ever asked me about this, from which I assume that no one's ever looked at the code they're using. That's kind of scary, because the comment isn't in there just to annoy people, you really could build a rather nasty backdoor in there. There may actually be products out there which are released in binary-only form where the vendor has built in a backdoor at that point, although I saw a posting from foo@anon.org in alt.2600 saying he'd looked at the product and it was fine, so it must be OK."*

That is, many eyes do not help if they are all looking at something else.

The most important contribution, therefore, is the fact that discoveries are shared and can, in some domains (such as networking code), influence commercial code whether these influences are visible or not.

## The Marketplace

Concurrent with the emergence of open source has been a drive by some portions of the U.S. Government (notably the U.S. Department of Defense) to develop and/or procure a "trusted" operating system. A major problem with modified commercial operating systems has been the difference in priorities between the marketplace and a knowledgeable, specialized consumer such as the U.S. Government. In particular, the security features and development processes and documentation required have resulted, when the vendors have been engaged, in multiple development efforts—one driven by commercial considerations and the other(s) driven by specific considerations such as security, an audit process, etc.

Separate development of the secure version inevitably results in a **TOAD** (**T**echnically **O**bsolete **A**t **D**elivery) version of the operating system, since the audit process, among other factors, inhibits introduction of new features while underway. The obvious and only cost-effective way to solve these problems is to ensure that no separation occurs, requiring that security considerations be "mainstreamed."

As open source systems are developed by volunteers and often driven by aesthetics (such as a desire for a "secure" system) rather than market considerations, a potential opportunity was identified by author Douglas Maughan of the U.S. Defense Advanced Research Projects Agency and embodied in a smallish (by DARPA standards) program called Composable High Assurance Trusted Systems (CHATS). The goal, at a high level, is to introduce required security features into open source operating systems such as Linux, FreeBSD, and OpenBSD such that they will be in whatever mainstream version exists and that they will be present in commercially supported versions of these operating systems, allowing their procurement by governments and other interested parties.

The initial goals of the DARPA Composable High Assurance Trusted Systems program included adding new security functionality to existing open source operating systems, as well as the political/community effect of demonstrating the value of useful security and analysis tools and techniques to the open source community. This approach by DARPA to work "directly" with the open source community was seen as a risky endeavor by both parties. The open source community was leery of DARPA's commitment to open source, and DARPA was unsure of this new role of research partner and the uncertainty of product delivery. However, DARPA felt that these open source technologies are critical for systems of the future to be protected from imminent attack. The CHATS program has focused on developing the tools and technology that enable core information infrastructure systems and network services to protect themselves from the introduction and execution of malicious code and other attack techniques and methods (Sullivan & Dubik, 1994). These tools and technologies are intended to provide the high assurance trusted operating systems need to achieve comprehensive, secure, highly distributed, mission critical information systems. The CHATS program intended to fundamentally change the existing approach to development and acquisition of high assurance trusted operating systems technology by dramatically improving the state of assurance in current open source operating systems and, further, developing an architectural framework for future trusted operating systems. Such technologies have broad applicability to many programs within DARPA and the DoD (MITRE, 2003).

A most important consequence of the CHATS approach is that technologies developed under the program are demonstrated and evaluated on a large number of open source system platforms, for all to see and use. The open source development model provides a conduit for technology transition directly into products and services that will employ and support trusted operating system technology.

# POSSE: TOWARD AN OPEN SOURCE SECURITY COMMUNITY

The Portable Open Source Security Elements (POSSE) Project at the University of Pennsylvania is an example of a DARPA Composable High Assurance Trusted Systems (CHATS) project. In this section, we will describe the goals of the POSSE project (such as supporting widespread availability of high quality cryptographic systems) and the project organization we have used to accomplish these goals. The project organization has generally worked, although several challenges have arisen over time. Nonetheless, as we detail here, the project has been successful both in its internal goals and in its goals of influencing both other open source projects and commercial vendors.

A major goal of POSSE is the development of a (growing) community of individuals interested in and capable of enhancing the security of operating systems. Open source systems serve three purposes towards achieving this goal:

1. They provide a natural diversity, avoiding the "single point of failure" noted above.
2. They provide a basis through which a community of developers can express their knowledge about secure systems.
3. The "open source" characteristic of the software allows the knowledge to be freely shared, even with those who might not themselves choose to share knowledge.

Our model is illustrated in Figure 1. What the model shows is that the POSSE project not only generates its own portable security technologies, but takes a stronger social engineering stance than the "chuck wagon" approach of putting the technologies out and shouting "come and get it." Rather, meetings of developers (at the

*Figure 1: The POSSE synchronize and synthesize process model.*

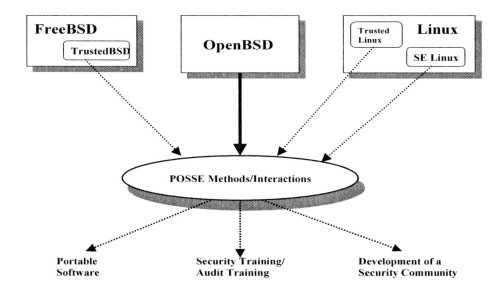

"waist" of the diagram) build up the strengths of the security community, cutting across project boundaries, and raise all boats on the same tide.

## POSSE Project Goals

An abstract view of the overarching project goal is to create and grow a community of open source developers with security as a major focus. Without getting into debates of software engineering "religion," our team studied open source operating projects and found that the OpenBSD project had many of the properties we desired. In particular, it had a very strong focus on security issues, had a small but extremely capable group of developers—several of whom were extremely interested in the technical contributions we wanted to make—and the project leader, Theo de Raadt, was interested in the basic proposition of community building.

Much of the project focus beyond the technological developments has, in fact, been on community building. Important sub-goals have been:

a.  Propagating technologies such as the OpenSSH secure shell, which is now distributed with, among other platforms, the Apple Macintosh OS-X, as well as maintaining the multiplatform portability of the OpenBSD system itself (see OpenBSD.org).

b. Exporting methodologies such as OpenBSD audit to multi-OS security infrastructures such as OpenSSL, and investigating the strength of tool-based versus expert audit in this task.
c. Collaborating with other open source and free software efforts on security projects of common interest, such as an Extended Attribute File System with TrustedBSD (part of FreeBSD), an open source secure bootstrap with the University of Maryland, and an IPSEC for Linux (Keromytis, Ioannidis, & Smith, 1997).
d. Large face-to-face developer meetings, typically before or after major conferences that attract developers such as USENIX. These meetings have proven surprisingly successful, resulting in, for example, a new packet-filtering firewall for OpenBSD, called "pf."
e. Collaborating with security hardware vendors to rapidly generate support software for their devices, such as cryptographic acceleration hardware.

While we will say more in the section in this chapter on POSSE outcomes, in the Spring of 2003 we feel that, on the whole, these goals have been and continue to be met.

## POSSE Project Organization

One of the first questions is how one would organize such a project. While the usual challenges of distributed organizations were all present (decision-making, personnel changes, control of resources, etc.) some particular challenges we faced were raised by the combination of goals and the fact that the CHATS program was funded by DARPA, an agency that is part of the United States Department of Defense.

a. Many of the OpenBSD volunteers were working on their own time but were employed by commercial enterprises.
b. The work we envisioned for POSSE demanded essentially full-time commitments for the OpenBSD and OpenSSL developers responsible for certain sub-projects.
c. Many of the OpenBSD and OpenSSL participants are non-U.S. nationals.
d. Open Source projects do not have a corporate or non-profit corporate structure with which contracts can be negotiated.

We have worked out a solution that has largely been successful. The University of Pennsylvania has contracted to DARPA to perform the items in a statement of work more or less covering the POSSE goals, with some more details as laid out here in the section on POSSE Project Management Challenges. Several U.S.-based, OpenBSD developers became Penn employees. Subcontracts were used for one other U.S.-based developer, and subcontracts were created for Columbia University, as well as subcontracts in Canada and the U.K.

Universities in general, and the University of Pennsylvania in particular, provide an ideal structure with which to carry out such arrangements, since it is a U.S. entity with a structure capable of contracting, has many modes and methods for employing and contracting, and has intellectual property policies for software that are extremely attractive for a funded open source project. DARPA's only request has been an acknowledgement that DARPA funding was used to create the software; the BSD license rights are completely preserved. It is interesting to note how frequently DARPA is acknowledged in the OpenBSD source tree—many of the acknowledgments are in the original Berkeley source, but more and more (53 in OpenBSD 3.3) are showing the POSSE agreement number!

Our project takes a broader view of what we must do than technology alone. We see that the important tech transition is first among the small number of individuals in each open source effort who are security-focused and second among the core teams of each effort. While these groups are one and the same in the OpenBSD effort, and it is unique in this respect, the important intellectual "customers" are developers who should have their "security" thinking caps stapled to their "developer" thinking caps, so that security is a first-class consideration in every open source effort. Our effort to document the OpenSSL auditing process, to involve many people in development activities, and our aggressive outreach to other projects, enabled by the DARPA resources, raised everyone's standards by several notches.

## POSSE Project Management Challenges and Solutions

We outline here four major challenges we faced and our approaches.

1. **Decentralized development.** The OpenBSD and OpenSSL development communities are worldwide and mainly volunteer. POSSE hired two developers (authors Rahn and Wright) at Penn as senior software engineers, residing in the Midwest and Middle Atlantic regions of the U.S., and structured a subcontract with AL Digital, Ltd., a UK firm through which Ben Laurie's services (Laurie is an OpenSSL developer and an author of this chapter) were made available. Such geographic distribution means there must be good communication channels (for example, Internet Relay Chat and Instant Messaging), people must be familiar with and trust each other (frequent communication for trivial matters can be annoying), and tasks must be neatly separated so people can work independently as much as possible.

2. **Integration with existing working methods.** There are already cultural mechanisms and protocols to build consensus among members of the developer community. These mechanisms and protocols can be leveraged by using developers who are already aware of the processes and culture (e.g., Keromytis, Rahn, and Wright), although a certainly degree of friction will always occur because of potentially conflicting goals—this is the overhead of developing a consensus.

3. **Minimize administrative overheads.** We used the structure and specialized skills effectively. In particular, the university has significant resources for purchasing, sub-contracting, and reporting. As academics must typically both perform and report on research, it was natural for the academics on the project (Smith, Greenwald, and Keromytis) to write quarterly reports, aggressively report on technical progress in the academic literature, and inject scientific rigor where appropriate. This had the benefit of focusing the developer's attention on development.
4. **"Light-touch" management.** From the start of POSSE, we worked very hard to identify capable and highly motivated people and gave them interesting problems to work on. Not surprisingly, they have implemented clever solutions with a great degree of autonomy.

Management of the project, as anticipated, has been challenging. As we noted in the original POSSE proposal,[1] there are the challenges of distributed management, strong personalities, and knitting together of sometimes quite distinct development cultures. For example, the OpenSSL system must work across many operating systems and its collaboration is much looser, less structured, etc., than the OpenBSD development team, which is tightly integrated and led by Theo de Raadt. In some ways, the development culture of OpenBSD resembles the "Surgical Team" model of hierarchy developed by Brooks (1975), while the OpenSSL development model is more analogous to the "Programming Group" model of Weinberg (1974). The OpenBSD methodology is driven by biannual releases that incorporate whatever software is ready for "prime time," while the OpenSSL releases are more event-driven than periodic release-driven. Thus, the OpenBSD model for what OpenSSL should look like and when it should look that way is clear, while achieving larger scale consensus for OpenSSL took more time, leading to some tension.

In particular, one focus of our work had been the support of hardware cryptographic acceleration, as discussed in the next section, and, further, its integration with SSL to accelerate use of cryptography. Our belief was that cryptographic operations should be perceived by users to be fast (as we have recounted elsewhere—see Miltchev, Ioannidis, and Keromytis, 2002, for example), as this would encourage their use. OpenSSL modifications were necessary to accommodate some of these changes and, based on discussions at an early developer meeting, these changes were undertaken by the OpenSSL community. However, the pace and development style of the two teams clashed, as the OpenSSL release and consensus model did not mesh smoothly with the aggressive release cycles of OpenBSD, and some tempers flared, with many telephone exchanges to and from the University of Pennsylvania people acting as intermediaries.

The seriousness of the culture clash should not be underestimated, and dealing with such potential clashes must be dealt with in any management plan intending to meld multiple open source projects. During the lifetime of the POSSE project, an unhealthy and somewhat permanent rift opened up between the OpenBSD and OpenSSL communities. Major management effort was required to prevent a "fork" of OpenSSL (one for OpenBSD and one for the rest of the world), and this effort was and continues to be successful. Within the POSSE project, this rift was smoothed by project-level successes. These included the many OpenSSL fixes, patches, and enhancements that have emerged from both the OpenSSL auditing efforts and the OpenBSD cryptographic framework as well as cryptographic accelerator support, and modifications to OpenSSL to run on OpenBSD. These features are discussed next.

# POSSE: OUTCOMES AND SUCCESS EXAMPLES

The next three sections provide examples of the progress made as a result of the POSSE project. The first of these covers the cryptographic framework, the basis of hardware cryptography support in OpenSSL. The second covers the extended attribute file system, intended to provide controls similar to those of security enhanced Linux (www.nsa.gov/selinux). The third covers audit of the OpenSSL system and some of the important consequences for Internet security.

## Hardware Cryptography Support

The OpenBSD cryptographic framework (OCF) (Keromytis, Wright, & de Raadt, 2003) uses a service virtualization model that provides access to cryptographic services while hiding details of specific cryptographic hardware accelerator cards (cryptographic providers) behind a kernel-internal API. User-level applications such as the OpenSSL library or the SSH daemon can access the hardware through the */dev/crypto* device, which acts as another kernel application to the framework. While the implementation details of the framework are outside the scope of this chapter, we provide sufficient detail to both understand the measurement methodology and at least to first order, reproduce our experiments.

Inside the operating system kernel, the framework presents two interfaces: one to device drivers, which register with the framework and specify what algorithms and modes of operations they support; and one to applications (e.g., IPsec or */dev/crypto*), which create "sessions." Sessions create context in specific driver instances selected by the framework based on a best-match basis with respect to the algorithms used. Applications queue requests on sessions, and the cryptographic framework, running as a kernel thread and periodically processing all requests, routes them to the appropriate driver. Once the request has been processed, a callback function provided by the application is invoked that continues processing. A software pseudo-driver registers with the framework as a default when no hardware acceleration is available. Public key operations are modeled in the same way, although no session is created.

In summary, the framework provides asynchronous operation, load balancing, application and cryptographic provider independence, and support for both symmetric and public key operations. For our discussion, the most important attribute of the framework is that it provides an *identical common path* to the cryptographic providers available in the system, regardless of their nature (hardware vs. software) or other characteristics (performance, details of the card interface, etc.).

The framework is implemented and has been in use with IPsec since OpenBSD 2.8, although it continues to evolve in response to new requirements. Public key support and the */dev/crypto* API were introduced in subsequent versions of OpenBSD. The OpenSSL crypto library uses this API by default since OpenBSD version 3.1. The OCF has also been ported to FreeBSD, and we are working on Windows and Linux versions.

## Extended Attribute File System

The Extended Attributes work from TrustedBSD is an extension to the BSD UFS layer that allows new meta-data to be persistently associated with filesystem objects (files and directories). These meta-data are arbitrary *(name, value)* pairs and can be used to implement Access Control Lists, Sensitivity Labels, POSIX process capabilities, SubOS user IDs (Ioannidis, Bellovin, & Smith, 2002), etc. Besides the obvious extensions to UFS, there are API modifications to accommodate handling of Extended Attributes, as well as the necessary userland tools to manage them.

This work was introduced for TrustedBSD (Watson, 2000), but given the similarity of the kernels, it was believed to be fairly straightforward to import it to OpenBSD and integrate it with the rest of our security architecture. The combination of the /dev/policy interface (Ioannidis, Keromytis, Bellovin, & Smith, 2000), Security-enhanced Linux features, and Extended Attributes should result in a very flexible security enforcement mechanism.

The enhanced file system has been designed and implemented by author Dale Rahn in concert with Robert Watson of NAI Labs/TrustedBSD. The implementation effort has been kept completely synchronized with that of TrustedBSD.

The /dev/policy policy device has been implemented for OpenBSD and continues to be refined. As a major goal of this work was support for SE-Linux, we also undertook an effort (by Tom Langan of Penn) to provide SE-Linux features. For example, the extensions included checking permission on every I/O system call related to files, networks, etc. Conventional BSD systems check just once on open or equivalent. This extension was successful and is available, but not in the OpenBSD release. The /dev/policy notions, including the use of advanced policy specification languages, were applied directly.

## OpenSSL Audit

OpenSSL is used as a technical building block of the secure Apache (Laurie & Laurie, 1999) web server. Web servers are, with considerable accuracy, considered the operating systems of the WWW. Apache is the dominant web server, widely used by commercial and industry sites, and has a greater than 70% market share. Apache provides an operating environment for concurrent transaction processing, script execution, and any other requests that arrive on an HTTP (80) or HTTPS (443) port. The server keeps multiple threads running concurrently to overcome disk and other latencies and provide high performance. A number of services are provided, such as perl scripting, that can help process client PUTs and GETs. When secure Apache is used, the SSL protocol ensures that the transactions with the server are authenticated and encrypted; this behavior is selected, for sites which support it, by prefacing the site name with https: to indicate that the security features are to be used.

The OpenSSL Project is a collaborative effort to develop a robust, commercial-grade, full-featured, and open source toolkit implementing the *Secure Sockets Layer (SSL v2/v3)* and *Transport Layer Security (TLS v1)* protocols as well as a full-strength general purpose cryptography library. The project is managed by a worldwide community of volunteers that uses the Internet to communicate, plan, and develop the OpenSSL toolkit and its related documentation.

A major research issue addressed by POSSE was the portability of the effective OpenBSD audit methodology to other open source efforts. As an experiment, applying the audit methodology to OpenSSL seemed appropriate, given the importance of the OpenSSL software and Apache to electronic commerce. OpenSSL had never been audited, had accreted code from many programmers, and had many patches, and thus was an ideal candidate for the careful scrutiny of a code audit. The strategy we chose was to start the audit with tools, to see what "low-hanging fruit" could be picked by these tools and eliminated in the code base. For example, John Viega's RATS tool (Viega & McGraw, 2001) can help with fixed-size buffers and detected over 500 instances of fixed-size buffers (which can be exploited for buffer overflow attacks). After some poor initial experiences with RATS, we found that creating search patterns was reasonably powerful. While we looked at Splint (Larochelle & Evans, 2001), we did not end up using it. We were able to detect some errors using a tool supplied by David Wagner (Wagner, Foster, Brewer, & Aiken, 2000).

An important observation about the OpenSSL auditing process is that publicized holes in other systems (for example, on security mailing lists) suggested analogous code in OpenSSL to check, and a variety of problems were identified in this fashion. This suggests that *experience* can play a large role in code auditing, since problematic code will often follow a pattern, which can both be exploited by an experienced attacker and repaired by an experienced auditor. The conclusion at this stage is that tools are an effective way of both pruning low-hanging fruit and identifying chunks of code that need attention. However, many problems still require insight and experience in the auditor.

The OpenSSL audit discovered and fixed holes in OpenSSL identified on the Internet. The holes in OpenSSL were fixed just before Defcon and were totally due to CHATS funding. A patch of over 3,000 lines of code secures a host of problems of lesser severity and generally hardens OpenSSL against future attacks. The OpenSSL audit was largely performed by Ben Laurie of AL Digital, Ltd. AL Digital's auditing efforts proved prescient; the fixes in OpenSSL illustrated a potential hole in other systems that was exploited to write the Sapphire/Slammer worm.[2] The worm exploited people who had failed to patch a persistent problem with available security updates (Arbaugh, Fithen, & McHugh, 2000).

A large body of auditing notes and an outline of a book on the OpenSSL audit have been produced, but it is unclear what the final disposition of this information will be. Our operating assumption is that some cleanup and publication on the WWW would extract the maximum value from these unique notes describing both the use of audit tools (such as John Viega's RATS) as well as the manual audit.

## Discussion

Almost 37,000 lines of new code are directly attributable to this project (as measured by a scan of the OpenBSD 3.2 source tree), and the POSSE project has directly contributed to the 3.0, 3.1, 3.2 and 3.3 releases of OpenBSD.

In addition, a variety of creative new work has been done. An example of this is the W^X (for Write XOR eXecute) project. This goal of this project is to modify the executable and shared library layout so that the a typical program had no regions of memory that were *both* writable and executable.

This change prevents one of the common attacks where a buffer overflow is used to write code into the address space of a program, then execute that code. This change was introduced with OpenBSD 3.3 on several architectures: alpha, sparc, sparc64. Changes are in progress to add support for this protection to i386 and macppc (PowerPC) architectures with OpenBSD 3.4.

In addition, a modification to GCC called ProPolice written by Etoh was integrated. ProPolice rewrites the layout of stack allocated data including a logical "canary" to detect buffer overruns. This change, coupled with W^X mappings and a randomized stack gap, greatly reduces the chance of a buffer overrun attack being successful.

## CONCLUSIONS

The freedom of open source development has led to a plethora of UNIX-derived and UNIX-like source trees. Each tree has, at best, partially instantiated security features, although OpenBSD has the advantage of audited code. Inadequate resources, insufficient motivation for portable solutions, and too few security experts for all trees have been major barriers. POSSE helps to surmount these barriers and more closely match the resources to the requirements.

We have created a project that has been having a substantial impact on the open source community, and beneficiaries have included a variety of commercial vendors who examine or incorporate features from open source systems directly in their own systems. For example, many security appliance vendors use OpenBSD or a minimized version of OpenBSD as the platform for their systems. OpenSSH is shipped with Apple's machines and is extremely widely used.

This has proven a challenging project to manage. There are distributed developers, many distractions, and strong personalities. Nonetheless, we continue to believe that the university is an ideal model for a management entity for this type of effort. By design, its "loose coupling" and open style of discourse provide an easy means by which the long-term goals addressed in the CHATS program can be effectively addressed. Producing a new generation of security-conscious (and capable) developers is a natural enterprise for a university.

## ACKNOWLEDGMENTS

This work was supported by the Defense Advanced Research Projects Agency (DARPA) and Air Force Research Laboratory, Air Force Materiel Command, USAF, under agreement number F30602-01-2-0537. Statements made herein are neither explicit nor implied positions of the U.S. Government.

The authors thank Theo de Raadt, the founder and leader of the OpenBSD Project, for his persistence and technical vision.

## REFERENCES

Arbaugh, W. A., Fithen, W. L., & McHugh, J. (2000). Windows of vulnerability: A case study analysis. *IEEE Computer*, 33(12): 52-59.

Bach, M. J. (1986). *The design of the UNIX operating system*. Englewood Cliffs, NJ: Prentice Hall.

Brooks, F. P. (1975). *The mythical man-month*. Reading, MA: Addison-Wesley.

Clark, D. D. (1988). The design philosophy of the DARPA Internet protocols. In *Proceedings of SIGCOMM 1988*, 106-114.

Cusumano, M. A. & Selby, R. W. (1997). How Microsoft builds software. *Communications of the ACM*, 40(6): 53-61.

Daley, R. C. & Dennis, J. B. (1968). Virtual memory processes and sharing in MULTICS. *Communications of the ACM*, (5): 306-312.

Fano, R. M. & David, E. E. (1965). On the social implications of accessible computing. In *AFIPS Conference Proceedings 27*, 243-247.

Ioannidis, S., Keromytis, A., Bellovin, S., & Smith, J. (2000). Implementing a distributed firewall. In *Proceedings of Computer and Communications Security (CCS) 2000*, 190-199.

Ioannidis, S., Bellovin, S., & Smith, J. M. (2002). Sub-operating systems: A new approach to application security. In *Proceedings of the 10th SIGOPS European Workshop*, pp. 108-115.

Keromytis, A., Wright, J., & de Raadt, T. (2003). The design of the OpenBSD cryptographic framework. In *Proceedings of the USENIX Conference*, pp. 181-196.

Keromytis, A. D., Ioannidis, J., & Smith, J. M. (1997). Implementing IPsec. In *Proceedings of Global Internet (GlobeCom) '97*, 1948-1952.

Larochelle, D. & Evans, D. (2001). Statically detecting likely buffer overflow vulnerabilities. In *Proceedings of the 2001 USENIX Security Symposium*.

Laurie, B. & Laurie, P. (1999). *Apache: The definitive guide*. Sebastopol, CA: O'Reilly.

Lions, J. (1977a). *A commentary on the UNIX operating system*. Bell Laboratories.

Lions, J. (1977b). *UNIX operating system source code, Level Six*. Bell Laboratories.

McKusick, M. K., Bostic, K., Karels, M. J., & Quarterman, J. S. (1996). *The design and implementation of the 4.4 BSD operating system*. Reading, MA: Addison-Wesley.

Miltchev, S., Ioannidis, S., & Keromytis, A. (2002). A study of the relative costs of network security protocols. In *Proceedings of USENIX Annual Technical Conference (Freenix track)*, 41-48.

MITRE (2003). *Use of Free and Open-Source Software (FOSS) in the U.S. Department of Defense*. MITRE Report MP 02 W0000101, Version 1.2.04.

Organick, E. I. (1972). *The MULTICS system*. Cambridge, MA: MIT Press.

Raymond, E. S. (1999). *The cathedral and the bazaar: Musings on Linux and open source by an accidental revolutionary*. Sebastopol, CA: O'Reilly and Associates.

Ritchie, D. & Thompson, K. (1974). The UNIX operating system. *Communications of the ACM*, 17: 365-375.

Ritchie, D. M. & Thompson, K. L. (1978). The UNIX Time-Sharing System. *The Bell System Technical Journal*, 57(6): 1905-1930.

Saltzer, J. H., Reed, D. P., & Clark, D. D. (1984). End-to-end arguments in system design. *ACM Transactions on Computer Systems*, 2(4): 277-288.

Schroder, M. D., Clark, D. D., & Saltzer, J. H. (1977). The MULTICS kernel design project. In *Proceedings of the 6th ACM SOSP*, 43-56.

Schroeder, M. D. (1975). Engineering a security kernel for MULTICS. In *Proceedings of the 5th ACM SOSP*, 125-132.

Sullivan, G. R. & Dubik, J. M. (1994). War in the information age. U.S. Army War College: Strategic Studies Institute (SSI), (23 pages).

Thompson, K. (1978). UNIX Implementation. *The Bell System Technical Journal*, 57(6): 1931–1946.

Viega, J. & McGraw, G. (2001). *Building secure software*. Reading, MA: Addison-Wesley.

Wagner, D., Foster, J. S., Brewer, E. A., & Aiken, A. (2000). A first step towards automated detection of buffer overrun vulnerabilities. In *Proceedings of the Symposium on Network and Distributed Systems Security*, 3-17.

Watson, R. (2000). Introducing supporting infrastructure for trusted operating system support in FreeBSD. In the *Proceedings of BSDCon 2000*.

Weinberg, G. (1974). *The psychology of computer programming*. New York: Van Nostrand.

# ENDNOTES

[1] See http://www.cis.upenn.edu/~dsl/POSSE/
[2] See http://www.cs.berkeley.edu/~nweaver/sapphire/

# SECTION VI:

# Implications of the F/OSS Development Model – "The Broad Picture"

## Chapter XII

# The Impact of Open Source Development on the Social Construction of Intellectual Property

Bernd Carsten Stahl, De Montfort University, UK

## ABSTRACT

*This chapter discusses the impact that open source software has on our perception and use of intellectual property. The theoretical foundation of the paper is constructionist in that it holds intellectual property to be a social construction that is created and legitimized by narratives. In a first step, the chapter recounts the narratives that are usually found in the literature to justify the creation and protection of intellectual property. The two most important streams of narratives are the utilitarian and the natural rights arguments. In a second step, the paper proceeds to the impact that the use of information and communication technology (ICT) has on the narratives of intellectual property. From there, the chapter progresses to a discussion of the impact of open source software on these narratives. It will be argued that open source software changes our perception of intellectual property because it offers evidence that some of the classical narratives are simplistic. At the same time it will become clear that open source is not a frontal assault on intellectual property because it is partly based on ownership of intellectual artefacts. The conclusion discusses how this change of narratives caused by open source software may reflect on our institutions, laws, and regulations of intellectual property.*

# INTRODUCTION

There can be little doubt that the way we regulate intellectual property is of high importance for the oft-cited information society. Intellectual property regulations affect the way we do business on the Web, but they also go the heart of other developments in fields such as education, recreation, or government. In this chapter, I will take a narrative approach to intellectual property, with the aim of studying how new developments such as open source software impact on it. The central idea is that intellectual property is a social construction that is based on the acceptance of narratives. These narratives form the basis of the regulations that societies adopt. The main thesis is that open source software brings with it new narratives that conflict with the established ones and that this will lead to consequences in the way we perceive, regulate, and enforce intellectual property.

In order to render this hypothesis plausible, the chapter will start out by recounting the narratives that are used to justify and legitimize property in general. It will proceed to the stories that are used to constitute intellectual property and how these differ from physical property narratives. The subsequent section will discuss the influence that information and communication technologies have on intellectual property, and it will give an account of the development of intellectual property regulations based on these narratives. The following section will then discuss the impact that open source software has on this debate. It will relate how the stories upon which open source is based conform to or contradict the traditional justifications of intellectual property. The conclusion will then attempt an outlook on how these changing narratives may be reflected in intellectual property regulations.

# THE NARRATIVE OF INTELLECTUAL PROPERTY

The plausibility of this chapter hinges on the acceptance of the hypothesis that intellectual property is a set of rules that are based on narratives. It should be clear that the stories that we associate with social norms and their believability determine the effectiveness of these norms. Only if this starting point is accepted will it make sense to tell these stories, which will be done for the remainder of this section.

What are stories, what are narratives, which narratives are good, which are not, who tells them, who receives them? All of these are questions that cannot be answered exhaustively, especially not in one chapter. Narratives are those stories that we use to make sense of the world on an individual as well as a collective level (Ricoeur, 1994). These stories are transmitted by a multitude of channels, they can be contradictory, they change over time and between geographical areas, and they are very hard to pin down. Nevertheless, they are the stuff of which our culture and identity are made (Stahl, 2003). And like most concepts that have to do with culture, identity, personality, meaning, and understanding, they are highly fuzzy at the edges and therefore tend frustrate analytically minded scholars. This fuzziness need not concern us at this stage, however, because it is not the purpose of this chapter to prove, disprove, analyze, or scientifically validate narratives. This chapter aims to tell a narrative about narratives. It is a story about the stories that are told about intellectual property. As such, it cannot claim to be objective or true. At best, it will be plausible and believable to the reader. The point of the exercise is to look at the effects that a new narrative, namely, that of open source software, has on older narratives of intellectual property. The reason for doing so is to see how far these narratives are compatible and where they may need to change. This is a very broad exercise that can absolutely not do justice to all of the aspects involved. It is nevertheless useful because it may help us refocus on what intellectual property means to us, why we value it, and how we want to use it.

## The Story of Property

Since intellectual property is one aspect of a wider field of rights that are summarized under the heading of "property," it is a good starting point to look at the stories that justify property. First of all, property can be described as a "bundle of rights" (Donaldson & Preston, 1995). As such, property is always embedded in a framework of other rights and part of a social practice (DeGeorge, 1999). The bundle of rights that constitutes property contains several clearly specified rights. First, property confers on the owner the right to use something. This includes production, exchange, and consumption (Gauthier, 1986). Second, property gives the owner the right to exclude others from use (DeGeorge, 1999; Spinello, 2000).

There are two groups of narratives used to justify property: the natural rights and the utility narratives. Natural rights stories hold that property is something that originates in nature, some "intrinsic quality" (Warwick, 2001), that only needs to be expressed in human terms. Among the natural rights arguments, one can distinguish between two groups of such inalienable rights which are developed in Lockean labor arguments and Heglian personality arguments.

The labor arguments go back to Locke, who held that everybody has an inalienable right to their own labor and that therefore one has property rights in those things that are the result of one's labor. When one mixes one's labor with something, then one acquires property. This view has raised the reputation of labor to the high standard where it is now held in market economies (cf. Arendt, 1958).

The personality argument is similar and related to the labor argument but it carries a different emphasis. It is also based on the inalienable right everybody has to themselves, but instead of pointing out the necessity of compensation for onerous work, it sees property as an extension of one's body that represents an extension of one's rights to oneself (Höffe, 1996).

Nowadays, natural rights theories are quite difficult to sustain. While it is plausible that we all have a right to property in ourselves, it is hard to see what status this right would have. Is it a moral right or a legal one? Who will enforce natural rights and what happens when they are breached? Relatively few philosophers today would argue for natural rights, but they still have importance for approaches such as this one. In this chapter, natural rights are not accepted as naturally binding but as important stories that explain adherence, acceptance, and legitimacy of social rules.

Due to the inherent problems of natural rights, stories the second group of narratives about property, namely, the utilitarian stories, have gained prominence. These stories stress the fact that by instituting property rights overall utility is increased. There are several reasons why personal property is seen as useful. First, it appears to be one of the main motivators for people to work. It is thus the basis of effort, productivity, and consequently of social welfare (Donaldson & Dunfee, 1999). When people own means of production, they make careful use of them and treat them parsimoniously which leads to efficiency. At the same time, property that is linked to individuals ensures that owners feel responsible and can be held accountable (Nozick, 1974). Apart from these immediate advantages that aim at the improvement of production, property can also be said to have more far-reaching benefits. Among them there are economic and political arguments, such as Hayek's which views property as a necessary condition of individual freedom (Hayek, 1994).

## The Story of Intellectual Property

While traditional theories and justifications implicitly refer to the ownership of tangible physical goods, intellectual property aims at something different, namely, at "invention of the mind—original ideas, expressions, and their 'ownership'" (Lawrence, 1996). An important question is in which respects traditional and intellectual property differ (Ladd, 2000). Some authors hold that the old concepts of property cannot be applied to non-physical objects (Barlow, 1995). There are several reasons why the classical concept of property changes when applied to ideas. Chief among them one can find the fact that the sale or distribution of an idea does not affect the original creator's ability to use it (Johnson, 2001a; Kuflik, 1995; Weckert & Adeney, 1997). Also, it is much easier to reproduce or copy items that are subject to intellectual property (Johnson, 2001b), and it is consequently often hard to distinguish between the original and the copy, rendering the very distinction between original and copy meaningless (Hinman, 2002). The right to exclude others from use therefore takes on a new meaning and requires new justification. At the same time, the notion of intellectual property is continuously gaining in importance because it defines the way we deal with the most important resources of the knowledge society, with knowledge, ideas, information, and the results of creativity (Mason, 1986; Mason, 2000).

It is therefore important to understand how intellectual property can be justified. A look at the existing literature shows us that the same types of arguments are used that we have already encountered with regard to physical property. On the one hand, there are the natural rights approaches drawing on the labor or personality argument. In terms of software, for example, this means that a programmer has the right to the product of her efforts (Johnson, 2001a). At the same time the program can be seen as an extension of the self of the programmer and thus naturally hers (McFarland, 2001; Nissenbaum, 1995). On the other hand, there are utility-based justifications. The creation of exclusive rights to intellectual creations is supposed to advance social utility by opening a source of revenue to the authors and thus motivating them to create things of social use (Weckert & Adeney, 1997). This line of thought can be summarized by stating that the institution of intellectual property produces more utility—be it in the form of money, knowledge, art, or anything else—than its absence.

## Limits of Intellectual Property

It should be noted that independent of the stories used to justify it, there are always limits to intellectual property. The exact form of these limits, however, depends on the original justification. Generally, property ends where it collides with equal rights. In the case of natural rights this can refer to other rights that are also perceived to be natural, such as the right to life, the pursuit of happiness, or freedom (Halévy, 1995). The limits of property can refer to the entire bundle of rights. A property owner may not have the right to do certain things with her property or she may not have the right to exclude others from use. Similarly, in the case of utility-based justifications, intellectual property finds its limits in those situations where it no longer produces utility (cf. Boyle, 2001; Forester & Morrison, 1994; Ladd, 2000; Snapper, 1995).

The limits of copyright as the probably most important expression of intellectual property are often discussed under the heading of "fair use." According to fair use, copyright can often be broken for educational or non-commercial uses. Patents as another expression of intellectual property also know exceptions. Exceptions to intellectual property are important in our context because they indicate the point to which legitimizing narratives are deemed acceptable and where their limits are.

# INTELLECTUAL PROPERTY AND ICT

One of the reasons why questions of intellectual property have been hotly debated during the last few years is the impact that information and communication technology (ICT) had on it. In this section, I will briefly review what the specific consequences of the introduction of ICT on intellectual property are, the developments in the area of intellectual property protection, and where they seem to be heading. Toward the end of the section, I will discuss some arguments that are critical of these developments.

## Narratives of Intellectual Property in ICT

While ICT does not pose fundamentally new problems regarding intellectual property, it nevertheless changes our way of dealing with items of intellectual property in a basic way. Computers and networks allow the dissemination and copying of material, they allow new forms of collecting and collating ideas and texts, they allow new types of analysis and use of data and information. Computers have "greased" (Moor, 2000) information and accentuated those areas where intellectual property produced problems before. Original and copy have become indistinguishable. At the same time, intellectual property gains in importance as a basis of the knowledge-based society, as a commodity and also as a social resource. The financial value of intellectual property is already in the hundreds of billions of dollars and continues to grow (Boyle, 2001; Delong, 2000). This has propelled the question of the justification of intellectual property in the information age to the forefront of legal, political, and social debates.

The classical justifications of property discussed in the last section can be found again in this area. For the following discussion, I will limit the debate to intellectual property in computer programs in order to facilitate the contrast with open source software, but it should be noted that similar arguments can be found on both sides regarding the other big problematic area, namely, that of content such as texts, music, or films.

Ownership in computer programs can be justified by looking at natural rights, and one can find the argument that programs belong to the programmer because of a Lockean labor theory or because of a Hegelian personality theory. In the one case, the emphasis is on the work the programmer invested; in the other case, it is on the fact that the program is some kind of extension of the programmer's mind (Nissenbaum, 1995).

More common than natural rights arguments, however, are arguments that justify ownership in computer programs by pointing at the utility aspect. Ownership in computer programs is supposed to "stimulate creativity, innovation and entrepreneurship" (Mason, 2000), to motivate programmers (cf. Stallman, 1995) and thereby improve the well-being of society.

## The Development of Intellectual Property Protection in the Knowledge Society

On the basis of the stories recounted so far, a strong consensus has developed that intellectual property is important in modern societies and that it needs to be protected by legal measures. At the same time, the threats to intellectual property are becoming more pronounced and clear in the public perception. The ease of access and reproduction of digitized material facilitates breaches of intellectual property by the average individual as well as the professional. Such breaches have been called "piracy," an interesting term that shows how serious these problems are deemed to be. It also shows that breaches of intellectual property are seen to be equivalent to breaches of normal property rights, such as theft or maybe even robbery. At the same time that such strong feelings about intellectual property develop, the legal protection is becoming more and more difficult. This is caused, on the one hand, by technical problems and, on the other hand, by the international nature of modern information technology which defies the national character of legal rights (O'Rourke, 2001).

The result of this is that attempts to protect intellectual property are increasing in scope and scale. This development is mainly driven by those who have a strong commercial interest in intellectual property, mostly the big media and software companies (Lipinski & Rice, 2002). For example, it has been argued that the position of intellectual property rights holders has continuously been strengthened by extending legal constructs to software (cf. de Laat, 2002; Syme & Camp, 2002) and creating new ones such as the American Digital Millennium Copyright Act.

It should be noted that this development, albeit initiated and lobbied for by very strong groups, does not proceed without protest. Most of the voices that have been raised against it say that the strengthening of intellectual property protections goes against the grain of the moral narratives that were used to justify it in the first place (Birrer, 2001).

The strongest opposition to the strengthening of intellectual property protection argues that it is a form of economic dominance of business interests over other legitimate stakeholders in society. Big companies are seen to form legislation to their own interests (Benkler, 2001; McFarland, 2001; Smiers, 2001). This can be seen as a victory of the law over ethics (DeGeorge, 1998) or, worse, as an expression of cultural imperialism (Weckert, 2000). The strengthening of intellectual property rights eliminates recognized exceptions and fair use (Lowe-Petraske, 2002) which are an integral part of its legitimacy.

More fundamentally, many authors doubt the moral narratives used to justify the strengthening of intellectual property. Many of them believe that software "piracy" does not produce social damages because most pirated software would never have been bought anyway (Weckert & Adeney, 1997) or because using software without permission does not deprive the original owner of its use (Siponen & Vartiainen, 2002). Finally, there are arguments that refer to competing narratives such as that regarding freedom of information, and one can find the contention that from an ethical point of view freedom of information is a higher good than the protection of intellectual property (Ladd, 2000). This argument can again be supported by different ethical narratives. This clash of narratives—supporting either the strengthening or weakening of intellectual property protection—is where open source software becomes interesting for the debate.

# OPEN SOURCE SOFTWARE AND NARRATIVES OF INTELLECTUAL PROPERTY

In this section, I will first outline what open source software is in order to then analyze how it impacts on intellectual property narratives.

## Open Source Software

Open source software can be defined as software for which users have access to the source code that distinguishes it from most commercially published software that allows users only access to the object code (Madey, Freeh, & Tynan, 2002). Apart from the accessibility of the source code, open source software is also characterized by the organization of its development, which can be described as self-organizing, collaboration, and social networks. It is based on a different paradigm of development when compared with proprietary software. It is based on the principle of continuous improvement through frequent releases, collaboration among developers and users, and adherence to open standards through open source licenses (Mishra, Prasad, & Raghunathan, 2002).

Open source software should not be misunderstood to stand for a complete lack of ownership. In fact, it tends to be distributed with licenses. Unlike proprietary licenses, these OSS licenses allow users to modify and improve the source code, to further distribute code, whether modified or not. Within the OSS realm there are different types of licenses that allow users different degrees of freedom (de Laat, 2002). Some licenses, such as those promoted by the Open Source Initiative, follow ideological goals and are described as a "bill of rights for the computer user" or are used for outright political goals (Syme & Camp, 2002). Other forms of open source software licenses are meant to be clear countermeasures to copyrighted software. The example here is "copyleft" that originates with Richard Stallman's Free Software Foundation and which explicitly aims at keeping software in the free domain.

Open source software is also sometimes called "free software" and some authors argue that it should be made available free of charge (Syme & Camp, 2002). Most authors agree, however, that the term "free software" ("libre software") means that the user is free to read the source code, to modify it, and to use the software according to her own devices (Feller & Fitzgerald, 2000). This means that open source software and products that are based on it can be bought and sold. For most of the OSS available today, the latter view seems to apply. OSS licenses are not meant to preclude programmers and companies from making money, but they aim to keep the knowledge behind the software in the public realm. For our topic it is interesting to note that OSS does not fundamentally question the idea of intellectual property. In fact, it is based on the assumption that one can determine the use of the fruits of one's mental efforts. This is of course one of the central ideas of intellectual property. Even such far-reaching attacks against intellectual property as the copyleft movement are only possible on the assumption that the copyleft license is binding and that the authors of source code have the right to determine that it remain open.

## Open Source Challenges to Intellectual Property Narratives

The development and unexpected success of OSS have led to several challenges to traditional intellectual property narratives. One of the central stories used to defend intellectual property protection is that it is necessary to provide immediate financial incentives so that able individuals will dedicate their time to producing valuable ideas. This story has simply been proven wrong by open source software. Programmers in OSS projects typically do not get paid directly for their participation. So why do they do it? Basically, one can distinguish two answers to this question, an altruistic one and an egoistic one. The altruists hold that programmers participate in OSS for the greater good of things. They are in it for the knowledge, for the understanding, in order to provide a better service. Not surprisingly, some authors therefore stress the importance of shared norms or ideologies as the basis of OSS (Stewart & Gosain, 2001). As a consequence, OSS has the image of being "more ethical" than other software (cf. Faldetta, 2002; Garfinkel, 2003).

On the other hand, one can also find narratives that explain the lack of financial rewards for programmers in less altruistic economic terms. According to those narratives, programmers participate because they see a chance of increasing their personal market value, their human capital. Given that OSS is often thought to be as good if not better than proprietary software, the social processes (such as peer review) that allow programmers to participate are such that they only admit the best. This means that programming in a prestigious OSS project can be taken as a sign of high competence and aptitude. Indeed, empirical research has shown that participation in OSS is correlated to higher salary in the programmer's normal job (Hann, Roberts, Slaughter, & Fielding, 2002). OSS can thus have the function of a signal that programmers use to demonstrate to potential employers that they are talented (Mishra et al., 2002).

OSS also threatens other utility-based narratives. While intellectual property defenders hold that intellectual property protection is necessary to produce good and reliable software, albeit at a steep price, OSS defenders contend that open sources projects are not only cheaper, they are also more productive. Stallman (1995) names four reason why open source software has a higher productivity than proprietary software: the programs find a wider usage, they can be adapted and customization need not start from scratch, programmers get a better education, and efforts do not have to be duplicated. The self-referential process of OSS development also finds it easier in many cases to employ large amounts of resources to the solution of complex problems such as programming or, maybe more importantly, debugging (cf. Raymond, 2001).

This implies that another point in favor of OSS is its quality. When compared with proprietary software, OSS seems to be at least as good. It is of course a difficult endeavour to make general statements comparing the quality of different types of software. However, research suggests that neither OSS nor proprietary software is unequivocally better than the other (Mishra et al., 2002). Feller and Fitzgerald (2000) observe that the market share that OSS has achieved is a good indicator of its quality. Especially the flagships of OSS such as Linux or Apache have been successfully established in the market, which can best be explained by superior quality.

One can summarize that the success of OSS renders the utility-based defenses of intellectual property problematic. What about the other group of narratives, those based on natural rights? Due to the problems associated with them, they play a less prominent role. While natural rights are seldom the focus of OSS stories, they nevertheless play an implicit role because they are usually taken for granted. We have seen that OSS is based on the idea that creators have rights over their creation, even though they may waive the right to use them for immediate financial advantages. According to the economic explanation, programmers who participate in OSS projects in order to increase their human capital rely on being recognizable as contributors to the joint effort. That means that their contribution must be identifiable, which in turn requires a clear and unalterable relationship between creator and creation. While programmers may renounce the copyright to their work, they clearly retain what is sometimes called "moral rights," the right to be identified as an author (Warwick, 2001). These rights are part of the natural rights tradition because they do not seem to require any utility justification and are generally accepted.

## CONCLUSIONS

Due to the complexity of the subject and space restraints, this chapter could not do justice to all of the narratives and all of the aspects involved in the discussion concerning intellectual property. For example, it did not distinguish sufficiently between different types of intellectual property that are based on the same narratives but have developed their own narratives building on these. It may similarly have missed some of the finer points of OSS. Nevertheless, it claims to mirror the most important aspects of the current debate as well as their narrative history.

The reason why this development of stories may be interesting to the reader despite its limitations is that these narratives have very strong and manifest consequences in our social world, most importantly in the development of intellectual property regulation, be it national or international. These rules and regulations are of central importance for the development of the knowledge society, and they will have a huge influence on how we will be able to access data and information, on how we teach and learn, on how we entertain ourselves, and, more generally, on how we interact.

It is part of the nature of the approached chosen in this chapter that no hard and fast conclusions in the sense of managerial strategies or governmental policy recommendations can result. This chapter has portrayed intellectual property as a collection of stories, and in this framework must itself be viewed as just one more story among many others. For the reader who found this story believable, however, some conclusions might be drawn. The narratives that are used to widen the scope of intellectual property protection have lost some of their credibility. Open source software shows that good software can come into existence without the strong measures proposed by many laws. Furthermore, there are plausible arguments that from a societal point of view open source software may be more desirable than proprietary software. Given that intellectual property is a social construction, societies may want to explore these possibilities and change the way they attribute property rights. This can of course only happen as the result of a huge national and international debate, which will take time. This debate will have to consider economic, political, social, ethical, and related matters and it will produced winners and losers. Lately, the winners seem to have been the owners of intellectual property, mostly the big software and entertainment companies. The success of open source software may be a good argument in the debate that will change the tide and see users and consumers as winners.

# REFERENCES

Arendt, H. (1958). *The human condition*. 2$^{nd}$ edition. Chicago, IL: University of Chicago Press.

Barlow, J. P. (1995). Coming into the country. In D.G. Johnson & H. Nissenbaum (Eds..) *Computers, ethics & social values*. 15-18. Upper Saddle River, NJ: Prentice Hall.

Benkler, Y. (2001). The battle over the institutional ecosystem in the digital environment. *Communications of the ACM* (44:2): 84-90.

Birrer, F. A. J. (2001). Applying ethical and moral concepts and theories to IT contexts: Some key problems and challenges. In R.A. Spinello & H.T. Tavani (Eds.), *Readings in cyberethics*. 91-97. Sudbury, MA: Jones and Bartlett.

Boyle, J. (2001). A politics of intellectual property: Environmentalism for the Net? In R.A. Spinello & H.T. Tavani (Eds.), *Readings in cyberethics*. 231-251. Sudbury, MA: Jones and Bartlett.

Burk, D. (2002). Lex genetica: The law and ethics of programming biological code. *Ethics and Information Technology* (4:2):109-121.

Currie, W. (2000). *The global information society*. Chichester, UL: John Wiley & Sons.

Davis, J.C. (2000). Protecting intellectual property in cyberspace. In R.M. Baird, R. Ramsower, & S.E. Rosenbaum (Eds.), *Cyberethics - Social and moral issues in the computer age*. 243-256. New York: Prometheus Books.

De George, R. T. (1999). *Business ethics*. 5$^{th}$ edition. Upper Saddle River, NJ: Prentice Hall.

de Laat, P. B. (2002). Open source networks in industry. In I. Alvarez et al. (Eds.), *The transformation of organisations in the information age: Social and ethical implications. Proceedings of the sixth ETHICOMP Conference*, 13 - 15 November 2002, Lisbon, Portugal. 403-415. Lisbon: Universidade Lusiada.

Delong, J. V. (2000). Mind over matter. In R.M. Baird, R. Ramsower, & S.E, Rosenbaum (Eds.), *Cyberethics - Social and moral issues in the computer age.* 234-242. New York: Prometheus Books.

Donaldson, T. & Dunfee, T. W. (1999). *Ties that bind: A social contracts approach to business ethics.* Boston, MA: Harvard Business School Press.

Donaldson, T. & Preston, L. E. (1995).The stakeholder theory of the corporation: Concepts, evidence, and implications. *Academy of Management Review* (20:1): 65–91.

Faldetta, G. (2002). The content of freedom in resources: The open source model. *Journal of Business Ethics* 39:179–188.

Feller, J. & Fitzgerald, B. (2000).A framework analysis of the open source software development paradigm. In *Proceedings of the International Conference on Information Systems.* 58-69.

Forester, T. & Morrison, P. (1994). *Computer ethics - Cautionary tales and ethical dilemmas in computing.* 2nd edition. Cambridge, MA / London: MIT Press.

Garfinkel, S. (2003). The free-software imperative. *Technology Review* (106:1), February: 30.

Gauthier, D. (1986). *Morals by agreement.* Oxford, UK: Clarendon.

Halévy, E. (1995). *La formation du radicalisme philosophique I - La jeunesse de Bentham, 1776 - 1789.* Paris: Presses universitaires de France.

Hann, I.-H, Roberts, J, Slaughter, S., & Fielding, R. (2002). Economic incentives for participating in open source software projects. In *Proceedings of the Twenty-Third International Conference on Information Systems.* 365-372.

Hayek, F. A. von (1994). *The Road to Serfdom.* 50th Anniversary Edition. Chicago, IL: University of Chicago Press.

Hinman, L., M. (2002). The impact of the Internet on our moral lives in academia. *Ethics and Information Technology* (4:1): 31–35.

Höffe, O. (1996). *Immanuel Kant.* 4. Auflage, München: Becksche Reihe Denker 506.

Johnson, D. G. (2001a). *Computer Ethics.* 3rd edition. Upper Saddle River, NJ: Prentice Hall.

Johnson, D. G. (2001b). Ethics on-line. In R.A. Spinello & H.T. Tavani, Herman (Eds.), *Readings in cyberethics.* 26-35. Sudbury, MA: Jones and Bartlett

Kuflik, A. (1995). Moral foundation of intellectual property rights. In D.G. Johnson & H. Nissenbaum (Eds.), *Computers, ethics & social values.* 169-180. Upper Saddle River, NJ: Prentice Hall.

Ladd, J. (2000). Ethics and the computer world – A new challenge for philosophers. In R.M. Baird, R. Ramsower, & S.E. Rosenbaum, (Eds.), *Cyberethics - Social and moral issues in the computer age.* 44-55. New York: Prometheus Books.

Lawrence, J. (1996). Intellectual property future: The paper club and the digital commons. In: C. Ess (Ed.), *Philosophical perspectives on computer-mediated communication.* 95-114. Albany, NY: State University of New York Press.

Lipinski, T. A. & Rice, D. A. (2002). Organizational and individual responses to legal paradigm shifts in the ownership of information in digital media. In I. Alvarez et al. (Eds.), *The transformation of organisations in the information age: Social and ethical implications. Proceedings of the sixth ETHICOMP Conference*, November 13-15, Lisbon, Portugal. 417-430. Lisbon: Universidade Lusiada.

Lowe-Petraske, A. (2002). CD copy-protection: Proprietary stealth and the ethics of the just war on piracy. In I. Alvarez et al. (Eds.), *The transformation of organisations in the information age: Social and ethical implications*. Proceedings of the sixth ETHICOMP Conference, 13 - 15 November 2002, Lisbon, Portugal. 433-444. Lisbon: Universidade Lusiada.

Madey, G., Freeh, V., & Tynan, R. (2002). The open source software development phenomenon: An analysis based on social network theory. In *Proceedings of the Eighth Americas Conference on Information Systems*. 1806-813.

Mason, R. O. (1986). Four ethical issues of the information age. *MIS Quarterly* 10: 5-12.

Mason, R. O. (2000). Intellectual property and open systems. In *Proceedings of the 33rd Hawaii International Conference on System Sciences*.

McFarland, M. C. (2001). Intellectual property, information, and the common good. In R.A. Spinello & H.T. Tavani, Herman (Eds.), *Readings in cyberethics*. 252-262. Sudbury, MA: Jones and Bartlett.

Mishra, B., Prasad, A., & Raghunathan, S. (2002). Quality and profits under open source versus closed source. In *Proceedings of the Twenty-Third International Conference on Information Systems*. 349–363.

Moor, J. H. (2000). Toward a theory of privacy in the information age. In R.M. Baird, R. Ramsower, & S.E. Rosenbaum, (Eds.), *Cyberethics - Social and moral issues in the computer age*. 200-212. New York: Prometheus Books.

Nissenbaum, H. (1995). Should I copy my neighbor's software? In D.G. Johnson & H. Nissenbaum (Eds.), *Computers, ethics & social values*. 201-213. Upper Saddle River, NJ: Prentice Hall.

Nozick, R. (1974). *Anarchy, state, and utopia*. New York: Basic Books.

O'Rourke, M. A. (2001). Is virtual trespass an apt analogy? *Communications of the ACM* (44:2): 98–103.

Raymond, E. (2001). The cathedral and the bazaar. In R.A. Spinello & H.T. Tavani, Herman (Eds.), *Readings in cyberethics*. 309-338. Sudbury, MA: Jones and Bartlett.

Ricoeur, P. (1994). Entretien avec Paul Ricoeur. In J.-C. Aeschlimann (Ed.), *Ethique et responsabilité - Paul Ricoeur*. 11-34. Boudry-Neuchâtel: Editions de la Baconnière.

Siponen, M. & Vartiainen, T. (2002). Teaching end-user ethics: Issues and a solution based on universalizability. *Communications of the Association for Information Systems* 8: 422–443.

Smiers, J. (2001). La propriété intellectuelle, c'est le vol! *Le Monde Diplomatique* 570, September: 3.

Snapper, J. W. (1995). Intellectual property protections for computer software. In D.G. Johnson & H. Nissenbaum (Eds.), *Computers, ethics & social values*. 181-190. Upper Saddle River, NJ: Prentice Hall.

Spinello, R. (2000). *Cyberethics: Morality and law in cyberspace*. London: Jones and Bartlett.

Spinello, R. A. & Tavani, Herman T. (2001). Note on the DeCSS trial. In R.A. Spinello & H.T. Tavani, Herman (Eds.), *Readings in cyberethics*. 226-230. Sudbury, MA: Jones and Bartlett.

Stahl, B. C. (2003). Cultural universality versus particularity in CMC. In *Proceedings of the Ninths Americas Conference on Information Systems*, Tampa, 04-06 August 2003.1018-1026.

Stallman, R. (1995). Why software should be free. In D.G. Johnson & H. Nissenbaum (Eds.), *Computers, ethics & social values*. 190-200. Upper Saddle River, NJ: Prentice Hall.

Stead, B. A. & Gilbert, J. (2001). Ethical issues in electronic commerce. *Journal of Business Ethics* 34: 75-85.

Stewart, K. & Gosain, S. (2001). An exploratory study of ideology and trust in open source development groups. In *Proceedings of the Twenty-Second International Conference on Information Systems*. 507-512.

Syme, S. & Camp, L. J. (2002). The governance of code: Open land vs. UCITA land. *Computers and Society* (32)3. Retrieved January 13, 2003, from: http://www.computersandsociety.com.

Velasquez, M. (1998). *Business ethics: Concepts and cases*. 4th edition. Upper Saddle River, NJ: Prentice Hall.

Warwick, S. (2001). Is copyright ethical? An examination of the theories, laws, and practices regarding the private ownership of intellectual work in the United States. In R.A. Spinello & H.T. Tavani, Herman (Eds.), *Readings in cyberethics*. 263-279. Sudbury, MA: Jones and Bartlett.

Weckert, J. (2000). What is new or unique about Internet activities? In D. Langford (Ed.), *Internet ethics*. 47-63. London: McMillan.

Weckert, J. & Adeney, D. (1997). *Computer and information ethics*. Westport, CT/London: Greenwood Press.

## Chapter XIII

# The Social Production of Ethics in Debian and Free Software Communities:
## Anthropological Lessons for Vocational Ethics

E. Gabriella Coleman, University of Chicago, USA

Benjamin Hill, Debian Project, USA

## ABSTRACT

*This chapter examines the way that participation in Free software projects increases commitments to information freedom among participants. With the Debian project as its core case study, it argues that in Free and Open Source software communities, ethics are reinforced through the sustained collaborative development of code and discussions and decisions around Free software licenses and project policy. In the final section, the chapter draws on the ethnographic analysis of ethical cultivation in Debian to describe a model of ethical volunteerism based on institutional independence, volunteer labor, and networks of trust that is applicable to a range of vocations.*

Copyright © 2005, Idea Group Inc. Copying or distributing in print or electronic forms without written permission of Idea Group Inc. is prohibited.

# INTRODUCTION

Free and Open Source software (F/OSS) development projects have been investigated from a number of perspectives with an emphasis on (1) F/OSS production techniques and legal codes, (2) developers' motivation to participate in projects, and (3) F/OSS' social and economic impact. We believe that one of the most novel dimensions to the networked and open production of code is its role as a socio-educational site for the cultivation of a more ethically dense practice of programming. In this way, we feel that F/OSS projects are institutions for ethical development through the practice and application of ethics in the same way that they are sites for the development of code. In our ethical examination of F/OSS developer communities, we move away from a typical account of "hacker ethics" toward a discussion of the rise and cultivation thereof. Participation in free software projects substantially deepens commitments to information freedom common among hackers. As a result, ethics are reinforced through the sustained collaborative development of code and, to a greater degree, through the discussions and decisions around free software licenses and project policy.

This piece utilizes an in-depth case study of the Debian project. Debian is one of the largest F/OSS projects, although it is also one that has received very little academic analysis to date. Debian is a non-commercial version of the GNU/Linux operating system maintained by over 900 volunteer developers who package[2] a wide variety of software applications. With the possible exception of the Free Software Foundation's (FSF) GNU project, Debian demonstrates the most overt commitment to the principles of free software as originally expressed by the FSF and the GNU General Public License (GPL) of any large F/OSS project. Debian developers have modified and extended these principles in their social contract and Debian Free Software Guidelines (DFSG).[3] While in some respects, Debian is unique in its explicit ethical codes, its uniqueness should not obscure its wider relevance. The type of moral cultivation that forms an important facet of the Debian social sphere is a component of many other F/OSS projects in less accentuated forms. Debian is ideal to analyze because it brings into clear relief the subtle yet significant processes of ethical socialization that occur in most F/OSS projects. Though we focus on the domain of F/OSS, this case study also serves as an example of the unique ways in which an ethical social order is built and sustained in the "immaterial" setting of the Internet.

We feel that an analysis of F/OSS communities' social lives is best served by a qualitative, ethnographic methodology. As a result, we have employed a "classic" anthropological model of participant-observation that combines audiotaped interviews with the examination of the "everyday" speech and discursive practices of Debian developers. This chapter is based on and references a number of sources, including text from Internet Relay Chat, mailing lists, published Debian documents, and Debian events and conferences. The chapter pulls from several years of data collection and fieldwork paired with survey data collected during interviews with

Debian developers from the U.S., Canada, Europe, and South America, including 45 audiotaped developer life histories and 12 email interviews. This ethnographic data is complimented by direct participation by one of the authors in each of these areas as a Debian developer, Debian applicant, and New Maintainer process application manager.

Through an application of our ethnographic knowledge of Debian and F/OSS, we posit a theoretical model of ethical volunteerism applicable to a wide range of vocations. We argue that institutional independence, volunteer labor, and networks of trust can act as key elements in facilitating moral development within occupational groups. With a deeper ethical practice, professionals can be more self-reflexive of their vocational values and, as a result, better situated to respond to external social, political, and economic forces. These responses, grounded in a more ethical form of labor, can provide an important foundation for political response and social change.

# THE SOCIAL LIFE OF ETHICS
## Approaching Ethics Socially

Western philosophy in the tradition of Kant approaches ethics as a realm of behavioral possibilities describing the ways that one ought to be behave given abstract, universal conditions.[4] This philosophy assumes that virtuous actions are determined by individuals through intellectual deliberation and adoption of ethical codes or laws. In contrast to this tradition, social theorists including M. M. Bakhtin[5] have rejected this model for the individualistic adoption of formal and universal rules and suggested that a "social life of ethics" is created and embodied in community-based social practice and lived experience.[6] For Bakhtin, socially conditioned life events and historical contexts frame ethical choices and give rise to social moral orders. Cultivation and adoption of ethical standards and behaviors arise and are under constant consideration, engagement, and reformulation through socialization, education, and lived experience. As a result, ethics are neither simply imposed by obligations determined through top-down structures nor individually chosen. Consistency of ethical form is achieved through the relative stability of social practices. However, there remains a possibility for transformation as "events" and their social context change over time. It is through this process of socially defined pedagogy and community-based self-fashioning that the sociological, historical, and economic context of ethics plays a far more decisive role than do static "codes" of conduct.

While the social cultivation of ethics in Free software communities is new academic terrain, a number of anthropological works on medicine and science have provided compelling studies of the way that individuals adopt values and make moral choices through embodied action influenced by institutions, technologies, and historically informed activities (Galison, 1997; Good, 1994; Gusterson, 1996;

Lurhmann, 2000; Rapp, 1999). In these studies, social experiences act to bolster or thwart ethical practice, shape the nature of ethical relationships between individuals, and frame a community's ethical orientation. We feel that this analysis is best served by de-emphasizing moral philosophy in favor of a short review of examples from sociological and anthropological literature that demonstrates both the way that ethics are adopted and the role that they play in shaping individual identity, behavior, and action.

The medical field is profoundly embedded in an extremely complex ethical terrain, as doctors routinely face difficult ethical situations. Doctors are expected to navigate these situations with the help of intensive training and a professed allegiance to the Hippocratic Oath. The Oath affirms the commitment doctors hold, both as individuals and as a vocational group, to do everything possible to heal and assuage the emotional and bodily suffering caused by disease. Morally saturated, the Oath unambiguously affirms the importance of linking care and compassion with skill and science in the management of illness.

Yet a number of analysts (Good, 1994; Lantos, 1997; Lurhmann, 2000; Scheper-Hughes and Lock, 1987) question the "moral competence" of American doctors. Claiming that medical school teaching has "a strong central element of institutionalized sadism," they argue that doctors' training hinders the "the art of medicine" by socializing students into an ethic of stoicism and detachment that ultimately leads to a deeply ambivalent patient-doctor relationship (Lantos, 1997, p. 77). Doctors' internship and residency periods cement this ethic, as overworked and poorly treated interns and residents see patients as "the source of physical exhaustion, danger, humiliation, and that doctors are superior and authoritative by virtue of their role" (Lurhmann, 2000, p. 91). Additionally, critics describe how many physicians come to perceive bodies as divorced from the personal and emotional traits that shape illness through an overemphasis on the biological and material dimensions of disease during training (Good, 1994; Lurhmann, 2000; Sheper-Hughes and Lock, 1987). The health care market erects even more "ethical barriers" by restricting the time doctors spend with patients and the types of services given. The Hippocratic Oath simply cannot undo years of training during which physicians are mistreated and overworked while the "person" drops from their patients bodies.

Gusterson's (1996) study of nuclear scientists at Berkeley's Lawrence Livermore Lab is a compelling ethnography of ethical contextualization and transformation. Gusterson demonstrates how politically liberal scientists, many of whom were uncomfortable with the existence of nuclear technologies before they were employed at Lawrence Livermore, became full-fledged and passionate supporters of nuclear weapons while working in the laboratory. These scientists came to see their work as meaningful and important contributions to the United States' nuclear defense program which, in turn, they came to see as essential to national defense and social welfare. Gusterson describes how scientists used collective joking and identification with machinery to eclipse and bypass some of the more morally uncomfortable issues surrounding nuclear weapons use and proliferation.

The experiences of doctors and nuclear physicists exemplify the way that ethical considerations can arise out of socialization, institutional conditions, technologies, class and ethnic backgrounds, and socio-economic forces. Though individuals ultimately hold ethical views and make moral choices, a wider set of factors shape the way that social actors come to hold and exercise ethics. In considering the ethics of "hacking," it is essential to leave behind the individualized view of ethics in favor of a more socially grounded analysis to examine the role that institutional conditions and social practices play in shaping ethics.

## Ethical Cultivation in F/OSS Projects

For many hackers, an explicit practice of ethics grows alongside direct participation in various hacker activities, such as attending conferences, reading interactive news websites, and attending local hacker meetings. While ethical social orders are the product and process of communities, they unfold dynamically according to individuals' life circumstances. As a result, there are a diverse number of ways through which ethics are considered and fostered by different hacker groups. These are often defined by an institutional context (the university, the BBS, "the underground," F/OSS projects, the business sphere), a particular hacking community (gaming, phone phreaking, security hackers, the hacker underground, UNIX), and historical factors (the arrest of certain hackers, specific battles over intellectual property, new technologies, the development of Linux, etc.).

One such hacker community is the F/OSS community formed around the production of high quality software whose source code is "freely" accessible.[7] Through everyday informal acts that include software development, debate, informal discussion, joking, trust building among developers, and the experience of sharing and learning, Free software projects sustain a vibrant ethic of information freedom.[8] The community is divided into many projects of varying sizes and has been lauded for bringing collaboration to new productive heights and introducing novel legal schemes to protect the free use, reuse, and distribution of knowledge. However, the importance of F/OSS projects as a site for the cultivation of ethics has been, up until this point, egregiously overlooked.

For F/OSS and for the Debian project in particular, these social and legal elements play an important role in contextualizing the cultivation of ethics. F/OSS developers' ethical and political aesthetic is defined in contrast to the predominant market-based styles of software production and intellectual property. F/OSS uses software licenses like the GNU GPL (the GPL is an example of what F/OSS developers call a "copyleft" license) to materially and symbolically reterritorialize knowledge and, as a result, eliminate the need to engage in illegal activity to gain total access to and control over source code or knowledge.

Copyleft licenses use law to make source code permanently accessible. As a result, no illegal act is necessary for those seeking to "hack on" a copylefted idea or piece of software. Richard Stallman, the father of and philosopher behind the Free software movement and ideology, devised a license, the GNU GPL, and a political organization, the FSF, to confront the issue of restricted knowledge and to create the conditions for a transparent domain for the creation and distribution of computer software.[9] Through participation in this realm, hackers contrast their experience with closed and proprietary development with the social and legal context of F/OSS. This contrast, paired with the ethical practices internal to many projects, are the means by which social actors come to adopt clearly understood and well-formulated ethical stances in favor of transparency and openness. Participation in F/OSS projects like Debian contributes to the solidification of a pre-existent ethical commitment to information freedom.

## *Ethical Cultivation in the Hacker Public Sphere*

While F/OSS hackers are most easily associated with an individual project, like Debian, they participate in a hacker sphere that serves to substantiate freedom as a concept with moral relevance beyond technological issues. This public sphere exists in a number of online and offline channels and consistently provides a space for rational argument and discussion around a multitude of political and legal issues. It often involves discussions framed by news articles, legal cases, editorials, and community-generated commentary. This nascent public sphere can be traced back to the 1980s in (Bulletin Board Systems (BBSs), the hacker conference scene, Usenet groups, and mailing lists. The rise of large-scale free software projects like Debian, the growth of large hacker conferences like USENIX, Defcon, HAL, and HOPE, the use of blogging for personal editorialization, and the central role that interactive web news sites like Slashdot and Kuro5hin have provided hacker communities with conditions for a more inclusive and dynamic public sphere. In these forums, discussions about censorship, politics, technology, intellectual property, and the media help define certain social trends as ethically important to hackers and bring social concerns into the realm of hacking.

Large hacker conferences and the widely read daily interactive news websites like Slashdot play particularly important roles in the constitution of a public sphere that allows for acute moral reflection in distinct and complementary ways. Interactive news sites are visited many times daily and help foster a constant interchange of information within F/OSS communities. Hacker conferences are infrequent but highly charged, allowing pleasure, politics, learning, and hacking to closely intermingle. More than in any other space, conferences allow hackers to see themselves as part of a wider moral, cultural, and increasingly political community. They explore the ways that their ethical goals extend beyond hacking, as conferences frame hacker ethics in relation to questions of freedom of speech, democracy, and scientific production during panels, talks, and informal conversations. At conferences, hackers often hear speeches by influential and charismatic political or cultural figures like Richard Stallman, Lawrence Lessig, or Jello Biafra, who overtly frame "hacker issues" as something with much larger political importance.

Online collaborations and interaction within a hacker public sphere constitutes a realm of "ethical doing" that establishes and furthers the value for information freedom. Developers might "practice ethics" by attempting to convince another F/OSS programmer to license his or her software under a DFSG-compliant license; others participate in a flame-war over the nature of transparency in the Debian project; others attend hacker conferences and informally discuss ethics; others write politically charged articles in local Linux publications. Even humor, such as the jokes over the existence of a "closed-door" cabal of leaders in Debian, contributes to the ethical consideration of transparency, openness, and universality. For most developers, these mundane acts combine over time to prompt ethical enlargement in a collective and informal fashion.

## Ethical Cultivation in Debian

F/OSS developers' attitudes toward freedom, set in broad terms in the larger F/OSS community and reinforced through the broad hacker public sphere, are particularized and reinforced in the context of individual F/OSS projects. By analyzing the particularities of ethical cultivation within the Debian project, we can gain insight into the ways in which these ethical social orders are built. Our data shows that over the course of participating in the Debian project, developers move toward a more vigorous and overt ethical stance toward the uniqueness of their project and the importance of free software than when first joining. This stance is captured in the following quote:

*We are hard core about being free. Red Hat will bundle non-free. What Debian throws into the mix is that we are free and we are serious about being free. Certainly, you don't have to have such a devotion to it but the fact is that there is a group of people that are so dedicated to freedom and openness.*

While the nature of this position is clear, the source of this devotion is less obvious. Many assume that this passionate adherence to freedom and openness is an ethical belief that developers bring to the project. Many people assume Debian self-selects those who are already extremely committed to freedom. Although most Debian developers approach the project with at least a fundamental version of this position, it is through participation over time that this ethical skeleton is given flesh. What was originally limited to a functional, "engineering ethic" of producing high quality software is transformed into a wider moral position as sharing and openness become ethical ends in themselves.

In this way, a general ethical consciousness permeates the underlying spirit of the project and is promulgated by developers through everyday acts of sociability. Debian acts as a moral refinery that enlarges the ethical inclinations of many developers in substantial ways; developers learn about legal issues and the differences between F/OSS licenses, as well as Debian social and technical policy in the process of their Debian work. Despite the variant ethical orientations that Debian developers hold prior to joining, participation usually contributes to the creation of a more explicit moral commitment to information freedom or substantially deepens existing commitments to freedom. For some, Debian's primarily engineering- and production-based ethic evolves to include other, non-technological issues related to freedom.

Prospective developers valuing information freedom are often attracted by Debian's reputation as the F/OSS project with the strongest and most explicit ethical commitment. This commitment is instantiated in Debian's Social Contract and in the DFSG and is implicit in mailing list and IRC (Internet Relay Chat) discussions. Other Debian developers are motivated more by pragmatic concerns including the desire to package a piece of software in Debian or the need to "give something back" to the Debian community whose software they use and enjoy. Despite these non-ethical orientations, and usually in combination with them, prospective Debian developers almost unanimously agree that open methods for software development produce better, higher quality software. Nonetheless, participation in Debian over time represents a form of ethical learning and socialization in which new values are adopted while others are refined and enlarged.

Developers' formal entry into Debian, The New Maintainer Process that is discussed in detail later in this chapter, marks a rite of passage into a project where ethics are made manifest through discussions, writings, and technical procedures. In this way, experiences that begin with the New Maintainer Process shape ethical sensibilities, influence the desire to continue participating, and form the basis for more overt forms of political engagement.

After entry into the project, ethical expansion continues through everyday acts that often include legal discussion on mailing lists and conversations on IRC. In interviews, many developers claim that although they were always committed to freedom, their knowledge of the legal issues surrounding software was bare until they began participating in Debian. Given the importance of licensing in the domain of F/OSS, and Debian in particular, it is not surprising that an ethical consciousness is instilled in Debian developers through the discussion of software licensing. This discussion is constant and complex; the debate engages with the fundamental meaning of software freedom and immerses it in a larger moral context. The following message from a mailing list demonstrates the way in which many developers equate and understand the role of free software as a meta-guarantor of freedom:

*Free software should create a sort of economy in which things are the way they would be if there were no copyrights at all. That's the intuition. In other words, when I write free software, I renounce the ability to control the behavior of the recipient as a condition of their [sic] making copies or modifying the software. The most obvious renunciation is that I don't get to demand money for copies. But I also don't get to demand that the person not be a racist; I don't get to demand that the person contribute to the Red Cross; I don't get to demand that the recipient contribute to free software. I renounce that little bit of control over the other person which the copyright law gives me, and in that way, I enhance their [sic] freedom. I enhance it to what it would be without the copyright law. You might say that public domain is good enough. But free software is about creating an economy of such freedom.*

The Debian developer IRC chat channel, where a series of topics are set each day, is another space where hundreds of developers connect. While many topics relate to technical issues, political and legal issues are almost equally frequent. These political topics often follow the form of the following example: "Topic for#debian-devel is SSL Patented? - http://www.theinquirer.net/?article=8029." Many developers follow the "topic link of the day," which becomes a shared concern for Debian developers.

While most developers contribute infrequently to mailing list discussions, many follow the higher traffic lists that pertain to Debian development, legal issues, and policy. On these lists, developers frequently raise questions about licensing and intellectual property in a way exemplified by the following message.

*Perhaps we need to be thinking about alternative ways to uphold the "protection of the moral and material interests resulting from...scientific, literary or artistic production[s]"? Surely existing copyright, trademark, and patent regimes, to say nothing of "work-for-hire," "paracopyright," and "trade secret" concepts, are not the only ways to give Article 27 force and meaning. In other words, I don't think it necessarily follows from Article 27 that we must have a global oligarchic hegemony of media corporations dictating to us what we shall and shall not read, watch, perceive, write, and share with our fellow human being[s].*

For this developer and for many others, F/OSS development blurs into a moral reflection on intellectual property. While the ethical strength and orientation of developers varies among individuals, ethical deepening plays a consistent role in many of the everyday acts that constitute participation in the project.

Licensing discussions are quite amenable to ethical reflection. However, it is not just "talk" about licensing, intellectual property, and transparency that forms the basis of ethical cultivation in Debian. The social experience of sharing and learning is another area where ethics are honed. As developers learn, they develop dedication to the organization and to its method of open collaboration. As they volunteer their time and skills, many feel that they gain tremendously through participation. These gains include material benefits such as free technical tools and more abstract social benefits like the satisfaction of building a quality product, peer respect and admiration, new collaborative abilities, technical skills and knowledge, and a sense of belonging in the Debian community. One long-time participant expressed this feeling during an interview in the following way:

*I've learned about the intricacies and history and every detail of the Debian distribution, how its disparate components fit together, how its packaging system works. I've learned all sorts of little oddities of technical lore, and I've picked up a few programming languages and a lot of general programming knowledge. I've learned how to collaborate with folks spread out over the world and across time zones. I've learned how to argue effectively online, and I've learned that even though I tend to shy away from arguments, there are things that are worth arguing for. I've learned how to think about the large effects work can have on a project, and how to take responsibility for and plan out those effects before hand.*

Given the penchant for learning and knowledge in the hacker community, it is not surprising that many developers come to value Free software projects as educational spaces. They feel that meaningful learning is grounded in the type of openness and transparency exemplified by Debian. Though many hackers are able to learn a tremendous amount on their own, mailing lists and IRC provide a means for developers to tap into the collective knowledge of their community. The experience of learning and sharing technical and non-technical skills contributes to the strong allegiance to Debian held by many developers and plays a role in the cultivation of ethics. During interviews, many developers claim that they have learned at least as many technical skills through developing free software as in more formal learning environments. In this educational space, the stereotypical geek qualities of elitism and bravado give way to a desire to help others. The fundamental hacker pursuit of knowledge becomes an endeavor that is recognized as a fully non-technical social process.[10] Although we can conceptually separate hackers' moral drive from their "engineering ethic," the two drives intermingle and interact to bolster each other. The technical success of free software projects and the personal gains from participation jointly reinforce the hacker belief in openness and information sharing with relevance and applicability in other domains of social and political life.

## The Debian New Maintainer Process: Trust in Virtuality

The incredible explosion of online communities, a broad term referring to grounds ranging from MUDs and MOOs to IRC, blogging, online activist organizing, and F/OSS, has been voraciously met by various theories and analyses about the "authenticity," "realness," and "nature" of such communities. The overarching goal of these analyses has been to determine whether Internet-based social orders, forms of identity, political engagements, and other types of community building are "really real" given the lack of bodily interaction and material presence of things and objects; theorists ask whether moral orders, empathetic care, political engagements, and meaningful relationships can exist when bodies rarely meet and text is the primary mode of communication and interaction.

Opinion and research on this subject varies greatly, although three major positions have emerged over the last 10 years. One is the opinion that social life on the Net reflects authentic forms of human identity, interaction, sociality, and political engagement (Doheny-Farina, 1996; Gulia & Wellman, 1999; Mitra, 2000; Negroponte, 1995; Rheingold, 1993). Supporters of this position believe that a real community is imagined to sustain emotional ties, be morally deep, or exhibit other forms of human closeness. Some of this work is Utopian in spirit and, in the opinion of skeptics, reflects a "need to authenticate," to "prove" that certain online spatial communities are meaningful and real. Some arguing this position go as far as claiming that online communities are more substantial than urban modern communities because they offer a form of emotional closeness and deep connections that are rare in modern urban settings (Rheingold, 1993).

In contrast is a more critical and, at times, even dystopian group that claims that online communities are socially thin and fake and can even act as a form of political misrecognition and domination. Some argue that without face-to-face interaction, true moral concern for others and communities, as a result, are impossible (Ostwald, 2000; Robins, 2000; Robins & Webster, 1999; Willson, 2000;). There are some in this group that argue that virtual spaces are too utterly synthetic and neutral to be authentic in a way that is not unique to cyberspace. They define virtual sociality as a manifestation of postmodern and hyper-capitalist polyesterization and compare it to extreme commodification and the proliferation of fake urban spaces best represented by Disneyworld and shopping malls (Otwald, 2000; Robins, 2000; Robins & Webster, 1999; Terranova, 1996).

Though these two sets of opposing positions differ dramatically in their description of online communities, both share certain key assumptions. They are concerned with what the authentic nature of community is (or should be) and are informed by a very primordialist sense of community in which a universal self psychology is assumed. A third position offers an alternative to such bi-polar treatments of online communities. Less concerned with what a community "ought to be," this literature calls for and offers a more refined analysis of the ways that sociality, politics, and community are sustained virtually (Danet, 2001; Fisher, 1999; Hakken, 1999; Hand & Sandywell, 2002; Kirshenblatt-Gimblett, 1996; Slater, 2002). Noting that the Internet is not a unitary space for interaction and politics (Hand & Sandywell, 2002; Miller & Slater, 2000), calling attention to historical conditions and the unit of analysis (Fisher, 1999; Hakken, 1999) and highlighting the ways that micro-sociological forms of interaction build materiality, relationships, and forms of play (Danet, 2001; Kirshenblatt-Gimblett, 1996; Miller & Slater, 2000; Slater, 2002), these studies provide a more compelling way to approach the subject of social interaction in virtual spaces. Authors arguing this position throw out the quest "to authenticate" (to prove the realness of interactions) and focus on how social orders are built and sustained and the ways that virtual and non-virtual interactions act in concert to produce unique forms of social expression and life.

Our research into the primarily virtual communities of Debian and F/OSS supports this final position. Comparing the reality of "virtual" ethical orders to offline orders is irrelevant in the context of our analysis. To gain insight into ethical cultivation in F/OSS, we must focus on the way that values are built and sustained through different social and technological practices within Debian and the hacker public sphere. One of the most relevant dimensions of the "collective" nature of ethical cultivation in F/OSS projects is the manner in which "embodied" ethical learning and the rise of social values occur in largely disembodied virtual spaces. As a prerequisite to the cultivations of ethics within any community, trust plays an essential role in Debian and in F/OSS more generally. Set in virtual space, the cultivation of trust must assume new forms and an increased importance.

A micro-sociological analysis of Debian's New Maintainer (NM) Process provides additional insight into the way that Debian attempts to facilitate trust building in "virtuality" and the way that trust helps facilitate the meaningful cultivation of ethics within the project. In our examination of the system, we focus on the creation of cryptographic trust and the role that explicit ethical engagement with the Social Contract and DFSG plays in the NM procedure. We highlight the way that virtual trust building retains connections to non-virtual interactions and the way that technical and social aspects of development intertwine to create socially productive forms of trust. Trust building in Debian demonstrates the flexible and innovative ways that social practices unfold in what are mostly virtual domains of practice.

When the Debian project was young, it was a close-knit community where most project members were familiar and interacted with a majority of other active members. As a result of this constant interaction, trust grew that enabled verification of identity, integration into Debian's ethical community, consistent familiarity with Debian project policy, and standardized technical competence. Prospective members needed only to informally demonstrate their technical competence and to claim knowledge and adherence to the project's social contract and policy before being admitted to the project. Socialization occurred organically and inter-developer trust was built almost wholly through personal and group-wide interactions.

During Debian's growth in size from about one hundred developers to nearly one thousand, the project found itself in a crisis. New members were admitted at rates faster than the project's ad hoc social systems could integrate them, and the group grew to a point where close interaction among many members was no longer possible. In reaction, Debian halted the acceptance of new maintainers until the project could develop a system to build trust—cryptographically, ethically, and technically—in a systemic and reliable manner.

The procedure developed was Debian's New Maintainer (NM) Process that aims to consistently facilitate and build trust in a virtual space among a very large group of developers from around the world. Sustaining trust among this large, culturally and linguistically diverse group poses unique challenges that the NM process is designed to remedy. The NM process creates a standard that all developers must meet - Richard Stallman, a Debian NM applicant, is held to the same standard as developers younger than the Free Software Movement. NM is structured not only as a test, but as a process for learning, mentoring, and integration into the project where prospective developers work closely with at least one older, "trusted" project member. While we emphasize some of the ethical and social elements of NM, it is important to note that it is just as much a method of displaying technical proficiency and a process of technical mentoring.

Before prospective developers even apply, they are first asked to identify the contributions they plan to make to the project. They are encouraged to demonstrate their commitment to Debian, to express why they want to join, and to display some level of technical proficiency. For most developers, this involves making a package and, because only existing developers can integrate a piece of software into the larger GNU/Linux distribution, they need to find an existing developer to "sponsor" their work. New maintainers work closely with their sponsors, who check their work for common errors and take partial responsibility for the new maintainer. This step is important because, in addition to technical skills, the new volunteer begins integrating into the social sphere of the project. Prospective developers are encouraged to join mailing lists and IRC channels that provide the medium for technical communication and an important piece of the Debian public sphere.

An NM's sponsor often acts as the new developer's advocate when the maintainer applies for membership in the project. Advocates are existing developers who vouch for new developers and their history of and potential for contributions to the community. After advocation comes the assignment of an application manager, an existing Debian developer, who handles the rest of the new maintainer process acting in a complex role that is mentor, teacher, examiner, and evaluator. The NM process that follows consists of three major steps. The three stages attempt to ascertain and confirm the new maintainer's identity, knowledge of and position on free software philosophy, and their technical expertise and knowledge. In terms of the social cultivation of trust and ethics, the first two steps are particularly significant within the Debian community.

The act of proving identity is accomplished through a complex hybridization of social and technical mechanisms involving cryptographic trust-building with Pretty Good Privacy (PGP) or, in Debian's case, its free software clone, the GNU Privacy Guard (GPG). This step is designed around the fact that each developer has control over a carefully controlled and fully unique cryptographic key attached to its owner's name and email address. These keys are used to generate "signatures" that can be used to verify that a particular message, text, or piece of software originated with the possessor of a particular key. When key owners meet in person, they prove their identity to each other by exchanging pieces of government-issued picture identification and identifying information about the key they use. Having traded this information, developers later place their unique cryptographic "signature" on each other keys to verify the fact that the developer signing has connected the key being signed with the name on the key with the individual in possession. In the first step of the new maintainer process, prospective developers use this cryptographic method to "prove" their identity to their application manager; to accomplish this, they must obtain the cryptographic signature of at least one existing developer.

As nearly every hacker within Debian has a key signed by at least one existing developer, and as many developers have keys signed by numerous others, nearly all maintainers are connected by what they call a cryptographic "web of trust." In the manner of the famous "six degrees of separation" model, Debian can use cryptographic algorithms to prove that—while it is now clearly impossible for every developer to have met each other developer—every developer can have met a developer, that has met a developer, that has met a developer, etc., until every developer is connected. Of the 963 keys in Debian official "keyring," 857 can trace a connection to every other developer through an average of less than five other developers; no key is more than an average of nine links away from any other key. Debian's administrative software depends heavily on these keys to identify users for the purposes of integrating software into the distribution, for controlling access to machines, for allowing access to a database with sensitive information on developers, and for restricting publication to announce-only email lists.

The development of this web of trust in the NM Process plays an essential role in Debian. A story illustrates this point: When it became clear that a developer who occupies an important technical position was unconnected to Debian's web of trust, a large number of developers expressed alarmed concern and anger on a Debian mailing list. Developers' strong and uniform reactions demonstrated the essential nature of these infrequent face-to-face interactions. Within three days, three nearby developers had driven to meet the individual in question and had succeeded in bringing him into the cryptographic web.

Integration into Debian's web of trust is an essential first step in new maintainers' integration into the Debian project. However, because this method of proving identity requires face-to-face networking with existing members of the project, it helps to foster the close-knit community feeling that the NM Process attempts to replicate or replace. These key-signing meetings, usually long social meetings over drinks or coffee and often with a number of Debian hackers, allow new maintainers to discuss Debian, its policies, its software, and its philosophies with other developers in a face-to-face manner and with a strong community feeling. This process connects and leads into the second, and often the most rigorous, part of the NM Process—philosophy.

During the "philosophy" step of the NM Process, application managers ask prospective developers a series of questions on Debian and Free Software philosophy. While general knowledge of the definition and philosophy related to F/OSS is essential, the questions usually revolve around Debian's Social Contract and the DFSG. New maintainers are asked a series of questions, which often vary substantially between application managers, to demonstrate their familiarity with the texts of these documents, their ability to apply and synthesize the concepts encapsulated within them, and to articulate their commitment and agreement. While each NM must agree to the social contract, the philosophy test is not used to ensure that all developers approach free software identically, but rather to ensure that all Debian developers are knowledgeable of, interested in, and committed to Free software discourse. Open-ended questions often turn into longer email conversations between application managers and NMs. While dynamic and heterogeneous, the process ensures both a common familiarity with ethical issues in F/OSS development and a consistent level of commitment to a set of ethical beliefs. The process aims to create consistency in developers' critical ability to dialog around a common philosophy, as opposed to prompting the individual adoption of a prescribed approach to Debian's moral code.

Many developers claim that writing answers to the philosophy section of their New Maintainer application was one of the first times that they coherently and explicitly compiled their ethical beliefs on software development. The act of externalizing ethics made some consider the wider social issues of freedom related to transparency, openness, and democracy. Although some developers focus on the moral implications of "freedom" in relation to technical questions, other developers use their NM application as an opportunity to enlarge the moral scope of questions of freedom. This is evidenced by the following excerpt from a New Maintainer application essay submitted by a prospective Debian developer:

*The Social Contract and the DFSG represent a very unique idea. In this day and age where society (at least in the U.S. and some other first-world countries) encourages individualism and tries to divide the people and control them, it is very refreshing to read the Debian Social Contract. Proprietary software made by commercial software companies/developers is exactly that, commercial. Those companies/developers are only about profit or advancing their agenda and will do what they need to in order to maximize that. Often this conflicts with doing the right thing for the user and here are some examples.*

In this short excerpt, it is clear that this developer reflects on the differences of moral code contained within the Debian Social Contract and those that are developed in society, as well as the differences between proprietary and Free software development. In contrast with other realms of hacker social life, including the corporate sphere, Debian's NM process represents an important nexus for the expression of ethics.

In the final step of the New Maintainer Process, applicants must demonstrate that they have the technical wherewithal to be trusted with the ability to integrate software into the Debian archive and to represent Debian to the world. This test is often filled with the presentation of a clean, policy-compliant, and bug-free example of the type of work the applicant aims to do within Debian (e.g., a package), although this is increasingly complemented by a series of technical questions.

In these ways, NM fulfills a community-forming function by establishing a common denominator of social, philosophical, and technical principles and by introducing the prospective member to the Debian hacker sphere and initiating the NM into Debian's critical dialog culture. Additionally, the process does not merely decide who can be trusted and who cannot. Rather, it establishes trust with the prospective Debian maintainer and integrates them in a social network of peers.

The significance of the F/OSS project as a site for ethical cultivation is not just that ethics "exist" but how they are made visible and accessible to wider groups of people. Ethics, much like other forms of cultural values, emerge through socialization and practice and are also shaped by socio-economic forces and considerations. Lived experience motivates action; spaces are needed to transform ethical inclinations into social realities. Free software demonstrates the importance and strength of cultivating ethics through everyday acts that constitute a sort of ethical habitus and prompts us to ask how ethical cultivation might be developed in other domains of professional or subcultural activity.

Understanding the internal dynamics of how to achieve ethical cultivation allows for more clear and realistic thinking on the creation of sustained ethical practices among other vocational groups where ethical actions have an immense social and political impact - like medicine and science. What F/OSS offers is not just new models for collaboration and novel ways to protect knowledge but a model for independent and volunteer associations in which ethics can be given fuller consideration among members of a vocational group.

# ETHICS AND THE CRISIS IN SCIENCE AND MEDICINE, NEW MODELS FOR VOCATIONAL ASSOCIATION

In a piece that asks whether F/OSS is like science, Chris Kelty (2001) concludes that F/OSS "is in fact part of science — and increasingly an essential part of it." In fact, as scientific knowledge and processes are increasingly affected by commercialization and privatization, Kelty implies that F/OSS may be more "scientific" than much of contemporary science.[11] We understand the second half of his article's title, "Free Science," as more than a description; it is an impassioned plea to sustain the standards, mores, and practices that guarantee openness and accessibility in science. Kelty warns that information sharing is essential, but not "intrinsic," to science and that it must be cultivated through a combination of social, literary, and moral techniques that include the open publication of work, conference participation, and active collaboration. Kelty (2001) concludes: "Openness can not be assumed, it must be asserted in order to be assured."

Like Science, the "art of medicine" has been affected by adverse institutional and economic conditions. As discussed earlier in this essay, the medical profession is marked by the problematic nature of socialization in modern medical training. This is compounded by expensive and controversial courting of physicians by pharmaceutical industry representatives. The enforcement of drug patents, especially in countries where average individual yearly salaries can not cover even a month of many pharmaceutical treatments, are particularly disconcerting. In the case of patents and drug costs, human lives are ultimately at stake; on the whole, physicians have been quiet on these issues.

Many doctors and scientists are alarmed in feeling that their professions have moved toward privatization and clinical stoicism with few vocal objections; the Hippocratic Oath in medicine and the traditions of openness and sharing in science are designed to protect against such shifts. In response, several scientists and doctors have attempted to raise awareness and response in their communities.[12] This response, however, has been limited and piecemeal. We believe that this silence signals the need for more active avenues of ethical socialization internal to medicine and science that form an ethical basis for engagement in collective responses to important political issues in their fields.

We can not attempt to offer a prescription for assuring openness in science and creating a more ethically active medical practice. We are aware of the number and complexity of considerations that must be taken into account in an exhaustive discussion of the democratization and ethicalization of science and medicine.[13] However, we feel that F/OSS' avenues for ethical socialization act as a model that provides one important step toward more ethically dense scientific and medical practices.

As scientists' opportunities to engage in "closed" forms of academic, corporate, and governmental science increase, and as medical and scientific production are increasingly affected by market pressures, it is increasingly important that scientists, doctors, and the full range of professions related to science and medicine are afforded opportunities to participate in vocational and volunteer spaces at least partially independent of academic, governmental, and corporate ties.[14] Meaningful change in a vocational field cannot be prescribed from the top-down or outside-in but must be instigated by the field's practitioners. It is the practitioners who are most familiar with the moral, political, and social issues relevant to their practices and with the ways that their activities might affect the societies in which they are situated. Their vocational knowledge, guided by an ethical vision, can form the basis for a more substantial political response.

F/OSS demonstrates how institutional independence, volunteer labor, and the cultivation of trust act as three fundamental pillars in the facilitation of ethically active vocations. In F/OSS, institutional independence is often framed in terms of "freedom"; while inherently ambiguous, its role is tremendously productive. In the scope of Free software licenses, "freedom" refers to individual rights as a producer, user, and distributor of software. In discussions among developers, the term morphs dynamically into a broad, but rich, social concept. During ethnographic interviews, Debian developers describe a rich diversity of approaches to freedom, yet, for most, freedom means that individuals should be able to create and learn without significant technical or legal barriers.

Ultimately, Free software is "free" because it allows hackers to creatively innovate and collectively create. It achieves a form of productive freedom that is rooted in conventions, values, techniques, and methods constituted around the act of writing code. Programmers in F/OSS projects have the freedom to collectively determine technical imperatives, styles of development, modes for the inclusion of members, ethical mores, and codes of behavior. Institutional independence allows for the collective body of social actors to decide what the nature of a certain activity should be and to practice the ethical codes and norms that are best suited to achieve the goals of such activity.

While absolute institutional independence is difficult, many F/OSS projects exhibit a high degree of productive independence for the internal cultivation of technical and ethical codes.[15] Additionally, the contrasting experience of working in both proprietary and F/OSS settings creates a space for a critical awareness of technical and ethical standards, the ways they function, and their relationship to social forces. With this critical awareness, F/OSS practitioners hold a clear sense of how social, economic, and legal conditions alter their vocation, understand how their own work might affect others, and feel strongly about what should or could be done as a result. It is only with critical reflection that a socially situated response to dynamic socio-economic and institutional conditions is garnered.

Associational independence already exists in organizations like the American Medical Association and other scientific and vocational groups. However, active, sustained volunteer work is rarely a feature of membership. This type of independence without labor dilutes the potential for active ethical practice among vocational practitioners. It is *labor* that realizes vocational goals — whether it is caring for the sick, making software, or building bridges — and it is *labor* that forms the basis for both the cultivation of standards and the corresponding ethical codes that uphold such activities. As we have argued in this chapter, action is motivated by lived experience and supportive spaces are essential to cultivate the social practice of ethical life. F/OSS demonstrates how the volunteer nature of certain vocational associations help to ensure independence, while labor allows for a more sustained ethical practice to emerge.

In addition, trust between participants is essential to realize the forms of collaborative labor needed to foster the application of ethical values; trust is dialectically reinforced through sharing, working, and learning. Even in F/OSS projects with clearly defined leaders, many decisions are made through a collective enterprise arising through a slow process of debate and discussion over mailing lists and IRC. Trust builds respect, understanding, and affinity in groups, which allows collaboration to flourish and makes consent in diverse groups possible. Trust turns fierce discussion and debate into a potential consensus-based decision-making process. This consensus facilitates groups' productive freedom to collaboratively determine the nature and content of their activities and underlying ethics.

F/OSS projects provide innovative examples of long-distance collaborations that have used the Internet to create a common space for the cultivation of ethics. The "location" of F/OSS volunteer associations within the technologies of the Internet makes sense among a vocational and cultural group whose life-world is intertwined with these technologies. While communication technologies important in F/OSS development facilitate long-distance labor, coordination, and identity building, they might be deployed differently to achieve other goals by other volunteer associations that may be more effectively situated in physical space.

In the wake of the widespread success of F/OSS software in recent years, F/OSS communities have garnered significant attention from groups outside the software development community. Researchers, activists, and corporations have been seduced by the wide applicability of F/OSS development methods, organizational models, collaborative models and licensing schemes to non-technical domains of activity. At the same time, F/OSS is confounding to many because classical economic models struggle with the complex and atypical nature of F/OSS' social practices, exchanges, and values. These same processes, when approached as a social practice, can reveal some of the diverse and culturally situated ways that humans collectively organize, create, share, and give.

It is in this spirit that we approach F/OSS as a domain of practice, the hacker public sphere and Debian as sites within this domain, and the establishment of trust through Debian's New Maintainer Process as a micro-sociological example of the mechanics of ethical cultivation in virtual F/OSS communities. It is in this way that we find insight into ethical cultivation as another important model of F/OSS activity applicable to the world beyond software.

# REFERENCES

Bakhtin, M. M. (1984). *Rabelais and his world.* Bloomington, IN: Indiana University Press. Translator: Helene Iswolksy.

Bakhtin, M. M. (1993). *Toward a philosophy of the act.* Austin, TX: University of Texas Press. Translator: Vadim Liapunov.

Danet, B. (2001). *Cyberplay: Communicating online.* Oxford, UK: Berg Publishers.

Dickson, D. (1988). *The new politics of science.* Chicago, IL: University of Chicago Press.

Doheny-Farina, S. (1996). *The wired neighborhood.* New Haven, CT: Yale University Press.

Fischer, M. (1999). Worlding cyberspace: Toward a critical ethnography in time, space, and theory. In G. Marcus (Ed.), *Critical anthropology now: Unexpected contexts, shifting constituencies, changing agendas.* Santa Fe, NM: SAR Press.

Galison, P. (1997). *Image and logic: A material culture of microphysics.* Chicago, IL: University of Chicago Press.

Good, B. J. (1994). *Medicine, rationality, and experience: An anthropological perspective.* Cambridge, UK: Cambridge University Press.

Gulia, M. & Welmman, B. (1999). Virtual communities as communities: Net surfers don't ride alone. In M. Smith & P. Kollack (Eds.), *Communities in cyberspace*. London and New York: Routledge.

Gusterson, H. (1996). *Nuclear rites*. Berkeley, CA: University of California Press.

Hakken, D. (1999). *Cyborgs@cyberspace*. New York and London: Routledge.

Hand, M. & Sandywell, B. (2002). E-topia as cosmopolis or citadel: On the democratizing and de-democratizing logics of the Internet, or toward a critique of the new technological fetishism. *Theory, Culture, and Society*, 19(1-2): 197225.

Kelty, C. (2001). Free software/free science. *First Monday*, 6(12). Available Online: http://www.firstmonday.dk/issues/issue6_12/kelty/.

Kirshenblatt-Gimblett, B. (1996). The electronic vernacular. In G. E. Marcus (Ed.), *Connected: Engagements with media*. Chicago, IL: University of Chicago Press, late edition: 3.

Lantos, J. D. (1999). *Do we still need doctors? A physician's personal account of practicing medicine today*. New York: Routledge.

Lurhmann, T. M. (2000). *Of two minds: The growing disorder in American psychiatry*. New York: Alfred A. Knopf.

Miller, D. & Slater, D. (2000). *The Internet: An ethnographic approach*. London: Berg.

Mirowski, P. & Sent, E.-M., Eds. (2002). *Science bought and sold: Essays in the economics of science*. Chicago, IL: University of Chicago Press.

Mitra, A. (2000). Virtual commonality: Looking for India on the Internet. In D. Bell & B.M. Kennedy, *The cybercultures reader*. New York and London: Routledge.

Moody, G. (2001). *Rebel code: Inside Linux and the open source revolution*. Cambridge, MA: Perseus Publishing.

Negroponte, N. (1995). *Being digital*. New York: Alfred A. Knopf.

Ostwald, M. (2000) Virtual urban spaces. In D. Bell & B.M. Kennedy, *The cybercultures reader*. New York and London: Routledge.

Rapp, R. (1999). *Testing women, testing the fetus: The social impact of amniocentesis in America*. New York: Routledge.

Rawls, J. (1971). *A theory of justice*. Cambridge, MA: Harvard University Press.

Rheingold, H. (1993). *The virtual community*. New York: HarperPerennial.

Robins, K. (2000). Cyberspace and the world we live. In D. Bell & B.M. Kennedy, *The cybercultures reader*. New York and London: Routledge.

Robins, K. & Webster, F. (1999). *Times of technoculture*. New York and London: Routledge.

Scheper-Hughes, N. & Lock, M. (1987). The mindful body: A prolegomena to future work in medical anthropology. *Medical Anthropology Quarterly*, 1(1): 641.

Slater, D. (2002). Making things real: Ethics and order on the internet. *Theory, Culture, and Society*, 19(5/6): 227245.

Terranova, T. (2000). Post-human unbounded: Artificial evolution and high-tech subcultures. In D. Bell & B.M. Kennedy, *The cybercultures reader*. New York and London: Routledge.

Wayner, P. (2000). *Free for all: How Linux and the free software movement undercut the high-tech titans*. New York: Harper Business.

Willson, M. (2000). Community in the abstract: A political and ethical dilemma? In D. Bell & B.M. Kennedy, *The cybercultures reader*. New York and London: Routledge.

# ENDNOTES

[1] Research for this chapter was made possible by a Social Science Research Council grant for the study of Philanthropy and the Nonprofit Sector and a National Science Foundation Grant (Award #0217470).

[2] Packaging is the systematic compartmentalization, customization, and standardization of existing software. Packaging forms the bulk of most Debian volunteers' work for the project.

[3] Debian's social contract and the DFSG are available online at http://www.debian.org/social_contract.

[4] This philosophical orientation has a long history in Western thought evident in such writers as Pascal and Descartes and is an approach that enjoys continued popularity in contemporary philosophical works on ethics and politics. John Rawls' theories of a just state in *A Theory of Justice* (1971) and Peter Singer's descriptions of a utilitarian, universalist stance towards life, death, and disability exemplify the continued salience of the treatment of ethics as abstract, deontological concepts created through individual choice and the imposition of codes of conduct.

[5] Various works of the Russian literary theorist Bakhtin engage with the question of ethics. Two of the most relevant are *Toward a Philosophy of the Act* (1993) and *Rabelais and His World* (1984). *Toward a Philosophy of the Act* deals explicitly with the question of morality and is written as a response to Kant's formulation of the categorical imperative. Avoiding absolutist or relativist views of ethics, Bakhtin argues that responsibility forms the particular situations that call for an ethical response. In *Rabelais and His World* Bakhtin situates the aesthetics and ethics of popular culture within particular historical periods, places, and events. In discussing medieval feasts, fairs, and markets, Bakhtin describes how medieval folk culture embodied and fostered certain genres of the grotesque and laughter that served to render visible, through inversion and parody, official and political ideologies and liberated people from other forms of moral condemnation.

⁶ There is a long tradition in Western social theory and philosophy, from Aristotle to Marx, Weber to Durkheim, and Bourdieu to Foucault, that has emphasized the way in which norms and values are cultivated through embodied social action within institutions and practice. We single out Bakhtin because he explicitly argues against the Kantian tradition in *The Philosophy of the Act*. His dialogical model of social creation is one in which social forms arise out of response, not rules, and is well suited to our analysis of F/OSS.

⁷ Many F/OSS hackers remind their users that they mean free as in "free speech" (libre), although often as in "free beer" (gratis) as well.

⁸ The concept of freedom within F/OSS communities takes on a broad, inclusive, and nuanced quality whose nature and history falls outside the scope of this chapter.

⁹ The copyleft and other such licenses have been integral for the free and open source production of software. However, many other social and technical conditions were necessary to really open up a vibrant space for the networked long distance and open collaboration of code. See Moody (2001) and Wayner (2000), among others, for accounts of the different factors that contributed to the constitution of this field of legal and technical production.

¹⁰ Granted, how one goes about learning from others is a highly stylized practice. There are culturally correct and incorrect ways by which one learns from other developers on mailing lists and IRC. We mention it to note that learning and sharing is uniquely deployed in hacker circles.

¹¹ See Dickson (1998) and Mirowski and Esther-Mijarm (2002) on the trend toward increased commercialization in science. For a comprehensive bibliographic list of empirical studies that examine how commercialization and sustained industry involvement have shifted the practices of science and medicine see *Integrity in Science's* selected bibliography at http://www.cspinet.org/integrity/bibliography.html.

¹² For example, No Free Lunch: http://www.nofreelunch.org, Doctors without Borders, Integrity in Science: http://www.cspinet.org/integrity/, Center for Science in the Public Interest: http://www.cspinet.org/ are all organizations who aim to shift the current ethical direction of their vocation. With the exception of Doctors without Borders, these organizations don't combine forms of active vocational labor with their politics.

¹³ See the last chapter in Dickson (1998) on some recommendations on how to democratize science in America. Though written in the 1980s, many of his insights are still relevant.

¹⁴ We are not claiming that F/OSS is fully independent from industry involvement or the government. Academic institutions such as Berkeley and MIT have historically played crucial roles in the development of key F/OSS applications; the Internet, where most development occurs, would not be possible without government funding and private research and development; many free

software developers have high paying jobs that give them the financial luxury to volunteer time; and corporations like I.B.M and Hewlett Packard have contributed significant funds to F/OSS projects. Despite these and many other connections (which we do not want to mystify as insignificant), it is arguable that F/OSS projects do have a degree of independence from the institutions and social structures that give them different forms of underlying technical and financial support. If HP files for bankruptcy, Debian will still exist as a project.

[15] Different F/OSS projects have different types of ties and relationships with governmental, corporate, and academic institutions that shape the degree and kind of independence. For example, many of the top kernel developers are employed by technology companies, and thus the Linux kernel exhibits a different type of relationship to the corporate sphere than that of other projects like Debian in which a much smaller percentage of its developers work on Debian for corporations.

# About the Editor

**Stefan Koch** is assistant professor of Information Business at the Vienna University of Economics and Business Administration in Austria. He received an M.B.A. in Management Information Systems from Vienna University and Vienna Technical University, and a Ph.D. from Vienna University of Economics and Business Administration. Currently, he is involved in the undergraduate and graduate teaching program, especially in software project management and ERP packages. His research interests include cost estimation for software projects, the open source development model, software process improvement, the evaluation of benefits from information systems and ERP systems.

# About the Authors

**L. Angelis** received a B.Sc. and a Ph.D. in Mathematics from the Aristotle University of Thessaloniki (AUTh), Greece. Since 1999, he has been a faculty member (currently he is a lecturer) in the Department of Informatics, AUTh. His research interests include statistical methods with applications to information systems, simulation methods, and algorithms for optimization problems.

**I.P. Antoniades** is a postdoctoral fellow at the Aristotle University of Thessaloniki, Department of Informatics, Greece. He received a B.A. in Physics from the University of Chicago, and a Ph.D. in numerical and simulation methods in Solid State Physics from the Aristotle University of Thessaloniki Physics Department (2000). He teaches undergraduate level courses in programming, mathematics, and theoretical informatics. His research interests include simulation methods in software development processes and quantum computing.

**Thomas Basset** is a student in sociology from the Ecole Normale Superieure de Chachan (France) and a doctoral student from the Centre de Sociologie des Organisations (Paris, France). After having completed a monograph of VideoLAN—an open source project—for his master's degree, he is working under Erhard Friedberg's supervision on a Ph.D. devoted to the social construction of a free software market. Thomas Basset has also been a student at the Institut d'Etudes Politiques in Paris. His email address is t.basset@cso.cnrs.fr and his website is http://thomasbasset.net.

**G.L. Bleris** is a full-time professor at the Aristotle University of Thessaloniki, Department of Informatics, Greece. He is head of the Programming Languages and Software Engineering Laboratory (PlaSE) of the department. He received a B.Sc. degree in Mathematics from Aristotle University of Thessaloniki and his Ph.D. degree in Solid State Physics from the Aristotle University of Thessaloniki Physics

Department. He teaches undergraduate level courses in mathematics, electronics, and theoretical informatics. His current research interests include simulation methods, empirical software measurement, evaluation, and management.

**Eva Castro-Barbero** earned an M.Sc. in Telecommunication Engineering from Universidad Politécnica de Madrid, Spain (1997). She is currently working toward a Ph.D. in Telecommunication Engineering at the same university. In 1997, she joined the ISABEL developer team. ISABEL is a distributed multimedia application that runs over heterogeneous networks and provides a customizable environment to define CSCW services. Her responsibilities included ISABEL application software design and implementation. Since 2002, she is a Teaching Assistant at the Rey Juan Carlos University, Madrid, Spain. Her current research interests include IPv6 at application level.

**José Centeno-González** teaches and researches at the Universidad Rey Juan Carlos, Móstoles, Spain. He joined the Computer Science Department of Carlos III University of Madrid in 1993, where he worked until 1999. His research interests include distributed systems programming, and protocols for distributed systems and mobility. He is also interested in the impact of Libre software on computer science teaching and industry. He graduated from Universidad Politécnica of Madrid with an M.S. in Electrical Engineering and expects to receive a Ph.D. in Computer Science from the same university in 2003.

**E. Gabriella Coleman** is an anthropology graduate student at the University of Chicago (USA) focusing on the anthropology of science, medicine, and technology. She has just recently completed her research on the ethics and politics of free software in the San Francisco Bay area. During this period, she also served on the board of the Online Policy Group and interned at the Electronic Frontier Foundation.

**Margaret S. Elliott** is a research specialist at the Institute for Software Research (ISR) and the University of California, Irvine, USA. She received her Ph.D. in Information and Computer Science (2000) and joined ISR in 2001. Prior to entry into graduate school, she worked for 10 years in software development for consulting firms and aerospace engineering, and in research and development for aerospace engineering. Her research interests include open source software development, virtual organizations, computer-supported cooperative work, occupational communities, organizational culture, court technology, and failures of large-scale software systems. She is an active researcher with more than 25 published papers.

**Vincent Freeh** is assistant professor of Computer Science at North Carolina State University (NCSU) (USA). He received his Ph.D. in 1996 from the University of Arizona. His research focus is system software, with concentrations in file systems, operating systems, and distributed systems. Professor Freeh has received an NSF

CAREER Award and an IBM Faculty Development Award. He was a captain in the U.S. Army Corps of Engineers before entering graduate school for his M.S. He worked at IBM in the Storage System Division until he returned to school to earn his Ph.D. Professor Freeh was on the faculty at the University of Notre Dame prior to coming to NCSU.

**Jesús M. González-Barahona** teaches and researches in Universidad Rey Juan Carlos, Madrid (Spain). He started to work in the promotion of Libre software in 1991, in PDSOFT (later Sobre). Since then, he has carried on several activities in this area, such as organization of seminars and courses, and the participation in working groups on Libre software. He participated in the Working Group on Libre Software promoted by the DG-INFO of the European Commission. Currently, he collaborates on several Libre software projects (including Debian), participates in, collaborates with associations like Hispalinux and EuroLinux, writes in several media about topics related to Libre software, and consults for companies on topics related to their strategy on these topics. His research interests include Libre software engineering and, in particular, quantitative measures of Libre software development and distributed tools for collaboration in libre software projects.

**Michael B. Greenwald** received an S.B. degree in Mathematics from MIT, and a Ph.D. in Computer Science from Stanford University. He worked at Symbolics Inc., at MIT's Laboratory for Computer Science (USA), and as an assistant professor in the Computer and Information Science Department of the University of Pennsylvania (USA). His main research areas are in networking, security, distributed systems, and performance evaluation.

**Michael Hahsler** is assistant professor in the Department of Information Business at the Vienna University of Economics and Business Administration, Austria. He teaches object-oriented programming and current topics in information business. His current research interests are object-oriented analysis and design methodologies and data mining.

**Jiayin Hang** has been an employee at Siemens Business Services GmbH & Co OHG C-LAB since September 2002, as a member of the "Business Challenges" team that was founded by Dr. Heidi Hohensohn in 2000. At the moment, she is working in the research project "NOW: Utilization of the open source concept in business and industry," funded by BMBF. Her major fields of work are also mobile services, information and collaboration, etc., with the objective to identify profitable use cases for technological topics. Before her employment, she studied business administration for four years with a focus on computer science while at the University of Applied Sciences in Berlin. There, she graduated with a diploma in Business Administration.

**Pedro de-las-Heras-Quirós** teaches at Universidad Rey Juan Carlos (Spain), and is a collaborator at BarraPunto.Com. Since the early 1990s, he has been a user of Libre software and has collaborated in the Sobre group, devoted to Libre software, since its foundation. He has worked in the organization of congresses, expos, and seminars related to Libre software, and has been editor and author for several publications on that topic.

**Benjamin Hill** is active in the Debian Project (USA) as a package maintainer, application manager for new maintainers, leading member of the Debian Non-Profit Custom Distribution, and as Debian's accountant and hardware donation manager. Additionally, he serves on the Board of Directors of Software in the Public Interest, Inc. and works as a Free/Open Source consultant for companies and NGOs in North America and Europe. Hill has authored a number of important documents on Free/Open Source Software development, including the Free Software Project Management HOWTO. He is an active researcher in the fields of intellectual property, communications studies, and science and technology studies.

**Heidi Hohensohn** is the head of Group "Business Development" of SBS C-LAB (Germany). Dr. Hohensohn has a Diploma in Business Administration (1992) and a Ph.D. (1997). She was the Scientific employee of the University of Paderborn at the Chair of Marketing (1993-1997), where she conducted several projects in market research and consumer-oriented communication for different companies and sectors of business. Since 1997, she has been responsible for Siemens C-LAB Marketing and Business Development, and, in 2000, founded the research team "Business Challenges" with the focus on concepts for IT-based services and solutions supporting changing market structures and processes.

**Jesper Holck** holds a Master of Science from the Danish Technical University (Denmark), with a Ph.D. in Computer Science from Roskilde University. After several years of teaching computer science at the Business College in Ballerup, he is currently an assistant professor at Copenhagen Business School, where his main research areas are information systems and systems development.

**Sotiris Ioannidis** is a Ph.D. candidate at the CIS Department, University of Pennsylvania, USA. He received his B.S. in Mathematics and his M.S. in Computer Science from the University of Crete, Greece (1994 and 1996, respectively). His main research areas are in operating system and network security and distributed systems.

**Niels Jørgensen** started as associate professor at Roskilde University in 2001, coming from Copenhagen Business School. Before that he was a software developer at Nokia. He completed his Ph.D. studies at Roskilde University, Denmark (1992). He has a master's degree in Math and Computer Science and a minor degree in Cultural Sociology. His main interests are open source, distributed systems (especially security), and optimized compilation.

**Angelos D. Keromytis** has been an assistant professor with the Department of Computer Science at Columbia University (USA) since 2001, and director of the Network Security Laboratory. He received his B.Sc. in Computer Science from the University of Crete, Greece, and his M.Sc. and Ph.D. from the Computer and Information Science (CIS) Department, University of Pennsylvania. His current research interests revolve around systems and network security, and cryptography. Previous research interests include active networks, trust management systems, and systems issues involving hardware cryptographic acceleration. His recent work has been on survivable system and network architectures.

**Kouichi Kishida** is director of SRA Key Technology Laboratory (Tokyo, Japan). His major interests are conceptual models of software development environments and the philosophical foundation of software engineering. He is now serving as the secretary general of the Software Engineering Association of Japan.

**Ben Laurie** is a director of the Apache Software Foundation and core team member of the OpenSSL team. He wrote Apache-SSL and "Apache: The Definitive Guide." He is also technical director of AL Digital, Ltd. (USA) and The Bunker, a secure hosting company. He specializes in security, privacy, and cryptography.

**Gregory Madey** is associate professor of Computer Science & Engineering at the University of Notre Dame (USA). His current research interests include free/open source software, agent-based modeling, chaos and complexity, data mining, e-Science, e-commerce, and neural computing. He has published in various journals including the *Communications of the ACM*, several IEEE journals, *The Journal of MIS*, *Decision Sciences*, *The European Journal of OR*, *Omega*, *The Journal of Systems Management*, *Expert Systems with Applications*, *International Journal of Electronic Commerce*, *Information Resources Management Journal*, and *Expert Systems*. He is a member of IEEE, ACM, Informs, and AIS.

**Vicente Matellán-Olivera** works as an assistant professor at Rey Juan Carlos University (Madrid, Spain). He has been involved in various activities related to "libre" software, among them the creation of OpenResources.com and BarraPunto.com.

**Douglas Maughan** is a program manager in the Advanced Technology Office (ATO) of the Defense Advanced Research Projects Agency (DARPA) in Arlington, Virginia (USA). His research interests are in the areas of networking and security. He manages research programs in active networks, fault tolerant networks, dynamic coalitions, trusted operating systems, and secure wireless networks. He has served on various program committees, including the Internet Society Network and Distributed System Security (NDSS) symposium and the International Conference for Distributed Computing Systems (ICDCS). While at DARPA, he has organized several large conferences to demonstrate DARPA technology

**Klaus Mayr** got his Diploma in 1990 and his Ph.D. at the Technical University of Munich in 1996. He currently works as a project leader at IFS IT GmbH (Germany), a company that develops specialized software for car diagnostics and automotive computing. Before he changed to IFS in 2003, he coordinated the research project NOW at 4Soft GmbH, which began in 2002. He worked as a programmer, software engineering expert, architect, and designer in several projects at sd&m from 1997 to 2001. His active engagement in open source originated with the development of "SHORE," an XML-based hypertext repository. The project started at sd&m and was finally donated to the open source community at www.openSHORE.org.

**Kumiyo Nakakoji** received a B.A. in Computer Science from Osaka University, Japan (1986) and an M.S. (1990) and Ph.D. (1993), both in Computer Science from the University of Colorado, Boulder. She is currently a professor at the Research Center for Advanced Science and Technology (RCAST), University of Tokyo, Japan, and directs the Knowledge Interaction Design Laboratory. Her research interests include human-computer interaction, specifically knowledge interaction design, which is a framework for the design and development of computational tools for creative knowledge work, and cognitive and social factors in software engineering.

**Alessandro Narduzzo** is assistant professor at the School of Economics, University of Bologna (Italy) and member of the ROCK (Research on Organizations, Coordination and Knowledge) Lab at the University of Trento. He has been a visiting scholar at the Organization Studies Department of Boston College, and studied at the Cognitive Science Department of the University of California, San Diego, and at the CREW (Collaboratory for Research on Electronic Work) of the University of Michigan. His research focuses on organizational routines, learning and artifacts design. In his empirical studies, he integrates both behavioral and cognitive analysis within an evolutionary perspective.

**Miguel Ortuño-Pérez** is a computer engineer and teaches in Universidad Rey Juan Carlos (Madrid, Spain), where he does research in mobile computing and Libre software. He has worked in Universidad de Oviedo on several projects related to distance learning systems.

**Dale Rahn** is an Open Source developer with OpenBSD. He is currently working for the DARPA-funded POSSE Project, adding security features to OpenBSD. These features include W^X (an effort to insure that no memory is both writable and executable at the same time) and dynamic loader changes to load libraries in random order and protect some vulnerable data sections of shared libraries. He also supports and develops portions of the PowerPC-based, OpenBSD/macppc platform. He has previously worked on cycle approximate simulations using ARM and other embedded processors at Motorola.

**Gregorio Robles** is a Ph.D. candidate at the Universidad Rey Juan Carlos in Madrid, Spain. His main research interest lies in Libre Software Engineering, focused on acquiring knowledge of Libre Software and its development through the study of quantitative data. As part of his work, he has developed or maintained many of the tools that are used for research in this area (CODD, codd-cluster, CVSStats, etc.) He formerly was involved in the FLOSS Project, an European Commission IST-program sponsored project, centered on researching Free/Libre/Open Source Software economic, social, and political consequences as well as gathering hard data about Libre Software developers (through a survey and automated authorship tracking in source code). In this area, he also took part in the WIDI survey during his stay at the Technical University in Berlin. He also worked for BerliOS at GMD (later Fraunhofer) FOKUS, a German Open Source mediator, where he did his diploma thesis. Gregorio is member of GNOME Hispano, the GNOME Foundation, and HispaLinux.

**Luis Rodero-Merino** teaches and does his Ph.D. work at the Universidad Rey Juan Carlos in Spain. His research interest lies basically in the study and implementation of peer-to-peer (p2p) networks. Luis has been researching JXTA, a p2p platform launched by SUN that offers many p2p services as well as a set of p2p protocols. Before joining GSyC, he worked at Ericsson España on services over telephone networks (Intelligent Networks) and in Telefonica I+D (Research + Development) on VoiceXML. Currently, he is also researching IP multimedia transmissions over multicast IP in mobile IP networks.

**Alessandro Rossi** is assistant professor at the Faculty of Economics, University of Trento (Italy) and a member of the ROCK (Research on Organizations, Coordination and Knowledge) Lab of the University of Trento. His research interests cover topics of coordination, division of labor, and management of innovation in the design and production of complex artifacts. He is also interested in applications of experimental economics as a tool for organizational design.

**I. Samoladas** holds a B.Sc. in Informatics from the Department of Informatics at the Aristotle University of Thessaloniki, Greece. Currently, he is a Ph.D. candidate at the same department, working in the area of Free/Open Source Software Engineering. Specifically, he focuses on the quality of the software produced in this way and various quantitative issues including empirical measurements of the process, modelling, and simulation. Other research interests include the analysis and design of e-Learning systems.

**Walt Scacchi** is a senior research scientist and research faculty member at the Institute for Software Research at the University of California, Irvine (USA). He received his Ph.D. in Information and Computer Science in 1981. He joined ISR in 1999, after serving on the faculty of the University of Southern California for 18 years. His interests include open source software development, software process engineering, software acquisition and electronic commerce, and organizational studies of system development. He is an active researcher with more than 100 research papers, and consults widely to clients in industry and government agencies.

**Jonathan M. Smith** is the Olga and Alberico Pompa professor of Engineering and Applied Science at the University of Pennsylvania (USA), and a professor in Penn's CIS Department. His research is centered on advanced communication and computer networking systems, with special interests in programmable networking, network security, and privacy. He is a member of ACM and Sigma Xi, a fellow of IEEE, and has consulted extensively for industry and government.

**Bernd Carsten Stahl** (Dr. rer. pol., Dipl.-Wi.-Ing., M.A., D.E.A) has studied mechanical engineering, business, economics, and philosophy in Hamburg, Hagen, Bordeaux, and Witten. From 1987 to 1997, he was an officer of the German Armed Forces. From 2000 to 2003, he lectured in the Department of MIS and the German Department of University College, Dublin, Ireland. Since 2003, he has been working as a Senior Lecturer in the Faculty of Computer Sciences and Engineering and as a Research Associate at the Centre for Computing and Social Responsibility of De Montfort University, Leicester, UK. His area of research consists of philosophical, more specifically of normative, questions arising from the use of information and communication technology. The emphasis in this area is on the notion of responsibility. He researches the application of such normative questions in economic organizations and also educational and governmental institutions. His second area of interest consists of epistemological questions in Information Systems research.

**I.G. Stamelos** is an assistant professor of Computer Science at the Aristotle University of Thessaloniki (Greece), Department of Informatics. He received a degree in Electrical Engineering from the Polytechnic School of Thessaloniki (1983) and a Ph.D. in Computer Science from the Aristotle University of Thessaloniki (1988). He teaches graduate level courses in language theory, object-oriented programming, software engineering, and enterprise information systems. He also teaches software project management at the post-graduate level. His research interests include empirical software measurement, evaluation and management, and open source software engineering. He is a member of the IEEE Computer Society.

**Renee Tynan** received her B.S. in Cognitive Science from the Massachusetts Institute of Technology, and her master's and Ph.D. in Social Psychology at Harvard University. She is assistant professor in the Department of Management in the

Mendoza College of Business at the University of Notre Dame (USA). Her research examines the role of identity threats and identity motivation in conflict, negotiation, and group behavior. She has published in various journals including the *Journal of Applied Social Psychology*, the *Journal of Applied Psychology*, and the *Journal of Communication*.

**Thomas Wieland** is professor for Telematics, Mobile Computing, and Computer Graphics at the University of Applied Sciences, Coburg, Germany, since 2002. After receiving his diploma in Mathematics (1994) and his Ph.D. (1996) from the University of Bayreuth, Germany, he worked for the German Aerospace Center (DLR) on large software systems for ground processing of satellite data and for Siemens Corporate Technology. At Siemans, he led the research team on flexible service networking and mobile computing. Moreover, Dr. Wieland was the founding editor-in-chief of the German magazine *Linux Enterprise* (2000-2001). He is author of four books on Windows and Linux programming, as well as of numerous papers and articles.

**Jason Wright** has been a developer with the OpenBSD project since 1997. He received a bachelor's degree in Computer Science from the University of North Carolina at Greensboro in 1999. Shortly after graduation, he went to work for Network Security Technologies, Inc. (NETSEC) as director of VPN Technologies, where the initial work for the OpenBSD Cryptography Framework (OCF) took place. He left NETSEC to devote full time to the POSSE Project and to continue working on drivers for cryptographic devices. He lives in Chantilly, Virginia, with his wife and two cats.

**Yasuhiro Yamamoto** received his Ph.D. in Computer Science from the Nara Institute of Science and Technology. He has been interested in information philosophy and interaction design. He has been working as an interaction designer for the last several years for a series of innovative systems supporting creative processes. His research interests in open source include developing a taxonomy for OSS communities based on the communication and interaction patterns among their members.

**Yunwen Ye** received his B.Sc. and M.S. in Computer Science from Fudan University, Shanghai, China, and his Ph.D. in Computer Science from University of Colorado at Boulder. He is currently a research associate at Center for LifeLong Learning (L3D), University of Colorado at Boulder (USA). He is also a Chief Researcher at SRA Key Technology Laboratory (Tokyo, Japan). His research interests include free and open-source software, software reuse, human-centered software development environments, and human-computer interaction.

# Index

## A

active developer 65
address space 254
advice 138
advice network 139
agent-based modeling 213
analysis of the CVS6 127
anarchy 2
Apache 253
Apple MacIntosh OS-X 247
appropriability 97
architecture 87
audit 245
audit process 245
authorities 138
authority score 139

## B

"Brooks' Law" 87
Bakhtin5 275
Baldwin 87
bazaar 127
bazaar style 73
betweenness centrality 141
broken-build 8
Brooks 87
buffer overflow 254

bug-tracking 7
bug reporter 66
business models 226

## C

case study 223
Cathedral 127
cathedral style 73
Clark 87
closed process 76
co-evolution of systems and communities 71
COCOMO model 47
collaboration networks 205
collegial organizations 128
committer 7, 13
community 152
community of practice 60
community structure 60
complex artifacts 85
computer-mediated communication 152
computer-supported cooperative 158
continuous integration 2
contributor 14
control 2
Conway's law 94
coordination 85

copyleft 277
core member 65
council style 75
cryptographic acceleration hardware 248
cryptography 250
CVS 6

## D

DARPA 245
Debian 274
Debian free software guidelines 274
decomposition 85
Defcon 254
design patterns 104
development cultures 250
development effort 48
development version 6
distributed denial of service (DDoS) 243
distribution of work 127
division of labor 85
doctors 276
don contre-don model 126

## E

Ecole Centrale Paris (ECP) 130
educational 282
electronic mail 155
Emacs 97
embedded 128
embeddedness 130
embedded relationships 131
empirical research 153
encryption 244
estimation 47
ethical development 274
ethical doing 279
ethical learning 280
ethnographic observation 130
ethnography 153
evolution 29
evolution of F/OSS systems 60
evolution of the associated F/OSS communities 60

evolution paths 68
exploration-oriented 60
extended attribute file system 248

## F

"free science" 288
firewall 248
forking 97
FreeBSD 1, 245
FreeBSD core team 12
FreeBSD project 85
Freenet 96
friendship 130

## G

GCC 254
GNU/Linux 29
GNU/Linux2 distributions 28
GNU/Linux kernel 85
GNU General Public License (GPL) 96, 277
GNU project 85
governance structure 96
Gutmann, Peter 244

## H

hacker meetings 277
hacker public sphere 278
hacking community 277
hierarchy 127
high assurance 246
Hippocratic Oath 276
hubs 138
HURD microkernel 85

## I

identity 285
individual incentives 126
information freedom 274
information hiding 85
institutional independence 289
integration 87
intellectual property 249, 260
interdependencies 88
interface 87

IPSEC 248
IRC 130

## J

"joining script" 62
Java 109

## L

layered structure of F/OSS communities 67
legitimate peripheral participation 70
libre software 28
license rights 249
lines of source code 33
Linus Torvalds 93
Linux 245

## M

mailing lists 9
management plan 250
many eyes 244
modularity 85
modularization 95
modular architectures 85, 94
monolith 94
Mozilla 1
Mozilla.org staff 12

## N

"no hiding" principle 96
narratives 260
natural product evolution 60
network 127
network analysis 130
network of advice 139
Newsgroups 9
new maintainer process 275
nuclear scientists 276

## O

online communities 282
OpenBSD 245
OpenSSH 247
OpenSSL 248
open development 76

open process 77
open release 76
open source 223
open source software 260
operating systems 243
organizational culture 152

## P

packages 30
Parnas 88
passive user 66
peripheral developer 65
philosophy 285
portability 253
preferential attachment 206
programming group 250
programming languages 30
project organization 248
ProPolice 254
public goods 126

## Q

qualitative research 154

## R

reader 66
Red Hat Linux 29
Red Hat Package Manager (RPM) 32
release 31
release cycles 250
release level process models 15
repositories 6
reviewers 13
role transformation 64

## S

"surgical team" 88
SE-Linux 252
secure sockets layer 253
security 242
security kernels 244
security policy 243
security practices 243
self-organizing system 204
service-oriented 60

sharing  279
signature  285
Simon  86
slammer worm  254
social construction  260
social contract  280
social movements  127
social network theory  205
social variables  131
software design  153
software development  223
SourceForge  204
Splint  253
sponsor  285
SSH  251
stakeholders  66
Stallman  243
super reviewers  13
surgical team  250

## T

technical expertise  128
technical skills  129
tech transition  249
tinderboxes  8
tinkering  94
tournament style  74
transaction costs  127
transparent process  76
transport layer security  253
trunk  6
trust  282
trusted  245
TrustedBSD  248

## U

U.S. Department of Defense  245
UNIX  85, 243
USENET  243
utility-oriented  60

## V

verification machines  9
version control data  106
virtual community  167

virtual organization  155
VLC  130
VLS  130
volunteer labor  289

## W

web server  253
WWW  253

## X

XEmacs  97

# IT Solutions Series – New Releases!

## Humanizing Information Technology: Advice from Experts

Authored by:
Shannon Schelin, PhD, North Carolina State University, USA
G. David Garson, PhD, North Carolina State University, USA

With the alarming rate of information technology changes over the past two decades, it is not unexpected that there is an evolution of the human side of IT that has forced many organizations to rethink their strategies in dealing with the human side of IT. People, just like computers, are main components of any information systems. And just as successful organizations must be willing to upgrade their equipment and facilities, they must also be alert to changing their viewpoints on various aspects of human behavior. New and emerging technologies result in human behavior responses, which must be addressed with a view toward developing better theories about people and IT. This book brings out a variety of views expressed by practitioners from corporate and public settings offer their experiences in dealing with the human byproduct of IT.

ISBN 1-59140-245-X (s/c) • US$29.95 • eISBN 1-59140-246-8 • 186 pages • Copyright © 2004

## Information Technology Security: Advice from Experts

Edited by:
Lawrence Oliva, PhD, Intelligent Decisions LLC, USA

As the value of the information portfolio has increased, IT security has changed from a product focus to a business management process. Today, IT security is not just about controlling internal access to data and systems but managing a portfolio of services including wireless networks, cyberterrorism protection and business continuity planning in case of disaster. With this new perspective, the role of IT executives has changed from protecting against external threats to building trusted security infrastructures linked to business processes driving financial returns. As technology continues to expand in complexity, databases increase in value, and as information privacy liability broadens exponentially, security processes developed during the last century will not work. IT leaders must prepare their organizations for previously unimagined situations. IT security has become both a necessary service and a business revenue opportunity. Balancing both perspectives requires a business portfolio approach to managing investment with income, user access with control, and trust with authentication. This book is a collection of interviews of corporate IT security practitioners offering various viewpoint on successes and failures in managing IT security in organizations.

ISBN 1-59140-247-6 (s/c) • US$29.95 • eISBN 1-59140-248-4 • 182 pages • Copyright © 2004

## Managing Data Mining: Advice from Experts

Edited by:
Stephan Kudyba, PhD, New Jersey Institute of Technology, USA
Foreword by Dr. Jim Goodnight, SAS Inc, USA

**Managing Data Mining: Advice from Experts** is a collection of leading business applications in the data mining and multivariate modeling spectrum provided by experts in the field at leading US corporations. Each contributor provides valued insights as to the importance quantitative modeling provides in helping their corresponding organizations manage risk, increase productivity and drive profits in the market in which they operate. Additionally, the expert contributors address other important areas which are involved in the utilization of data mining and multivariate modeling that include various aspects in the data management spectrum (e.g. data collection, cleansing and general organization).

ISBN 1-59140-243-3 (s/c) • US$29.95 • eISBN 1-59140-244-1 • 278 pages • Copyright © 2004

## E-Commerce Security: Advice from Experts

Edited by:
Mehdi Khosrow-Pour, D.B.A.,
Information Resources Management Association, USA

The e-commerce revolution has allowed many organizations around the world to become more effective and efficient in managing their resources. Through the use of e-commerce many businesses can now cut the cost of doing business with their customers in a speed that could only be imagined a decade ago. However, doing business on the Internet has opened up business to additional vulnerabilities and misuse. It has been estimated that the cost of misuse and criminal activities related to e-commerce now exceeds 10 billion dollars per year, and many experts predict that this number will increase in the future. This book provides insight and practical knowledge obtained from industry leaders regarding the overall successful management of e-commerce practices and solutions.

ISBN 1-59140-241-7 (s/c) • US$29.95 • eISBN 1-59140-242-5 • 194 pages • Copyright © 2004

---

**It's Easy to Order! Order online at www.cybertech-pub.com,
www.idea-group.com or call 717/533-8845 x10 —
Mon-Fri 8:30 am-5:00 pm (est) or fax 24 hours a day 717/533-8661**

 **CyberTech Publishing**

Hershey • London • Melbourne • Singapore

*Excellent additions to your library!*

# Advances in Software Maintenance Management:
## Technologies and Solutions

Macario Polo, Mario Piattini and Francisco Ruiz
University of Castilla–La Mancha, Spain

***Advances in Software Maintenance Management: Technologies and Solutions*** is a collection of proposals from some of the best researchers and practitioners in software maintenance, with the goal of exposing recent techniques and methods for helping in software maintenance. The chapters in this book are intended to be useful to a wide audience: project managers and programmers, IT auditors, consultants, as well as professors and students of Software Engineering, where software maintenances is a mandatory matter for study according to the most known manuals - the SWEBOK or the ACM Computer Curricula.

*ISBN 1-59140-047-3 (h/c) • US$79.95 • eISBN 1-59140-085-6*
*• 310 pages • Copyright © 2003*

"Software organizations still pay more attention to software development than to maintenance. In fact, most techniques, methods, and methodologies are devoted to the development of new software products, setting aside the maintenance of legacy ones."

*Macario Polo, Mario Piattini, and Francisco Ruiz*
*University of Castilla—La Mancha, Spain*

**It's Easy to Order!** Order online at www.idea-group.com or
call 717/533-8845 x10
Mon-Fri 8:30 am-5:00 pm (est) or fax 24 hours a day 717/533-8661

# Idea Group Publishing
Hershey • London • Melbourne • Singapore

*An excellent addition to your library!*

# Practicing Software Engineering in the 21st Century

Joan Peckham and Scott J. Lloyd
University of Rhode Island, USA

Over the last four decades, computer systems have required increasingly complex software development and maintenance support. The marriage of software engineering, the application of engineering principals to produce economical and reliable software, to software development tools and methods promised to simplify software development while improving accuracy and speed, tools have evolved that use computer graphics to represent concepts that generate code from integrated design specifications. ***Practicing Software Engineering in the 21st Century*** addresses the tools and techniques utilized when developing and implementing software engineering practices into computer systems.

*ISBN 1-931777-50-0 (s/c)* • *US$59.95* • *eISBN 1-931777-66-7*
• *300 pages* • *Copyright © 2003*

> "...during the past 30 years a generalized body of knowledge about design as other aspects of software engineering processes has emerged with some generally accepted standards."
>
> *Joan Peckham & Scott J. Lloyd*
> *University of Rhode Island, USA*

It's Easy to Order! Order online at www.idea-group.com or
call 717/533-8845 x10
Mon-Fri 8:30 am-5:00 pm (est) or fax 24 hours a day 717/533-8661

# IRM Press
Hershey • London • Melbourne • Singapore

*An excellent addition to your library!*